HOT
BIKES
Hugo Wilson

ARIEL

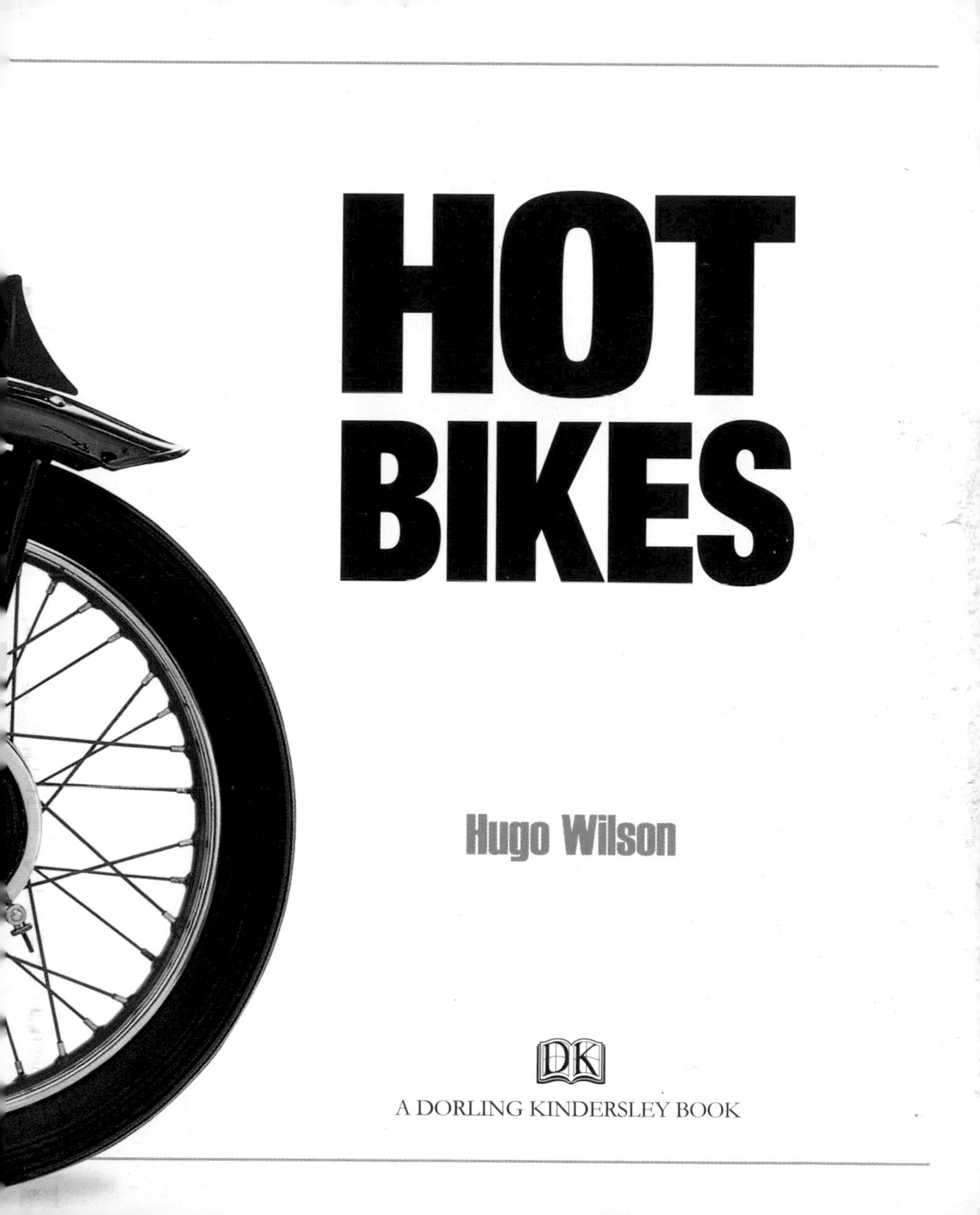

HOT BIKES

Hugo Wilson

DK

A DORLING KINDERSLEY BOOK

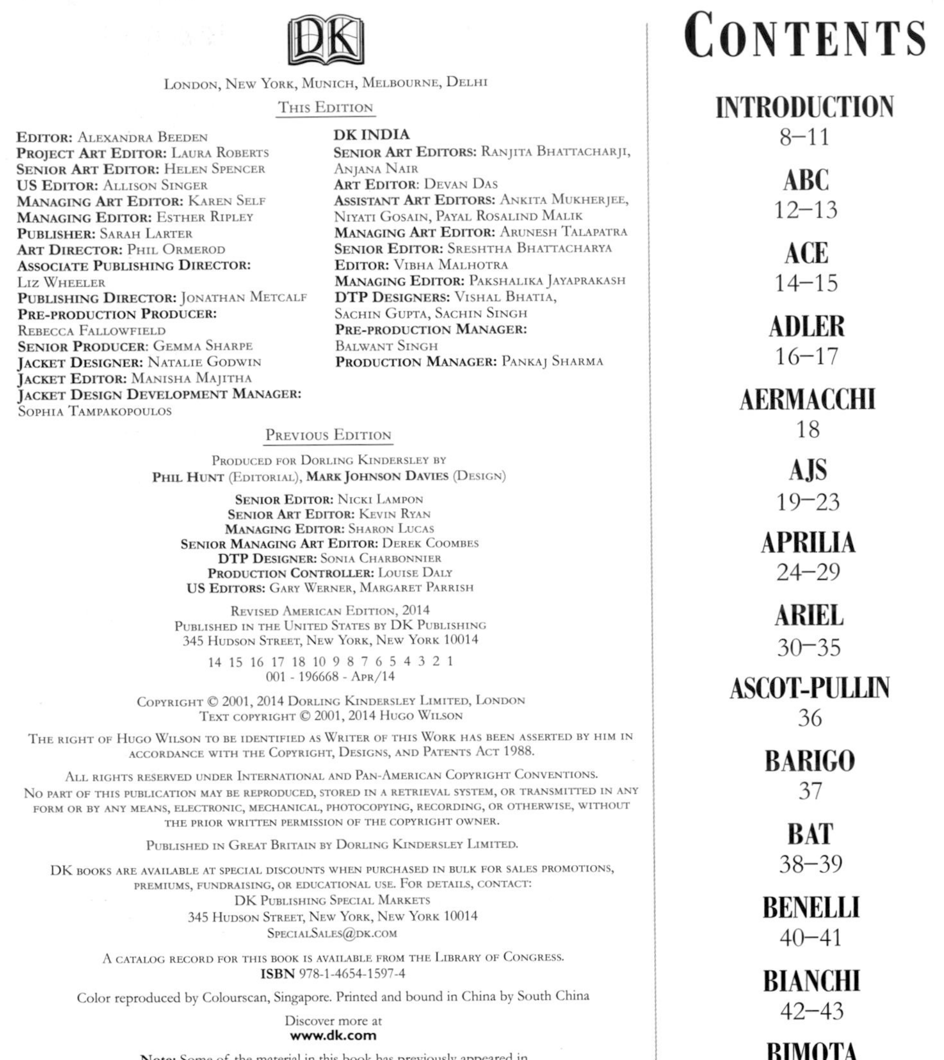

DK

London, New York, Munich, Melbourne, Delhi

This Edition

Editor: Alexandra Beeden
Project Art Editor: Laura Roberts
Senior Art Editor: Helen Spencer
US Editor: Allison Singer
Managing Art Editor: Karen Self
Managing Editor: Esther Ripley
Publisher: Sarah Larter
Art Director: Phil Ormerod
Associate Publishing Director: Liz Wheeler
Publishing Director: Jonathan Metcalf
Pre-production Producer: Rebecca Fallowfield
Senior Producer: Gemma Sharpe
Jacket Designer: Natalie Godwin
Jacket Editor: Manisha Majitha
Jacket Design Development Manager: Sophia Tampakopoulos

DK India
Senior Art Editors: Ranjita Bhattacharji, Anjana Nair
Art Editor: Devan Das
Assistant Art Editors: Ankita Mukherjee, Niyati Gosain, Payal Rosalind Malik
Managing Art Editor: Arunesh Talapatra
Senior Editor: Sreshtha Bhattacharya
Editor: Vibha Malhotra
Managing Editor: Pakshalika Jayaprakash
DTP Designers: Vishal Bhatia, Sachin Gupta, Sachin Singh
Pre-production Manager: Balwant Singh
Production Manager: Pankaj Sharma

Previous Edition

Produced for Dorling Kindersley by
Phil Hunt (Editorial), **Mark Johnson Davies** (Design)

Senior Editor: Nicki Lampon
Senior Art Editor: Kevin Ryan
Managing Editor: Sharon Lucas
Senior Managing Art Editor: Derek Coombes
DTP Designer: Sonia Charbonnier
Production Controller: Louise Daly
US Editors: Gary Werner, Margaret Parrish

Revised American Edition, 2014
Published in the United States by DK Publishing
345 Hudson Street, New York, New York 10014

14 15 16 17 18 10 9 8 7 6 5 4 3 2 1
001 - 196668 - Apr/14

Published in Great Britain by Dorling Kindersley Limited.

DK books are available at special discounts when purchased in bulk for sales promotions, premiums, fundraising, or educational use. For details, contact:
DK Publishing Special Markets
345 Hudson Street, New York, New York 10014
SpecialSales@dk.com

A catalog record for this book is available from the Library of Congress.
ISBN 978-1-4654-1597-4

Color reproduced by Colourscan, Singapore. Printed and bound in China by South China

Discover more at
www.dk.com

Note: Some of the material in this book has previously appeared in Dorling Kindersley's *The Encyclopedia of the Motorcycle, The Ultimate Motorcycle Book, Ultimate Harley-Davidson,* and *Classic Motorcycles* series.

Contents

INTRODUCTION

There are more than 300 bikes featured in this book—all of them different, and all wonderful in their own way. From the purposeful Aprilia RSV4 to the comedic Fantic Chopper; from the rugged KTM 990 Adventure to the stylish Indian Chief; from the bulbous Buell RR1000 to the spindly Weslake speedway bike—the only common factor is an engine and two wheels, although adding a third wheel in the form of a sidecar doesn't appear to affect the definition.

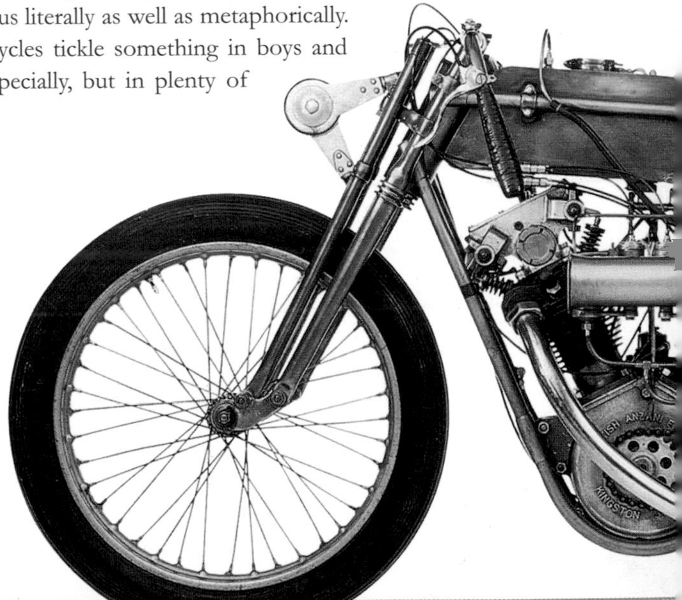

Motorcycles have always been more than just a means of transportation. From that moment in 1885 when Gottleib Daimler's son set off on the first ever motorcycle journey, a 6-mile (10-km) trip during which the rider's seat caught fire (definitely the first "hot bike"), motorcycles have moved us literally as well as metaphorically. Motorcycles tickle something in boys and men especially, but in plenty of girls and women, too. It is probably to do with escape, excitement, and technology—the combination of their physical form, their extraordinary ability, and the cultural meaning that they have acquired.

Throughout the history of the motorcycle, there has been a common quest from riders and engineers—performance. That is not just performance relating to power, but also performance relating to reliability, performance relating to handling, performance relating to safety, and performance relating to comfort. Motorcycle buyers have also had a more-than-passing interest in good looks; what's the point of a great performing bike if it's ugly? Along the way some people also demanded economy, but it was never a top priority for anyone who was interested in chasing his friends down the highway.

Racing and competition have always driven the development of the machines and the desires of the buyers. From that bizarre moment when two pioneer riders met on a dusty road and immediately began boasting of the performance of their bikes, there has been

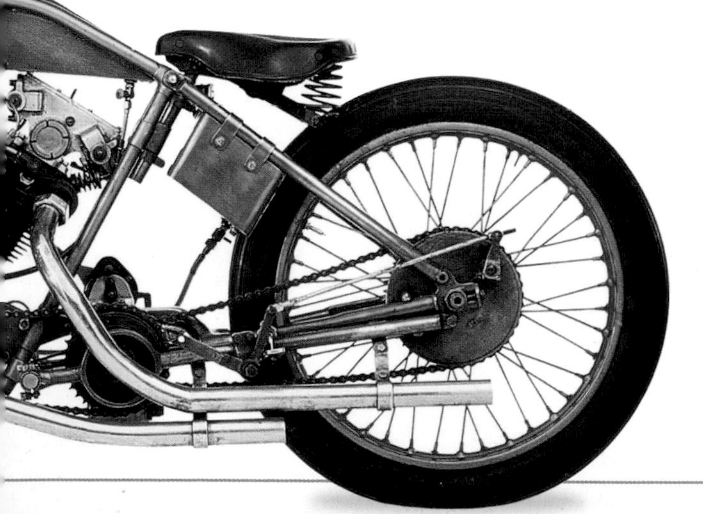

motorcycle racing. The direct result of that competition has been technological leaps driven by the need for racing success, but which soon transfer to sporting road bikes and eventually to the simplest moped. Sometimes those developments have been small, a change in the shape of a carburetor throat or a modification to the material used for the wheel spokes. Sometimes they have been huge—the appearance of disc brakes, float carburetors, or telescopic forks. In the 21st century, advances in electronic engine management systems have made engines cleaner, more economical, more powerful, and more user-friendly.

Sometimes the manufacturers didn't move fast enough. Owners would modify their bikes, improving performance, changing the looks, tailoring the bike to their needs whether for cosmetic or functional reasons. Few motorcyclists have been able to resist customizing their bikes in however small a way. And racers have always looked to make any allowable alterations to give them the competitive edge. Some people have built their own bikes. In the early years of the 20th century, with little competition, it was comparatively easy to become a motorcycle manufacturer. A selection of store-bought components thrown together with a unique name on the fuel tank was almost all you needed to qualify.

Now you need a unique vision to justify the development process and cost necessary to produce even a few bikes with your own name on the tank. Fortunately some people still have that drive and new names still crop up from around the world. Look at the amazing Britten—it is not just a functional race bike, but an amazing high-tech sculpture.

For most people, performance is not the only buying criterion. In fact, in the 21st century, when the performance of many machines exceeds the skill of some owners, it is not even the main consideration no matter what the buyer may say. Looks, and what they mean, are vitally important. People have always been concerned about what their choice says about them. What did it mean to buy Harley-Davidson compared to an Indian motorcycle in 1920s America? And what does it mean to buy a Moto Guzzi compared to a Suzuki in the 21st century? Psychologists can and do write theses on motorcycles and their owners.

The motorcycle as cultural icon has been influenced by films like *Easy Rider* and *The Wild One*, and although the bikes depicted in these films weren't really performance machines, they have affected our ideas about why we want a motorcycle and how we want that bike to look. Since this book was originally published, manufacturers have built faster, more powerful bikes, and some slower and more economical ones, too. Some brands have flourished and are now selling a wider variety of models, some brands have declined, and some, like the Norton, have been revived. For this updated edition of *Hot Bikes*, a selection of hot, new bikes have been added. The motorcycle is constantly evolving, with new technology, new designs, and new styles.

Every rider has a different motorcycling ideal, their own perfect bike, be it a Harley-Davidson or a Honda, Kawasaki or a KTM. Whatever your taste, you'll find your hot bike in this book.

ABC

WHEN THE PROTOTYPES OF THE ABC first appeared in 1919, they featured one of the most advanced specifications of any motorcycle up to that date. The ABC was equipped with front and rear suspension, front and rear drum brakes, a four-speed gearbox, a multiplate clutch unit, and a tubular cradle frame. The 398cc flat-twin engine was mounted transversely in the frame and had overhead valves. In many ways the ABC was a forerunner of the BMW transverse twin that appeared in Germany in 1923, though with chain rather than shaft drive. The light weight of the ABC promised good performance despite the small-capacity engine.

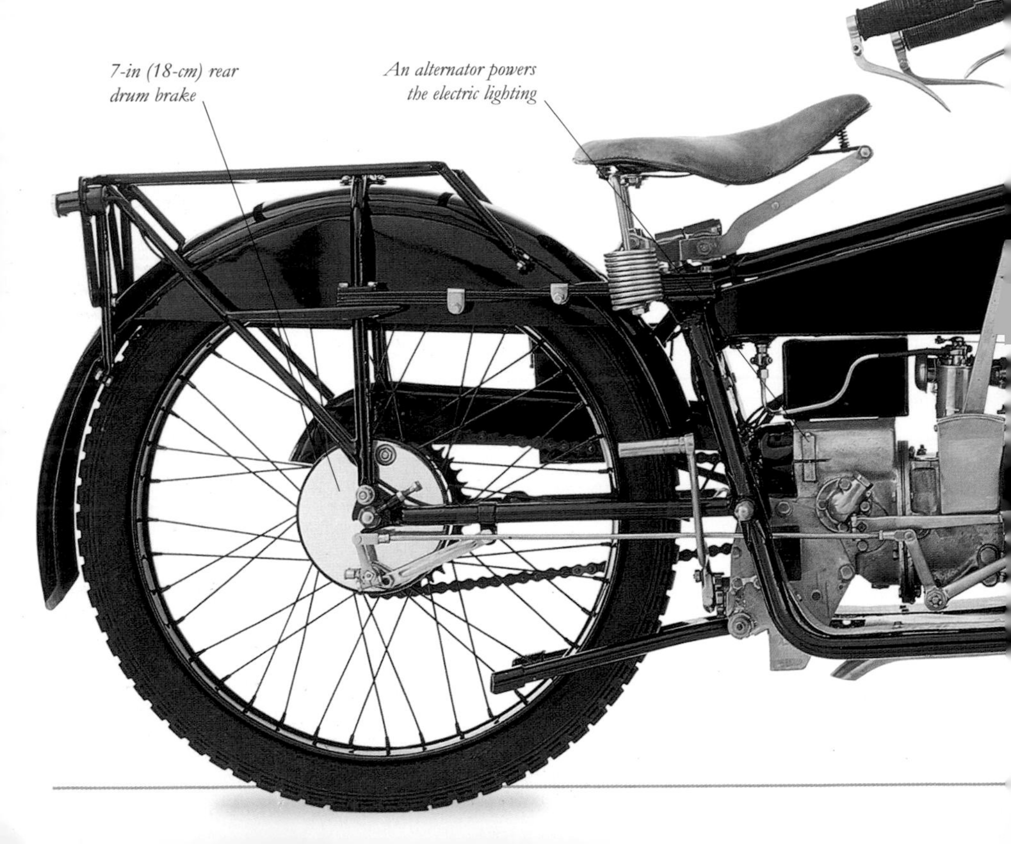

SPECIFICATIONS

MODEL ABC
CAPACITY 398cc
POWER OUTPUT Not known
WEIGHT 243 lb (110 kg)
TOP SPEED 60 mph (96 km/h)
COUNTRY OF ORIGIN UK

ADDED PROTECTION
On the ABC the cylinders are protected by the very widely splayed downtubes of the cradle frame. This also allows the use of an apron, giving some weather protection for the rider's legs.

Leg shields are attached within the frame rails

Leaf-sprung girder forks delivered a comfortable ride

Footboards are also mounted within the frame's perimeter

Originally made in England, the bikes were made under license in France between 1920 and 1924

ACE

THE ACE WAS DESIGNED by William G. Henderson after he left the Henderson company in 1919 *(see pp.184–86)*. The new machine followed the style of the Henderson, with the i.o.e. engine mounted in line with the wheels. There was a three-speed, hand-operated gearbox, and final drive was by chain. The front forks were a typical leading-link construction, but the Ace was considerably lighter than its opposition. Production models became available for 1920. Henderson was killed in an accident in 1922 and was succeeded as designer by Arthur O. Lemon. Unfortunately, production came to a stop late in 1924 when the firm hit financial difficulties and was bought by Indian; it marketed an Indian Ace until 1928.

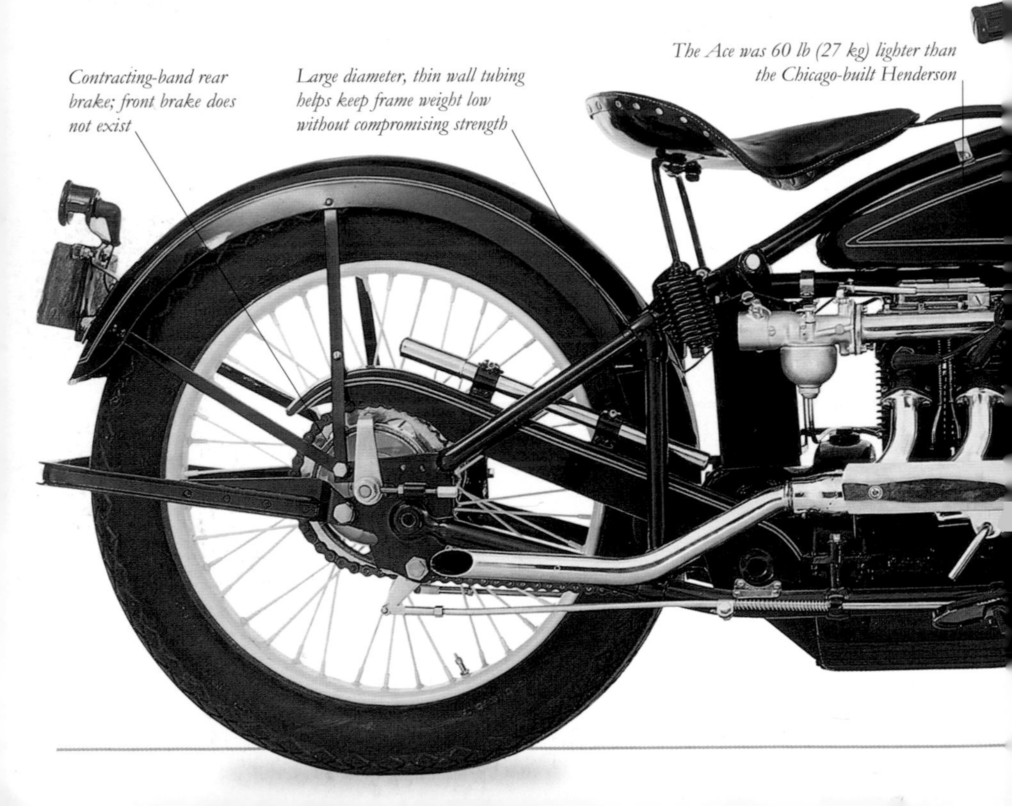

Contracting-band rear brake; front brake does not exist

Large diameter, thin wall tubing helps keep frame weight low without compromising strength

The Ace was 60 lb (27 kg) lighter than the Chicago-built Henderson

SPECIFICATIONS

MODEL Ace
CAPACITY 1229cc
POWER OUTPUT Not known
WEIGHT 365 lb (166 kg)
TOP SPEED Not known
COUNTRY OF ORIGIN US

MICHIGAN MACHINE
The motorcycle pictured here is one of the machines built by the Michigan Motors Corporation in 1926, before Indian acquired the manufacturing rights and production moved to Springfield, Massachusetts.

Leading-link front forks have a single, centrally mounted spring

Blue-and-cream paint finish was used on all Ace machines

ADLER *MB200*

ADLER REENTERED THE MOTORCYCLE market in 1949 after a gap of over 40 years. The new machines were lightweight, two-stroke singles. In 1951 a 195cc twin was added to its expanding line. The new model was the M200. A 250 version was also produced from 1952. To achieve the 247cc capacity, the bore was expanded to 54 mm, which resulted in the classic two-stroke dimensions of 54 x 54 mm. Designers in Britain and Japan were heavily influenced by the bike's design.

Leading-link front suspension with "clock" springs

Multiplate clutch is mounted on the crankshaft end

SPECIFICATIONS

MODEL Adler MB200
CAPACITY 195cc
POWER OUTPUT 16 bhp @ 5590 rpm
WEIGHT 297 lb (135 kg)
TOP SPEED 59 mph (95 km/h)
COUNTRY OF ORIGIN Germany

The first two-stroke twins made by two emerging Japanese producers, Suzuki and Yamaha, were based on the German machine

Horn positioned in a side panel

Denfeld dual seat was an optional extra; standard models came with a sprung solo saddle

CONFIGURATION

The twin had a 180° crankshaft with the clutch mounted on the left-hand end of the crank. Primary drive was by helical gear to a four-speed gearbox. The engine was mounted in a duplex cradle frame with plunger rear suspension.

Plunger rear suspension

AERMACCHI *Chimera 250*

THE FUTURISTIC AERMACCHI CHIMERA caused a sensation when it was launched at the 1956 Milan show. Under the easily removable alloy and pressed-steel bodywork was some slick engineering. The o.h.v. engine had a horizontal cylinder and was available first in 175 and then 250cc capacities. The frame was a tubular spine with rear suspension controlled by a single shock absorber mounted almost horizontally above the spine. Unfortunately the design was not a sales success, and it was only when the designer, Alfredo Bianchi, stripped away the bodywork and replaced the rear suspension with a conventional system that Aermacchi achieved sales and racing success.

SPECIFICATIONS

MODEL Aermacchi Chimera 250
CAPACITY 246cc
POWER OUTPUT 16 bhp
WEIGHT Not known
TOP SPEED 80 mph (130 km/h)
COUNTRY OF ORIGIN Italy

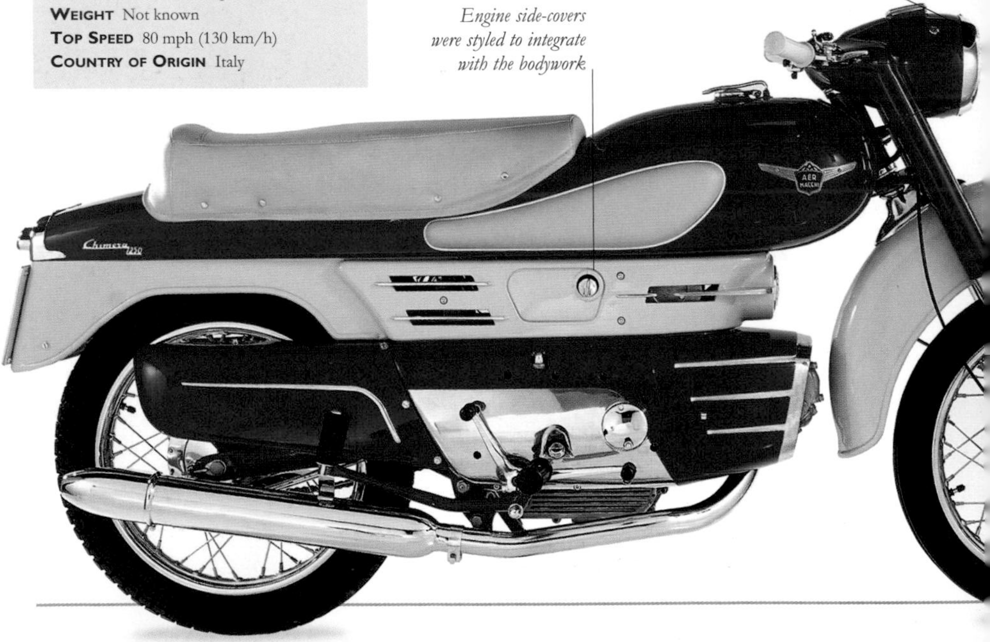

Alloy headlight binacle/top yoke

Engine side-covers were styled to integrate with the bodywork

Binks carburetor

Three-speed gearbox is AJS's own design

Exposed rocker gear has grease nipples for lubrication; valve clearance adjustment is on the pushrods

17-in (43-cm) wheels have chromed rims and full-width alloy brake

SPECIFICATIONS

MODEL AJS 350 G6
CAPACITY 349cc
POWER OUTPUT Not known
WEIGHT 210 lb (95 kg)
TOP SPEED 75 mph (121 km/h) (est.)
COUNTRY OF ORIGIN UK

AJS *350 G6*

IN THE EARLY 1920s, AJS achieved legendary sports success with its o.h.v. 350s, winning the Junior TT for four straight years from 1920. It also won the Senior in 1921, despite the 150cc capacity advantage of the opposition. The factory racing machines of 1922 were nicknamed "big ports" because of the exceptional size of the bike's exhaust port and the 2¼-in (57-mm) diameter exhaust pipe. Production models followed in 1923. The dimensions of the ports shrank for 1925 and 1926, as on this 1926 model, but returned to full size in 1927.

AJS *Supercharged V4*

DEVELOPED BY THE AMC race shop for the 1939 Senior TT, the formidable Supercharged V4 was built to respond to the supercharged Italian and German factory racers dominating the TT and Grand Prix circuits at the time. It was the first machine to lap a Grand Prix course at over 100 mph (161 km/h). The parallel-twin blocks were arranged in a 50° V-layout with chain-driven single overhead camshafts. The supercharger lay across the front of the magnesium crankcases. The outbreak of war prevented racing victory in 1940, and supercharging was banned in 1946 by the Auto Cycle Union (ACU).

SPECIFICATIONS

MODEL AJS Supercharged V4
CAPACITY 495cc
POWER OUTPUT 80 bhp (est.)
WEIGHT 405 lb (184 kg)
TOP SPEED 135 mph (217 km/h)
COUNTRY OF ORIGIN UK

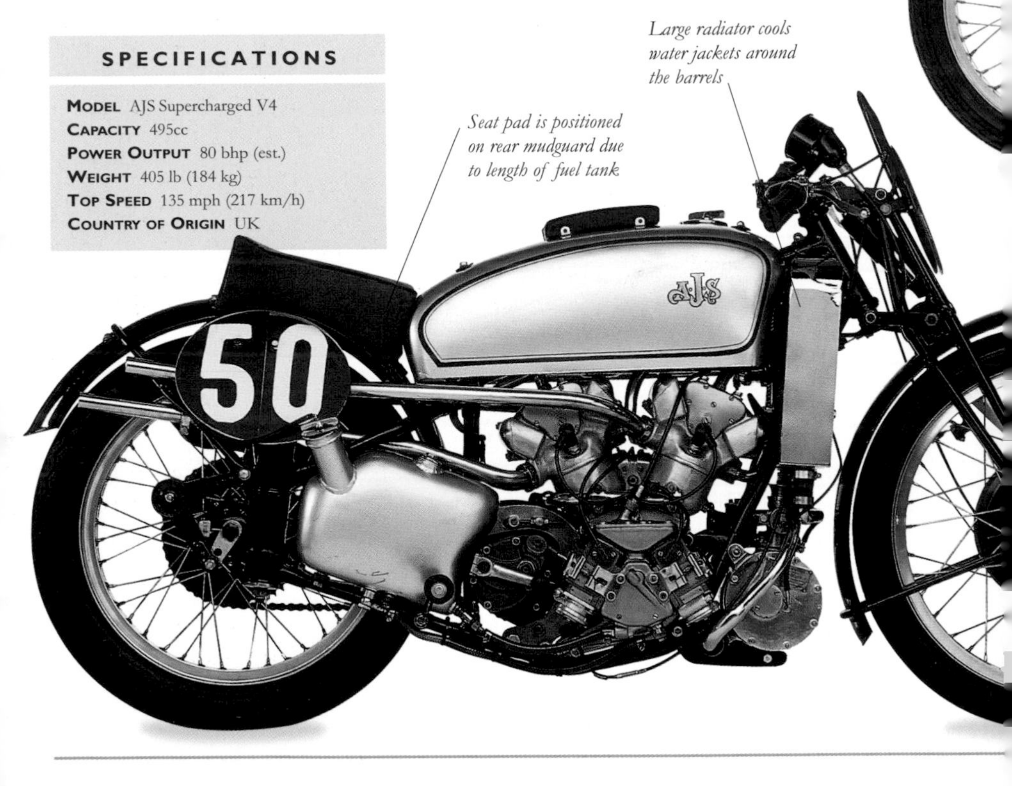

Large radiator cools water jackets around the barrels

Seat pad is positioned on rear mudguard due to length of fuel tank

Large-capacity pannier tank

Inclined engine drives racing gearbox through spur gears and a multiplate clutch

SPECIFICATIONS

Model AJS Porcupine
Capacity 500cc
Power Output 80 bhp @ 7600 rpm
Weight 143 lb (65 kg)
Top Speed Not known
Country of Origin UK

AJS *Porcupine*

The d.o.h.c., parallel-twin Porcupine first appeared in 1947. The nickname derived from the quill-like cylinder head finning between the twin cam boxes, although this feature had been deleted by 1954. Initial plans to cast the cylinder heads in silver for good thermal conductivity were abandoned due to cost. It should have had a supercharged engine, but the postwar ban on superchargers prevented this. Fed by twin carburetors with a single, remote-float chamber, the engine never realized its potential.

AJS *7R/3A*

FOR THE 1954 JUNIOR TT, race development chief Ike Hatch developed a three-valve version of the 348cc 7R. As well as changes to the cylinder head and cam box, the bore and stroke dimensions were revised to a near-square 75.5 x 78 mm. The 7R/3A, or "Triple Knocker," so called because of its three cam lobes, used the frame and running gear developed for the firm's Porcupine racers *(see p.21)*. Hatch went on to develop a shaft and bevel drive for the machine's triple-camshaft top end, but the project was shelved by the firm's accountants.

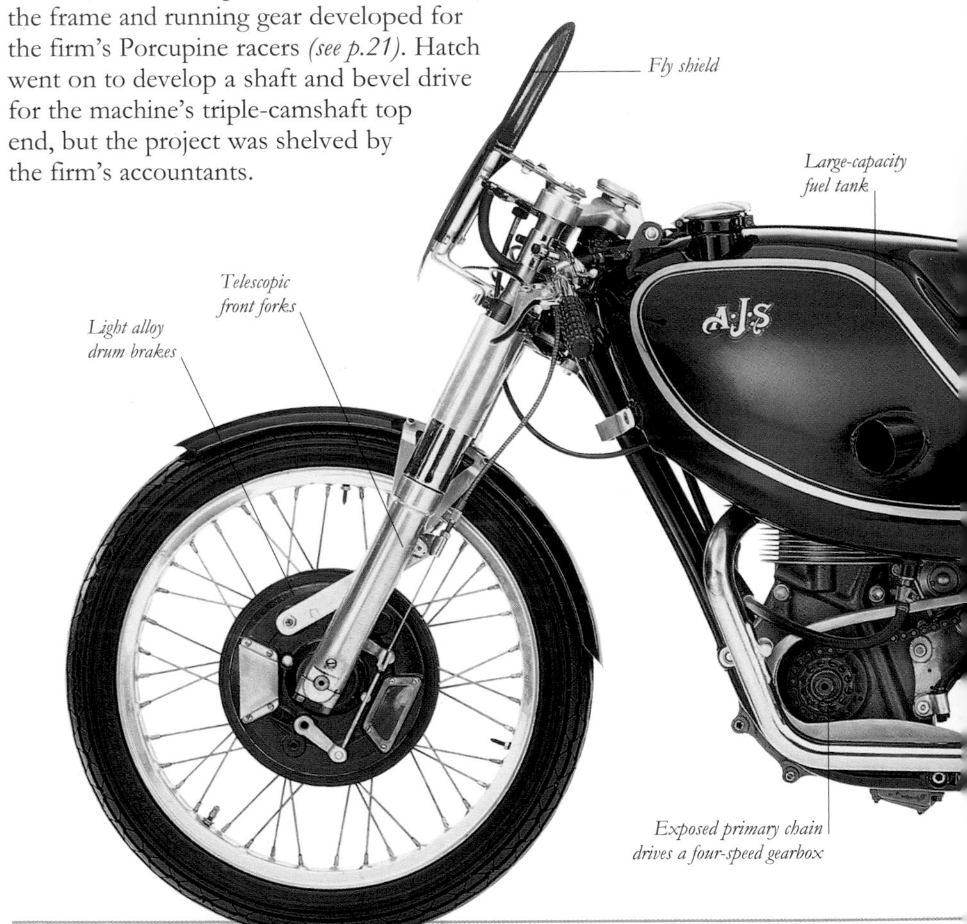

SPECIFICATIONS

Model AJS 7R/3A
Capacity 350cc
Power Output 40 bhp @ 8000 rpm
Weight Not known
Top Speed Not known
Country of Origin UK

Basic Front
The sparse front view of a racing bike from the early Fifties was not hidden behind a fairing. That was soon to change as aerodynamic "dustbin" fairings covered the front wheel of the bike as well as shielding the rider from the wind.

Race number plate

Road-race tire

Pannier-style fuel tank was originally developed for the Porcupine

Solo saddle

Light alloy wheel rim

Two exhaust pipes run from the twin-port cylinder head

APRILIA *RSV250*

THE YOUNG APRILIA COMPANY recognized competition success as an important way to enhance its image. Originally the firm produced motocross machines, but in the mid-1980s it began a campaign for Grand Prix success in the 250 and, later, 125 classes. The early racers used a Rotax tandem, twin-cylinder, disc-valve, two-stroke engine, but 1989's new machine had a V-formation engine layout. As well as being campaigned by the factory's riders, these bikes were also sold to private competitors. Max Biaggi won Aprilia's first 250 World Championship in 1994. The bike shown dates from 1990.

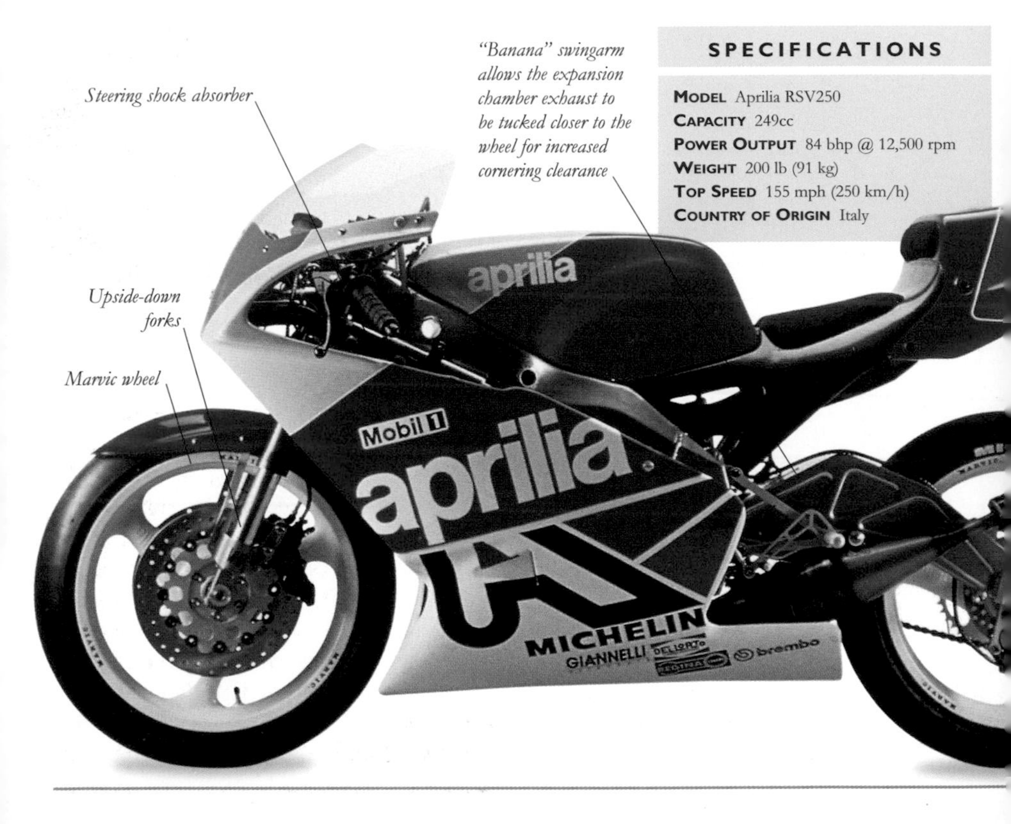

SPECIFICATIONS

MODEL Aprilia RSV250
CAPACITY 249cc
POWER OUTPUT 84 bhp @ 12,500 rpm
WEIGHT 200 lb (91 kg)
TOP SPEED 155 mph (250 km/h)
COUNTRY OF ORIGIN Italy

SPECIFICATIONS

MODEL Aprilia Pegaso 650
CAPACITY 652cc
POWER OUTPUT 49 bhp @ 7000 rpm
WEIGHT 345 lb (156 kg)
TOP SPEED 109 mph (175 km/h)
COUNTRY OF ORIGIN Italy

APRILIA *Pegaso 650*

CLOSE COOPERATION between Aprilia and Austrian engine-makers Rotax resulted, in 1993, in this high-tech trail bike. The engine has a five-valve cylinder head, with the valves arranged radially around a central spark plug. The valves are controlled by twin, chain-driven, overhead camshafts. The combination of technology, looks, price, and performance have made the Pegaso serious competition to Japanese domination of the trail bike market. Bikes like these were especially popular in Europe, where their ability to perform in urban situations, and on rough, twisting roads, was really appreciated. The Pegaso was updated over the following years and remained a contender for class honors despite the arrival of highly rated new competitors.

APRILIA *RSV Mille R*

STANDARD MILLE WAS THE FIRST large capacity superbike from the upstart Italian company Aprilia when it was introduced in mid-1998. The more expensive R version, equipped with better quality suspension, appeared alongside the basic model in 2000, and both were given a substantial makeover the following year. This is a 2001 model. The Mille is powered by a 60° V-twin engine, which is more compact than the 90° V-twin used by Ducati. This allows Aprilia to move the engine further forward for better weight distribution. A balancer shaft to reduce the vibration comes with the 60° configuration.

Single seat unit; the R has no provision for a passenger

Quirky Front

The distinctive three-reflector headlight is common to all Mille models, but the R version benefits from high-quality Öhlins suspension, lighter wheels, improved brakes, and an optional yellow color scheme.

Fairing slot exposes a glimpse of the coolant reservoir

Carbon-fiber mudguards and other components to reduce weight and keep up with fashion

Brembo Goldline four-piston brake calipers feature four separate pads in each caliper

Swedish-made Öhlins telescopic forks

Specifications

Model Aprilia RSV Mille R
Capacity 998cc
Power Output 112 bhp
Weight 403 lb (183 kg)
Top Speed 174 mph (280 km/h)
Country of Origin Italy

APRILIA *RSV4*

APRILIA INTRODUCED A NEW, surprisingly compact and purposeful 180 bhp 1000cc V-four sports bike in 2009. Initially this was produced as a high-specification machine to comply with the rules of World Superbike racing, and Italian rider Max Biaggi won the championship for Aprilia in 2010, and again in 2012. A slightly lower specification machine (shown here) was also produced. It was 10 percent cheaper but lacked some of the expensive, sophisticated, and ultra-lightweight components of the "Factory" model. For road riders, the absence of these parts was academic. It was still astonishingly fast with performance capabilities beyond the reach of all but the most talented riders.

Clutch has "slipper" action, and the bike has a racing-style "cassette" gearbox

Sachs rear shock absorber

Race Bike

Aprilia used a white finish for their basic R model. The more sophisticated and costly "Factory" model was finished in black and red, usually with gold wheels. With a chassis and fairing designed for racing success, this was a very compact bike.

Twistgrip controls the fuel injection via "fly-by-wire" electronics

Turn signal indicators are positioned in the rearview mirrors

Ducts below the headlights feed the engine's airbox

Front forks are made by Japanese supplier Showa

Front brakes use one-piece Brembo "monobloc" calipers

SPECIFICATIONS

Model Aprilia RSV4 2011
Capacity 999cc
Power Output 161 bhp @ 12,500 rpm
Weight 405 lb (184 kg)
Top Speed 186 mph (300 km/h)
Country of Origin Italy

ARIEL *Square-Four 1931*

AFTER TWO YEARS OF DEVELOPMENT, the Edward Turner–designed Square-Four was ready for the 1931 season. The new layout offered the advantages of four-cylinder power and smoothness in a compact unit. The disadvantages included high production costs and inadequate cooling. The Square-Four was essentially two parallel-twins sharing a common crankcase, cylinder block, and cylinder head with the overhung crankshafts geared together by central coupling gears. In the original design the coupling gear of the rear crankshaft also drove the three-speed gearbox, which was built into the unit with the engine. The engine was put into a frame very similar to that used on Ariel's 500cc single-cylinder model.

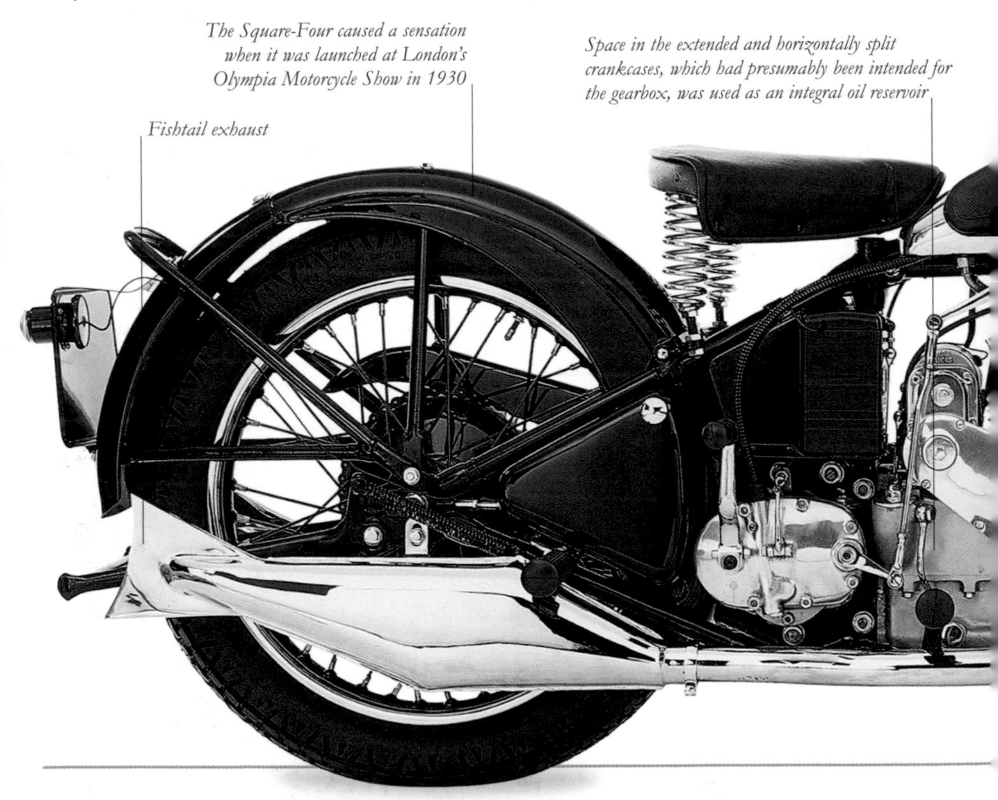

SPECIFICATIONS

MODEL Ariel Square-Four 1931
CAPACITY 597cc
POWER OUTPUT 24 bhp @ 6000 rpm
WEIGHT 413 lb (187 kg)
TOP SPEED 85 mph (137 km/h)
COUNTRY OF ORIGIN UK

SHAFT DRIVE

The overhead camshaft, magdyno, and double-gear oil pump were all driven via a half-speed shaft driven from the front crankshaft, and the end of the camshaft carried a distributor for the ignition system.

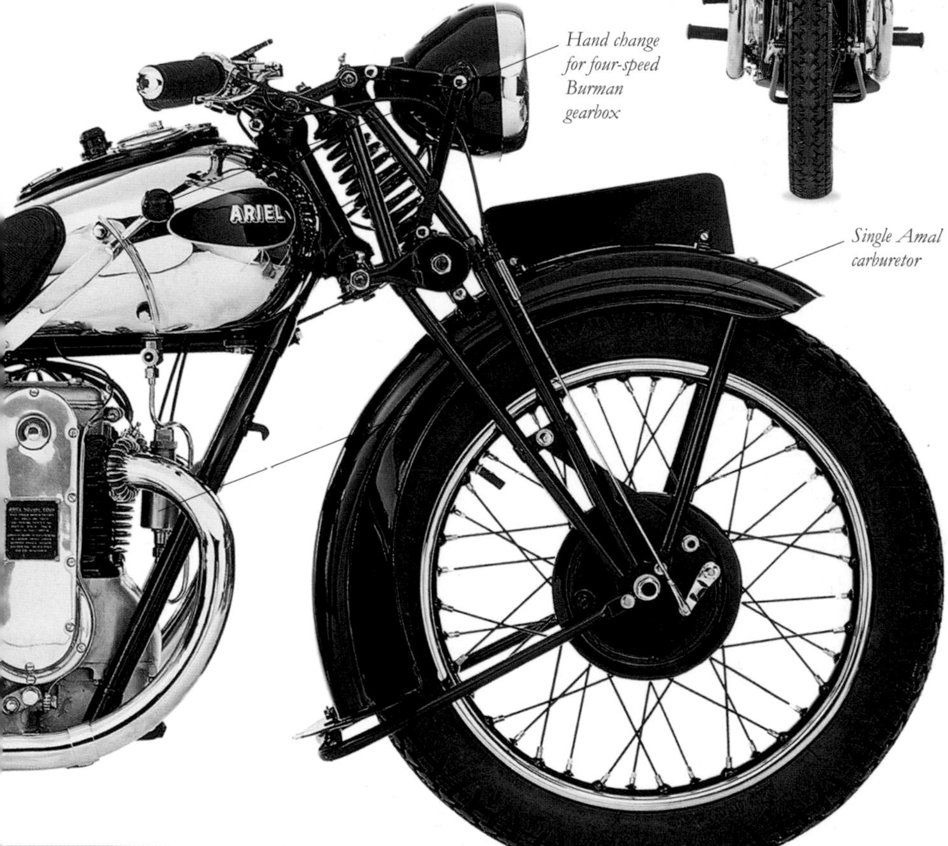

Hand change for four-speed Burman gearbox

Single Amal carburetor

Ariel *Red Hunter*

The pedigree of the Red Hunter goes back to designs produced by Val Page in 1926. The Red Hunter name was first used in 1932 on a Sports version of Ariel's o.h.v. 500 single. The Red Hunter line was soon expanded to include 250cc and 350cc models. From the start, the machine's merits were recognized by trials and grass-track riders, but it was not until Edward Turner embarked on a styling exercise that involved brighter colors and more chromium plating that it became popular with the general public. By the late 1930s, it was well established, with a loyal following until production of the model ceased in 1959. The bike illustrated on this page is a 350 model from 1937.

Specifications

Model Ariel Red Hunter
Capacity 347cc
Power Output 17 bhp @ 5600 rpm
Weight 320 lb (145 kg)
Top Speed 80 mph (129 km/h)
Country of Origin UK

Two high-level exhaust pipes are run from the twin port head

Instrument console positioned on the tank

All-alloy engine has four separate exhaust ports

Production of the Square-Four ceased in 1960

Single leading-shoe drum brake in 19-in (48-cm) front wheel

SPECIFICATIONS

MODEL Ariel Square-Four 1955
CAPACITY 997cc
POWER OUTPUT 40 bhp @ 5600 rpm
WEIGHT 425 lb (193 kg)
TOP SPEED 102 mph (164 km/h)
COUNTRY OF ORIGIN UK

ARIEL *Square-Four 1955*

THE SQUARE-FOUR of 1955 was almost the final development of the o.h.v. machine introduced in 1937. Over the years the machine had adopted telescopic forks, rear suspension, an all-alloy engine, and a four-pipe exhaust outlet. In this form, it represented the ultimate touring machine of the day with a performance equal to that of many sports machines. The engine was large and flexible, making the four-speed gearbox more of a luxury than a necessity.

ARIEL *Arrow Super Sports*

OFTEN KNOWN AS THE Golden Arrow, the Arrow Super Sports was introduced in 1961, the second sports derivative of the 1958 Leader design. It was inspired by Mike O'Rourke's performance in the 1960 Lightweight TT race: riding a modified version of the normal Arrow, he averaged a speed of 80 mph (129 km/h) around the course and finished a very respectable seventh. The Leader/Arrow family had always displayed excellent handling properties and, with only mild engine tuning and minor cosmetic and equipment changes, it captured the hearts of a generation. The bike shown dates from 1963.

18-in (46-cm) wheel

SPECIFICATIONS

MODEL Ariel Arrow Super Sports
CAPACITY 247cc
POWER OUTPUT 20 bhp @ 6650 rpm
WEIGHT 285 lb (129 kg)
TOP SPEED 80 mph (129 km/h)
COUNTRY OF ORIGIN UK

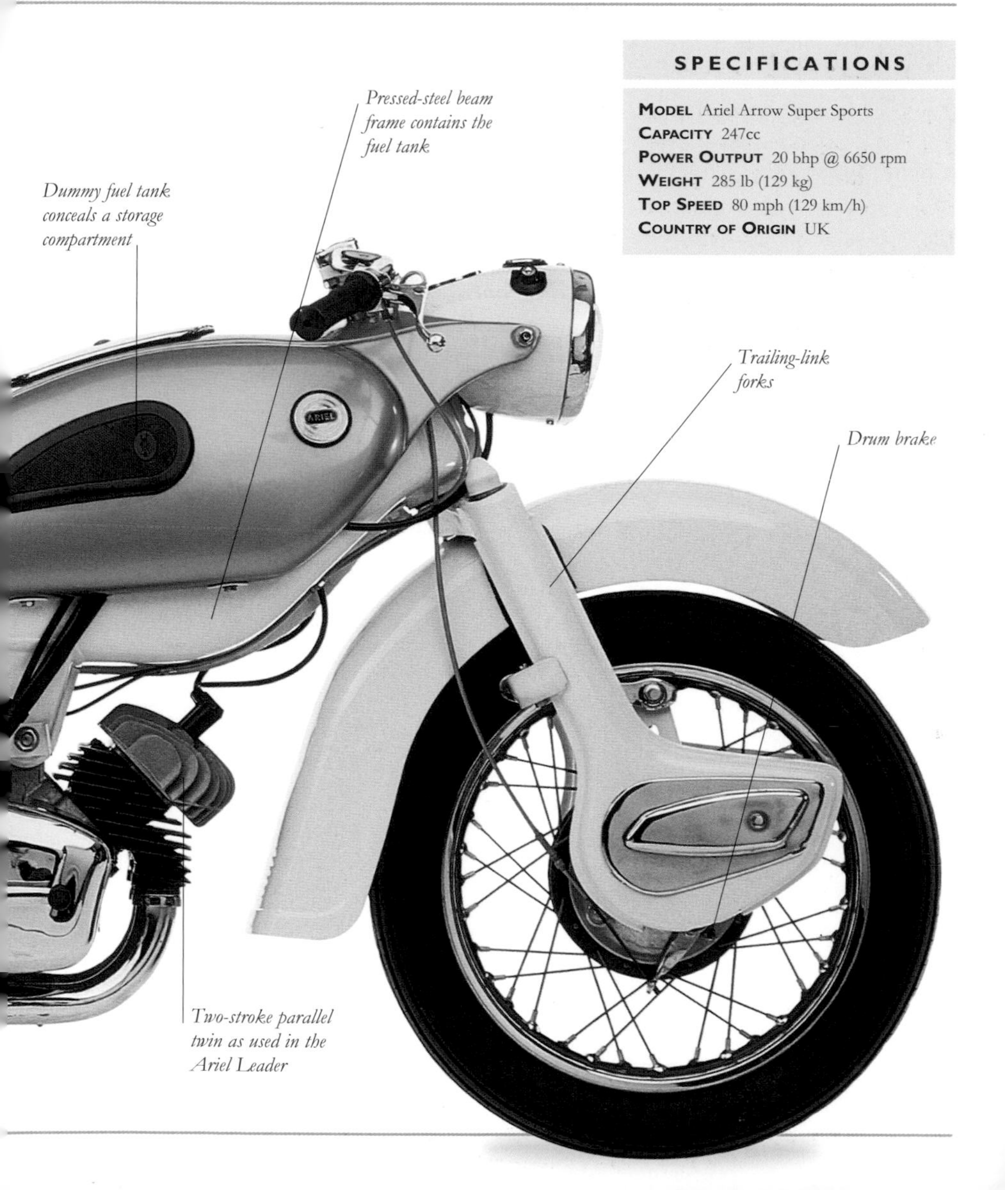

Ascot-Pullin *500cc*

THIS WAS A MACHINE embodying every worthwhile feature possible on a bike and lacking only one thing—success. The pressed-steel frame housed a unit-construction engine and gearbox, with the cylinder lying horizontally and the gearbox above the crankcase. All working parts, including the drive chain, were fully contained within a distinct enclosure, making the machine clean to ride and interesting to look at. The civilized image was enhanced by a neat instrument panel incorporated into the handlebar assembly, as well as optional leg shields and a windshield, with wiper available if required. Unfortunately, Britain was entering a recession at the time this model was produced and the bike did not survive for very long. The 500 illustrated here is a 1929 machine.

SPECIFICATIONS

Model Ascot-Pullin 500cc
Capacity 496cc
Power Output 17 bhp (est.)
Weight 330 lb (150 kg)
Top Speed 70 mph (113 km/h) (est.)
Country of Origin UK

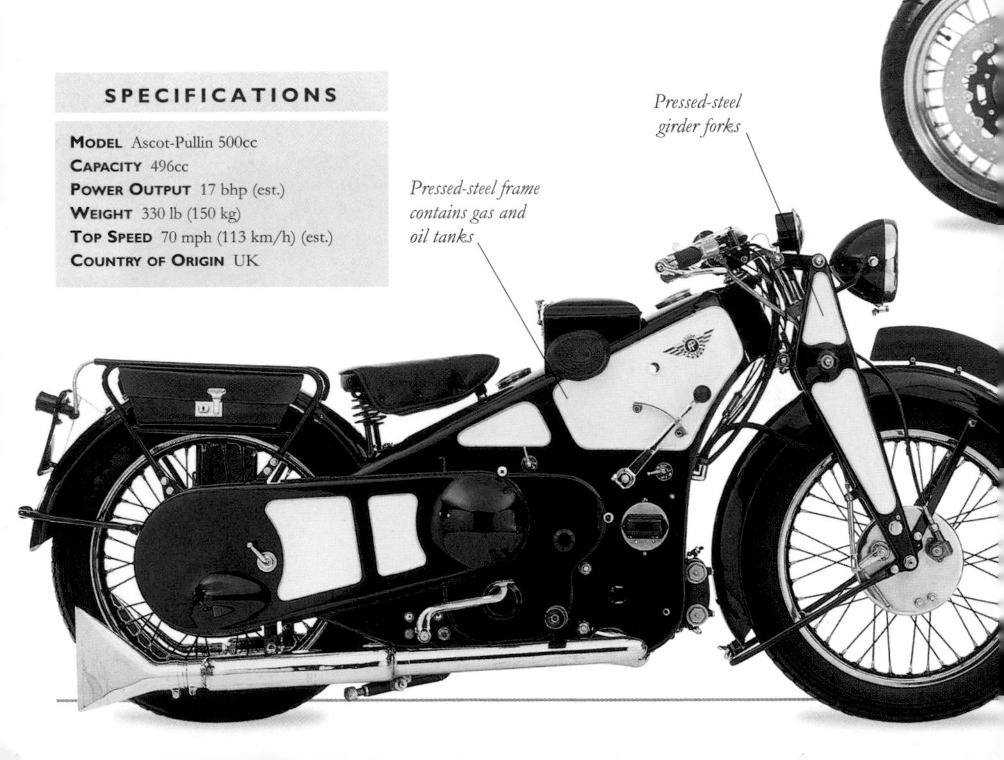

Pressed-steel girder forks

Pressed-steel frame contains gas and oil tanks

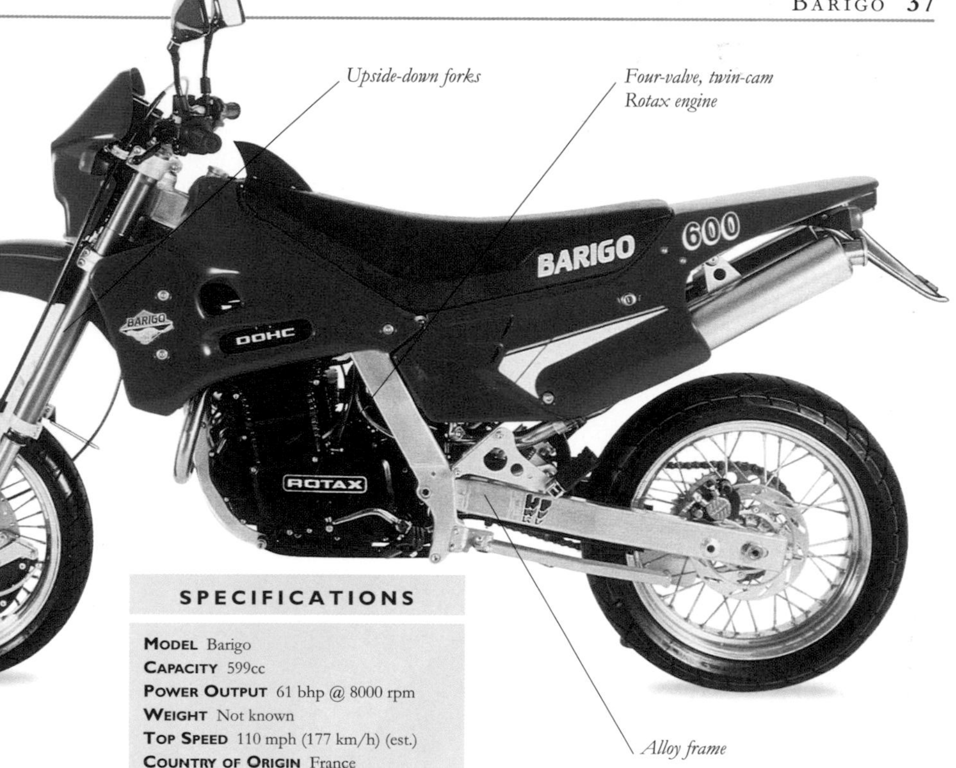

SPECIFICATIONS

MODEL Barigo
CAPACITY 599cc
POWER OUTPUT 61 bhp @ 8000 rpm
WEIGHT Not known
TOP SPEED 110 mph (177 km/h) (est.)
COUNTRY OF ORIGIN France

BARIGO

PATRICK BARIGO BUILT LIMITED numbers of Rotax powered bikes, until the business closed in 1997. This 1992 road bike is based on Barigo's Super Motard competition machines. Premium-quality suspension and brake components are used in its construction, and low weight, good handling, and a tractable engine offer impressive performance. Proven in the grueling Paris-Dakar rally and other trying events, Barigo was the only motorcycle manufacturer operating in France in the early 1990s.

BAT

THE FIRST BAT MOTORCYCLES were built by S.R. Batson in south London, England, in 1902; but by the time this machine was built in around 1904 the company was run by competition rider T.H. Tessier. He coined the slogan "Best after Test" (BAT) to advertise the proven ability of his machines, which ranged from 500 to 1000cc. This bike uses an a.i.v. engine produced by Harry Lawson's Motor Manufacturing Company (MMC), but DeDion engines were also used in early machines.

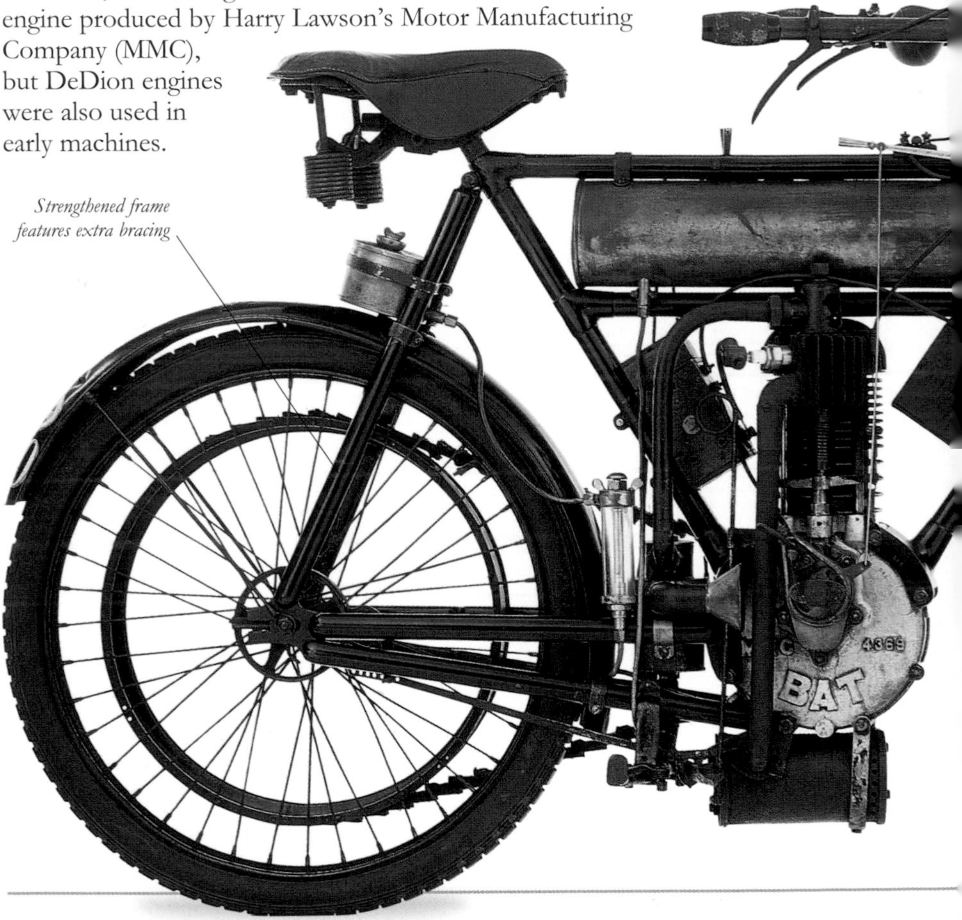

Strengthened frame features extra bracing

Engine Position

The vertically mounted power unit is bolted in the "new" Werner position, and the frame has additional stays running from the back-axle mounting lugs. For 1909, a sprung seat assembly was offered on the model.

Bulb horn

Acetylene lighting system

Bat regarded the use of pedals as an admission that engines were not powerful enough to climb hills; Bat riders had to dismount and push

Bicycle-style forks have additional bracing tubes

Bicycle-style stirrup brake

AH 130

SPECIFICATIONS

Model Bat
Capacity 500cc (est.)
Power Output Not known
Weight Not known
Top Speed 25 mph (40 km/h) (est.)
Country of Origin UK

Benelli *750 Sei*

The Sei was an attempt by Benelli's new owner, Alessandro De Tomaso, to better high Japanese standards. Ironically, the engine was an almost-perfect copy of a four-cylinder, 500cc Honda power unit, but with two extra cylinders grafted on. The alternator was mounted behind the cylinders in a rather futile attempt to reduce the engine's width. Although prototypes appeared as early as 1972, production models did not reach the market until 1974. This is a 1976 model.

Specifications

Model Benelli 750 Sei
Capacity 748cc
Power Output 71 bhp @ 8900 rpm
Weight 485 lb (220 kg)
Top Speed 118 mph (190 km/h)
Country of Origin Italy

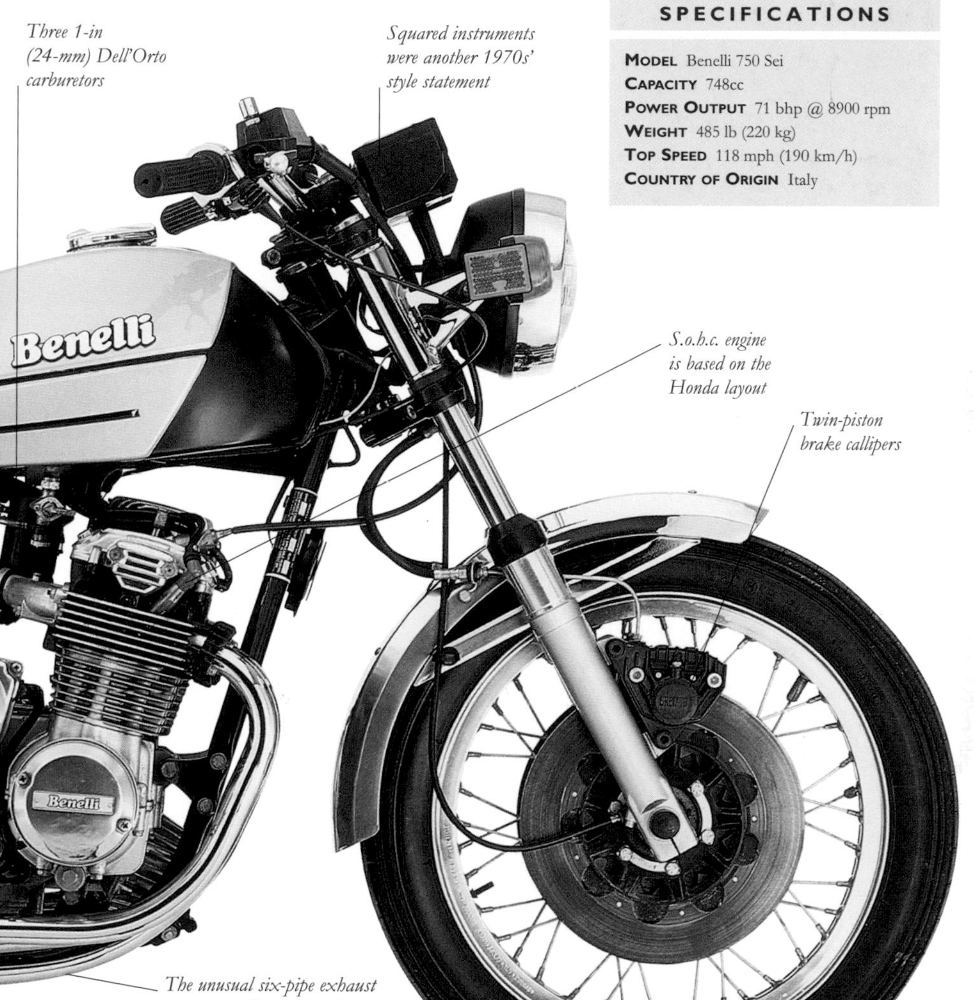

Three 1-in (24-mm) Dell'Orto carburetors

Squared instruments were another 1970s' style statement

S.o.h.c. engine is based on the Honda layout

Twin-piston brake callipers

The unusual six-pipe exhaust system was dropped on the 900cc versions that followed

BIANCHI *ES250/1*

THIS 1937 ES250/1 MODEL was typical of the quality machines produced by Bianchi. The firm was one of Italy's longest-lasting motorcycle companies, manufacturing machines from the turn of the 20th century right up until 1967. It branched out into producing airplane engines, cars, and trucks, and still exists as a bicycle manufacturer. This machine featured a shaft-driven overhead camshaft with exposed hairpin-valve springs. The cast-iron cylinder head had two exhaust ports with the pipes running down each side of the bike. The bottom of the engine featured an integral oil reservoir.

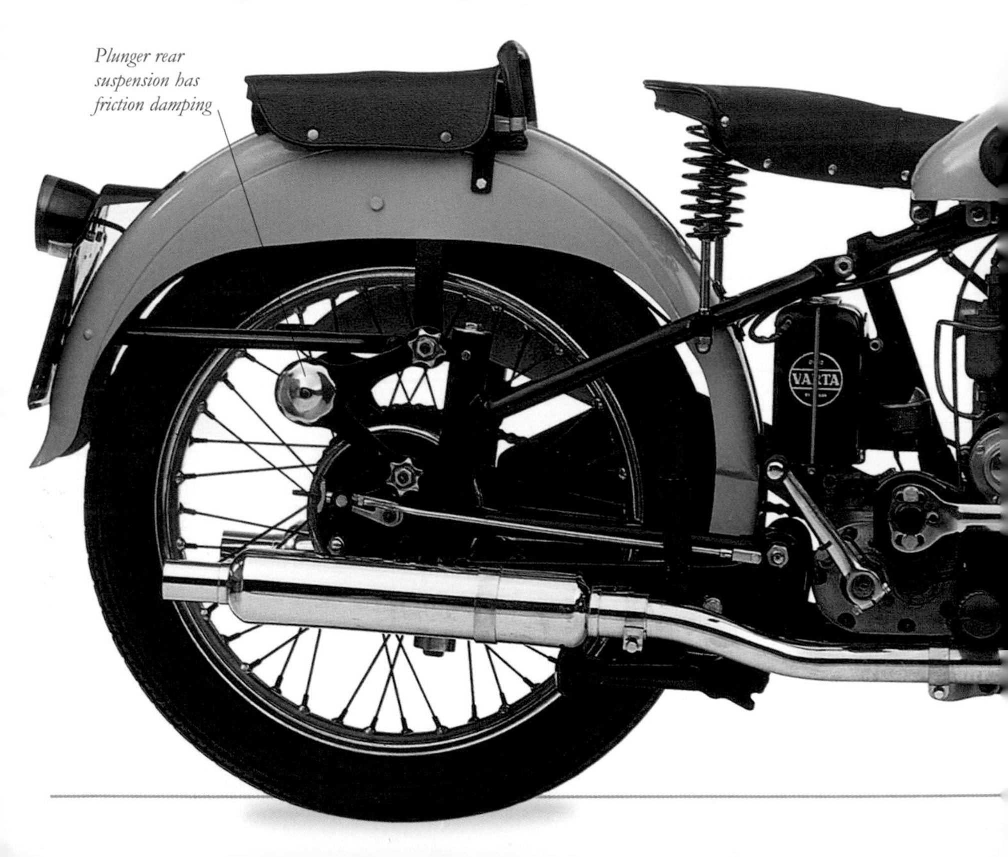

Plunger rear suspension has friction damping

SPECIFICATIONS

MODEL Bianchi ES250/1
CAPACITY 248cc
POWER OUTPUT 10 bhp @ 4800 rpm
WEIGHT 326 lb (148 kg)
TOP SPEED 65 mph (105 km/h)
COUNTRY OF ORIGIN Italy

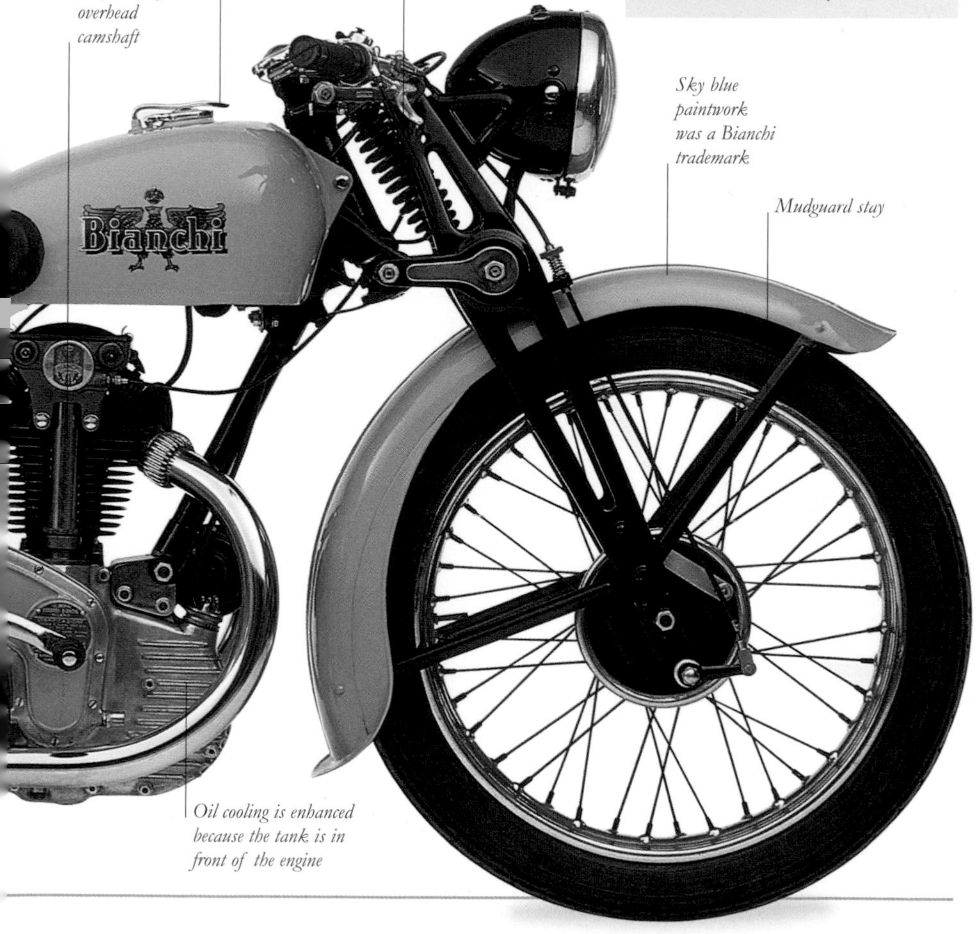

Pressed-steel girder forks

Gas cap retaining-clip

Shaft drive for overhead camshaft

Sky blue paintwork was a Bianchi trademark

Mudguard stay

Oil cooling is enhanced because the tank is in front of the engine

BIMOTA *Tesi 1D*

THE TINY ITALIAN BIMOTA company occupies an extraordinary position in the motorcycle marketplace. It builds superlative sports machines in very limited numbers and to very high standards, using the best available materials and components. The Tesi (Thesis) was first shown in prototype form at the Milan Show in 1982, and a racing prototype appeared in 1984. Early versions used V4 Honda engines in a carbon-fiber frame with the steering controlled by hydraulics.

The sports Tesi was only supplied with a solo seat

SPECIFICATIONS

MODEL Bimota Tesi 1D
CAPACITY 904cc
POWER OUTPUT 118 bhp @ 9000 rpm
WEIGHT 414 lb (188 kg)
TOP SPEED 165 mph (266 km/h)
COUNTRY OF ORIGIN Italy

NEAT DESIGN

When braking, a telescopic fork compresses and severe braking forces are transmitted high up in the frame so road shocks cannot be absorbed. On the Tesi the forces are passed backward directly to the frame.

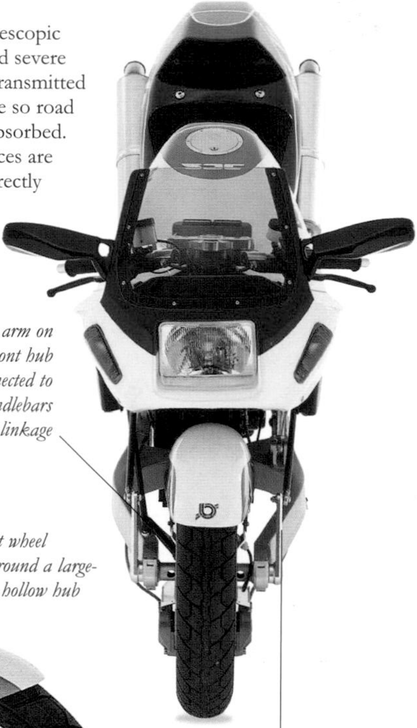

Steering arm on the front hub is connected to the handlebars by linkage

The front wheel rotates around a large-diameter hollow hub

Brake torque arm

Alloy swingarm

BMW *R32*

AT A TIME WHEN MOST MOTORCYCLES were still crude and impractical devices, the R32 was a revelation. Introduced at the 1923 Paris Show, it had a unitary engine/gearbox assembly and shaft final drive. The automatic lubrication system used oil stored in the engine's sump, and the combination of this system, shaft drive, and valanced mudguards made it an exceptionally clean and practical machine. The gearbox had three speeds and was mounted with the engine in a tubular cradle frame equipped with trailing-link, leaf-sprung front suspension. The quality of construction made the BMW considerably more expensive than its competitors, but the originality of the design guaranteed its success.

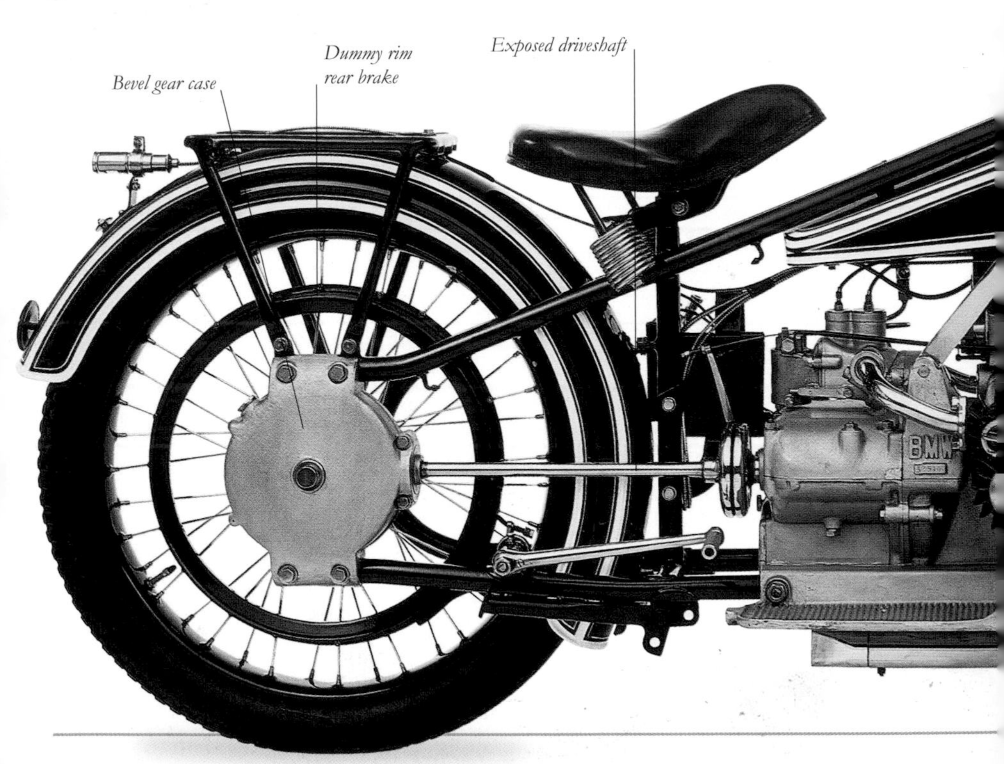

SPECIFICATIONS

Model BMW R32
Capacity 494cc
Power Output 8.5 bhp @ 3300 rpm
Weight 269 lb (122 kg)
Top Speed 53 mph (85 km/h)
Country of Origin Germany

Flat-twin Layout
The transverse flat-twin engine offers the advantages of excellent air cooling, with its cylinders protruding into the airflow. Unfortunately, it also risks cylinder damage in the event of an accident.

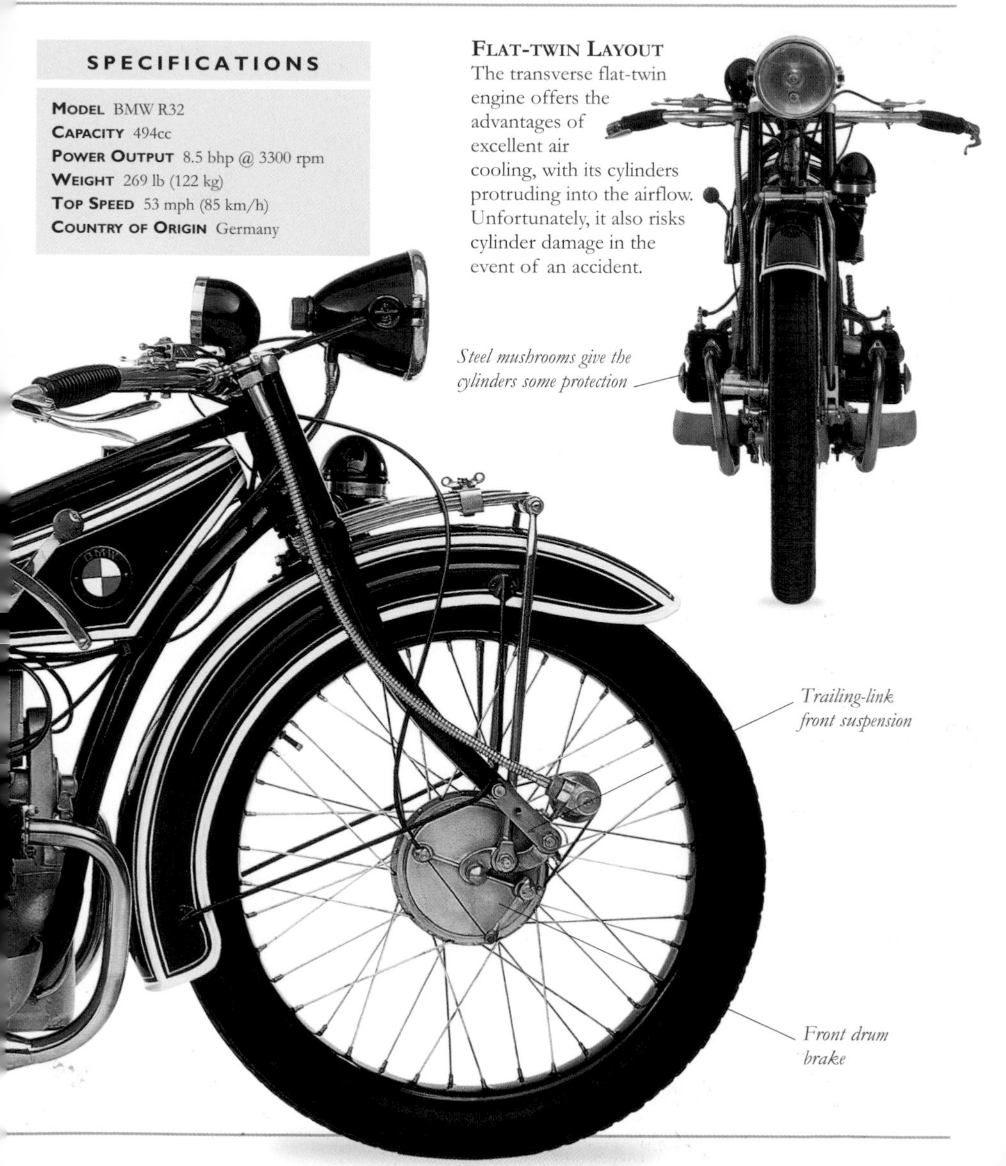

BMW *R12*

THE SIDE-VALVE R12 and its o.h.v. stablemate, the R17, built in 1935, were notable for being the first machines equipped with the BMW hydraulically damped telescopic fork, a leap forward that the rest of the world simply had to follow. Among the different types of front suspension offered over the preceding 30 years there had been a few telescopic types, but this was the first one to incorporate hydraulic damping as part of the design. Although the R12 retained some styling features from BMW's first bike, made back in 1923, it incorporated other new innovations, including a four-speed gearbox controlled via a handchange lever, and interchangeable wheels.

SPECIFICATIONS

MODEL BMW R12
CAPACITY 745cc
POWER OUTPUT 20 bhp @ 4000 rpm
WEIGHT 414 lb (188 kg)
TOP SPEED 75 mph (120 km/h)
COUNTRY OF ORIGIN Germany

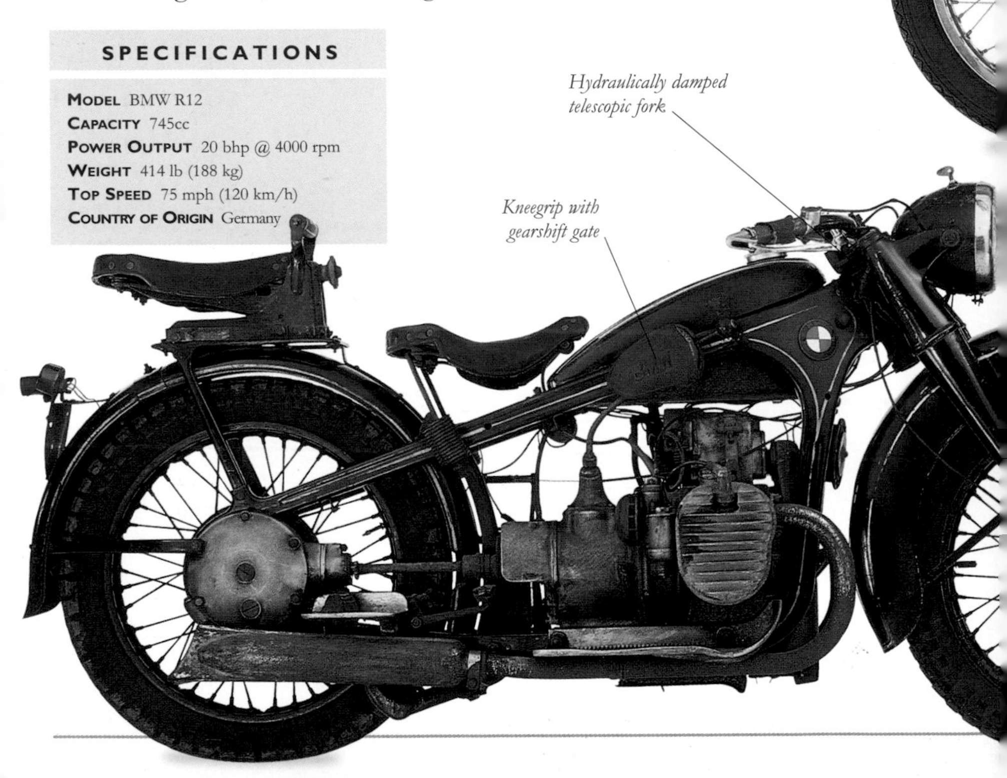

Aluminum cowling protects single carburetor

Linked brakes can be operated by handlebar lever or foot pedal

SPECIFICATIONS

Model BMW Kompressor
Capacity 492cc
Power Output 55 bhp @ 7000 rpm
Weight 302 lb (137 kg)
Top Speed 130 mph (210 km/h)
Country of Origin Germany

BMW *Kompressor*

Before World War II, German motorcycle manufacturers were encouraged by the government to compete in international races. In 1935, BMW, who had previously concentrated on off-road competition and speed-record attempts, produced this new 500cc road racer, which featured overhead camshafts and a supercharger. The power output on the Kompressor was about twice that of the British-built singles that had previously been the benchmark bikes.

BMW *Rennsport*

SUPERCHARGING WAS BANNED by the FIM (Federation of International Motorcyclists) after World War II, and BMW responded by introducing a new racing machine in 1954. While retaining the traditional BMW layout, the Rennsport was an all-new bike. It had a tubular cradle frame with swingarm rear suspension, and each cylinder had an overhead cam driven by shaft and bevel gears. The model was produced in both carburetor and fuel injection versions. While the bike never won a solo Grand Prix, in sidecar racing it proved almost invincible, winning 18 World Championships.

Earles forks were stronger than telescopics and ideal for sidecar events

SPECIFICATIONS

Model BMW Rennsport
Capacity 494cc
Power Output 48 bhp @ 8000 rpm
Weight 300 lb (136 kg)
Top Speed 201 mph (125 km/h)
Country of Origin Germany

Consistent Performer
Although the Rennsport never actually won a solo Grand Prix event, a series of consistently solid finishes by Walter Zeller in 1956 gave him the runner-up position on the 500cc championship table.

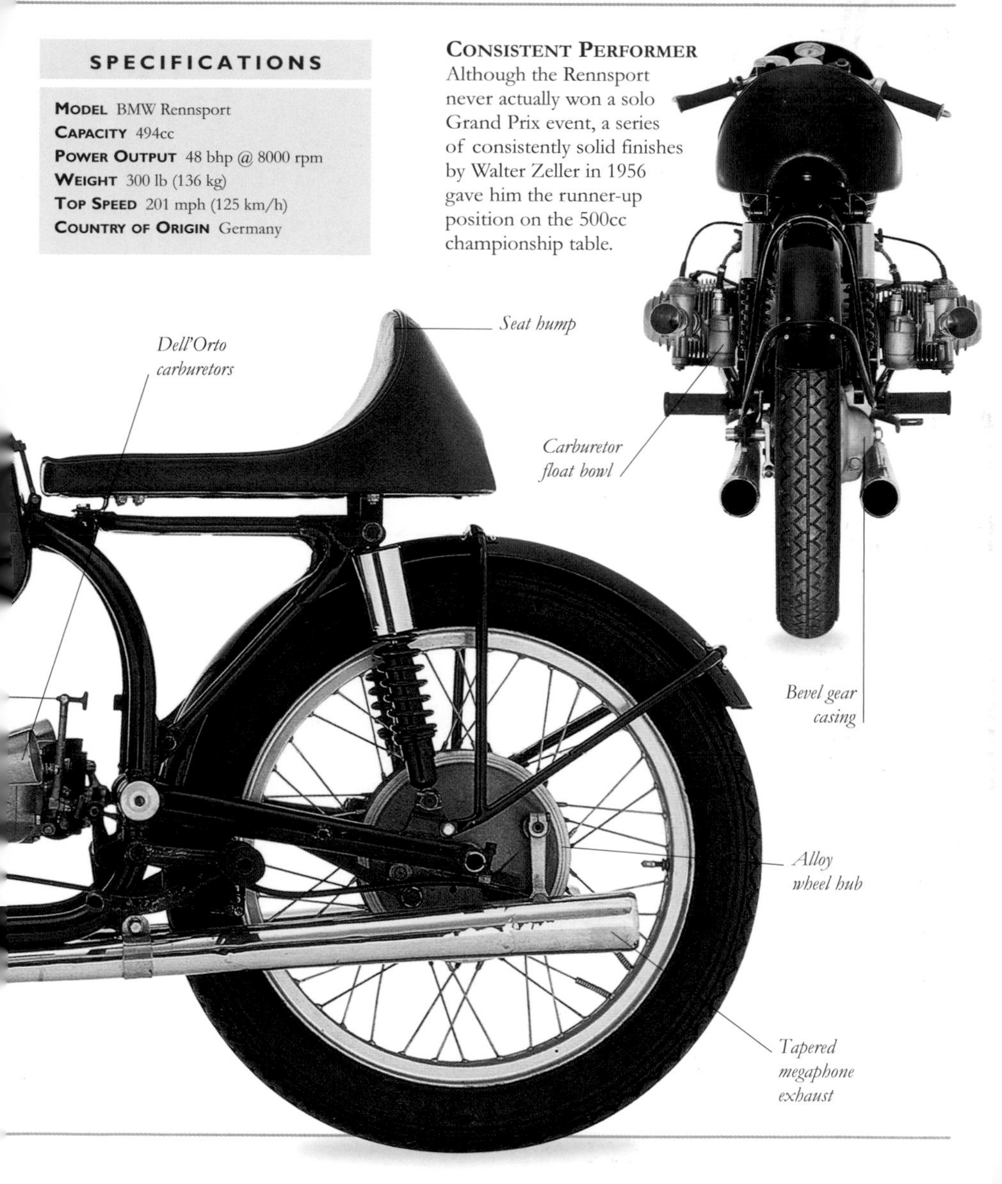

BMW *R60*

HAVING BEGUN POSTWAR production with prewar models, BMW's road bikes were comprehensively redesigned in 1955. The front and rear suspension on the new frames was provided by pivoting forks. The steering geometry and the strength of the new front suspension was ideal for sidecar use, and the BMW became the definitive sidecar machine. The combination shown here consists of a 1965 R60 linked to the classic German-built Steib sidecar. Special sidecar gearing was offered by most manufacturers, which helped the bikes pull the extra weight of the "chair," though performance was obviously reduced. Steib sidecars were built at Nürnberg from the late 1920s and were noted for their quality.

SPECIFICATIONS

MODEL BMW R60
CAPACITY 594cc
POWER OUTPUT 28 bhp
WEIGHT 445 lb (202 kg); S. car: 150 lb (68 kg)
TOP SPEED 75 mph (120 km/h)
COUNTRY OF ORIGIN Germany

Leading-link Earles forks were used on BMW road bikes from 1955 to 1969

After-market large-capacity fuel tank also contains oil and battery

Suspension adjustment lever

Frame around sidecar doubles as bumper bar

SPECIFICATIONS

MODEL BMW R69/S
CAPACITY 594cc
POWER OUTPUT 42 bhp
WEIGHT 445 lb (202 kg)
TOP SPEED 110 mph (177 km/h)
COUNTRY OF ORIGIN Germany

BMW *R69/S*

THE R69/S WAS A sports version of BMW's rather staid 600cc twin and was produced between 1959 and 1969. The power output was boosted by increasing the compression ratio to 9.5:1, and a close ratio gearbox was used. Cycle parts were similar to the standard R60 model. For high-speed touring the R69/S was the benchmark motorcycle of the late 1950s and early 1960s, and for some enthusiasts it represents the best BMW ever built.

BMW *R90/S*

HAVING SUCCESSFULLY REVISED its machines with the introduction of the /5 Series in 1969, BMW sought to improve the bikes further and change its conservative image with the /6 series, launched in 1973. Flagship of the new line was the 900cc R90/S. This was the fastest and most powerful road bike the company had ever built and, when finished in the smoke orange paint scheme shown here, the most colorful. Its increased power and performance helped to convince doubters that BMW could produce exciting machines. BMW used hydraulic disc brakes for the first time on these bikes. The bike illustrated here is a 1975 model.

Bikini fairing conceals instruments, clock, and voltmeter

IMPROVED STYLING
The machines that replaced the Earles-forked BMWs in 1969 had an altogether sportier and more modern appearance and were new in practically every respect. The stylist was now becoming as important as the engineer.

Fogged paintwork ensures that no two R90/Ss are identical

Drilled discs were intended to improve wet-weather braking

BMW's trademark horizontal cylinders

SPECIFICATIONS

MODEL BMW R90/S
CAPACITY 898cc
POWER OUTPUT 67 bhp @ 7000 rpm
WEIGHT 474 lb (215 kg)
TOP SPEED 125 mph (201 km/h)
COUNTRY OF ORIGIN Germany

BMW *K1*

STRICT NEW NOISE restrictions threatened BMW's old-fashioned air-cooled twins with extinction in the 1980s. The company responded in 1983 by introducing a line of three- and four-cylinder fuel-injected and water-cooled machines that retained the shaft drive and used a unique engine layout. The radical-looking K1 was introduced in 1990. Under the aerodynamic skin it had antilock brakes, paralever rear suspension, and a revised engine with four valves per cylinder.

Radiator cooling vent

Aerodynamic fairing designed in wind tunnel

Air duct

Electronic antilock braking system was exclusive to BMW at the time

SPECIFICATIONS

MODEL BMW K1
CAPACITY 988cc
POWER OUTPUT 100 bhp
WEIGHT 474 lb (215 kg)
TOP SPEED 125 mph (201 km/h)
COUNTRY OF ORIGIN Germany

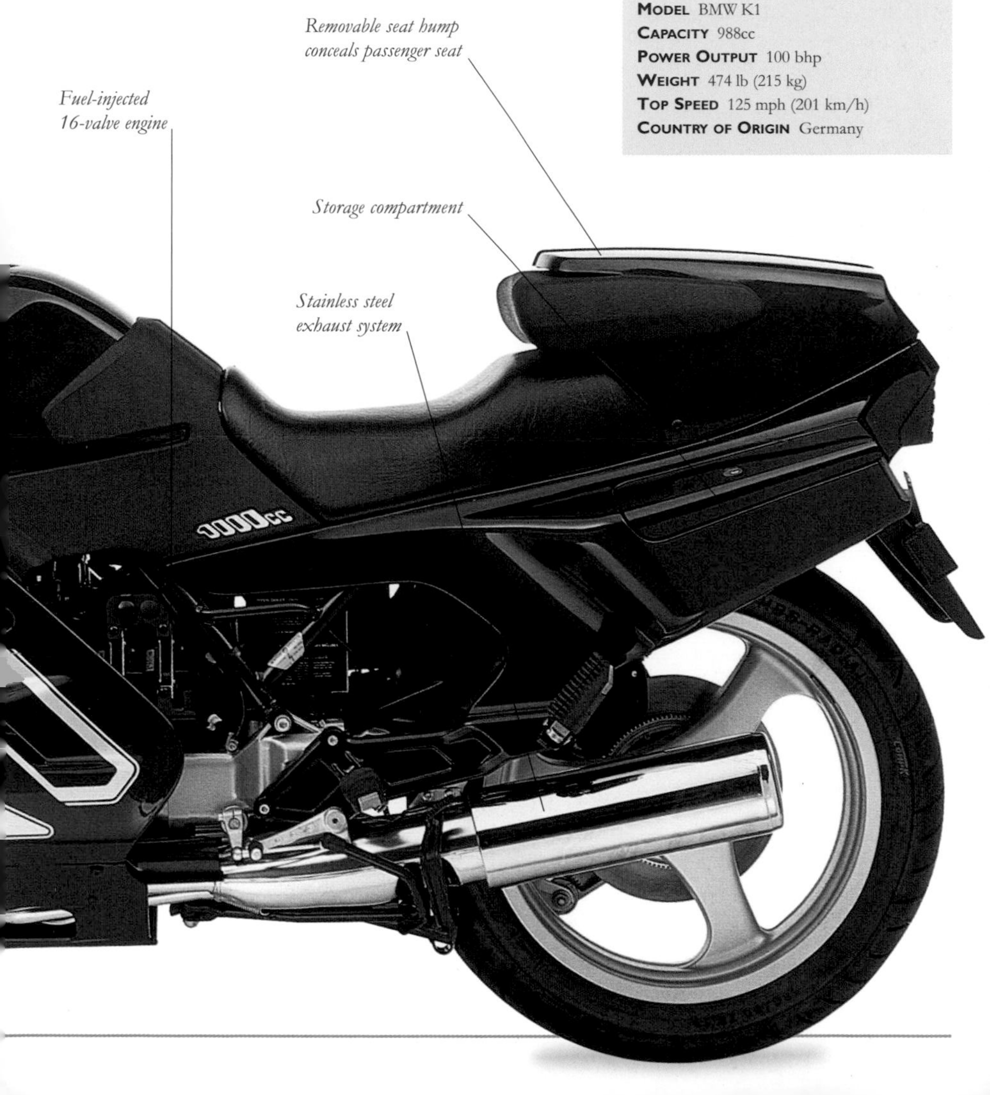
Removable seat hump conceals passenger seat
Fuel-injected 16-valve engine
Storage compartment
Stainless steel exhaust system
1000cc

BMW *R1100GS*

SEVENTY YEARS AFTER its first shaft drive flat-twin, BMW introduced a new line of radical machines based on that original layout. The new engines had eight valves and fuel injection, but the real innovation was in the chassis. Front suspension was by a new "Telelever" system that dispensed with a telescopic fork and much of the frame, now only made up of two minimalist subframes on which the steering head and the seat were mounted. Several models with 850 and 1100cc engines were available.

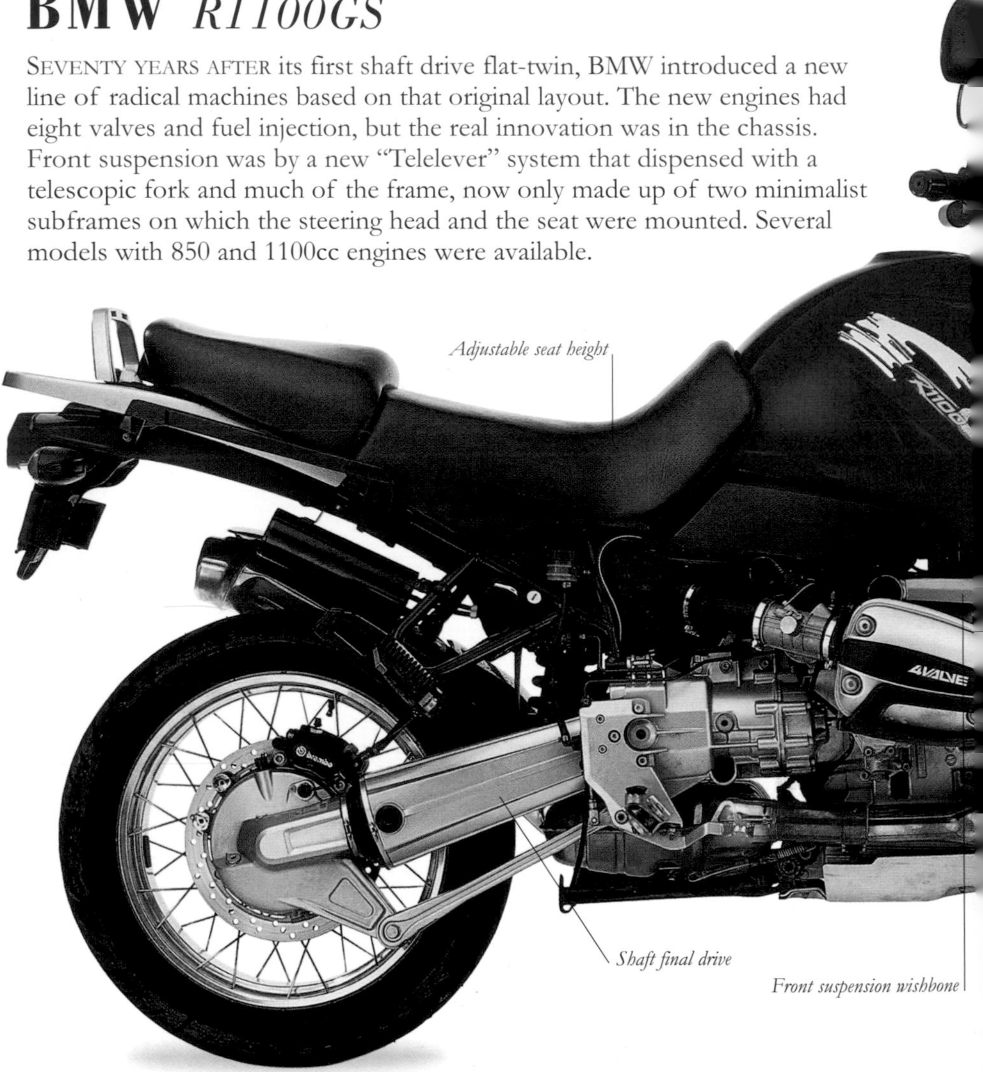

Adjustable seat height

Shaft final drive

Front suspension wishbone

Powertrain Details
The engine had a high-camshaft design that kept cylinder width minimal but gave the advantages of overhead cams. Fuel injection and a catalytic converter were used to keep emissions from the R1100GS to a minimum.

SPECIFICATIONS

Model BMW R1100GS
Capacity 1085cc
Power Output 80 bhp @ 6750 rpm
Weight 460 lb (209 kg)
Top Speed 133 mph (214 km/h)
Country of Origin Germany

BMW *F650ST*

THE F650 IS A TRULY European motorcycle. Conceived by BMW in Germany, it is assembled by Aprilia in Italy using a single-cylinder engine built in Austria by Rotax. This is a remarkable machine for BMW; it is the first single-cylinder BMW for 25 years and the first ever to feature chain drive. The F650 was launched in 1993 with the intention of attracting new buyers to the company, and its light, practical, fun composition makes it an ideal bike for inexperienced riders.

Cowling conceals radiator

TRUE ALL-ROUND BIKE
The F650 is not strictly intended for competition use, but is more an enduro-style bike equally at home on roads or tracks. The "F" in the model designation stands for "Funduro."

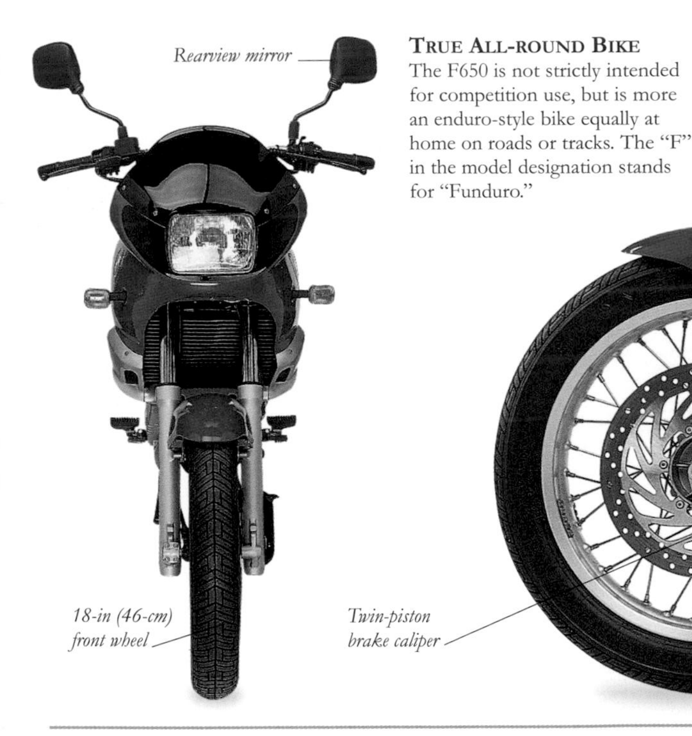

Rearview mirror

18-in (46-cm) front wheel

Twin-piston brake caliper

SPECIFICATIONS

Model BMW F650ST
Capacity 652cc
Power Output 48 bhp @ 6500 rpm
Weight 417 lb (189 kg)
Top Speed 109 mph (175 km/h)
Country of Origin Germany

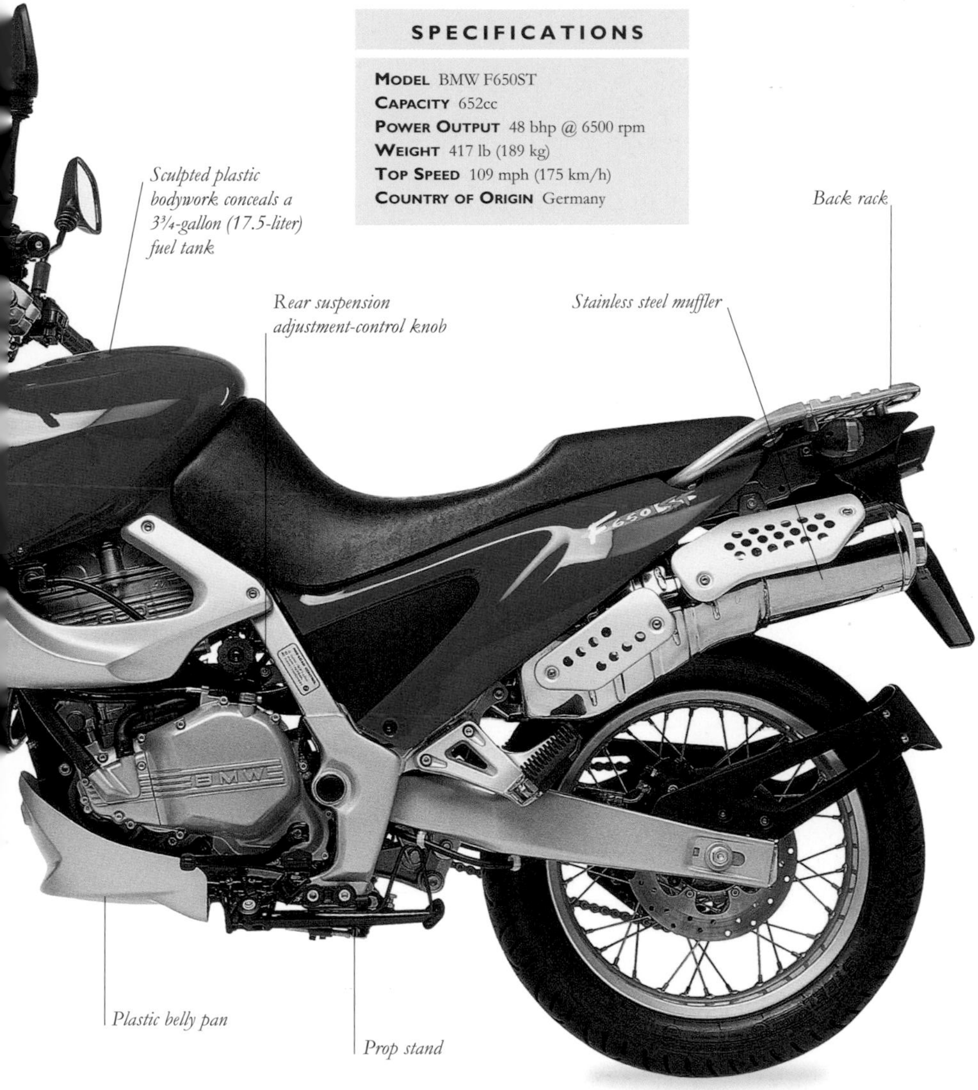

Sculpted plastic bodywork conceals a 3¾-gallon (17.5-liter) fuel tank

Back rack

Rear suspension adjustment-control knob

Stainless steel muffler

Plastic belly pan

Prop stand

BMW *K1600GT*

ALTHOUGH FAMOUSLY conservative in the past, by 2010 when the six-cylinder K1600GT was introduced, BMW had one of the most diverse ranges of any manufacturer. The new touring model was powerful, smooth, well equipped, and, despite considerable weight and size, surprisingly agile—at least on the move. Advanced technology included the use of "duolever" girder forks at the front, electronic "fly-by-wire" throttle control, and a headlight that kept its beam level no matter how the bike was pitched. As well as the traditional touring requirements of panniers and shaft-drive, the new bike also featured an adjustable windscreen for increased comfort. A more luxuriously equipped GTL model is also offered and there is an extensive range of accessories, including stereo, available for both bikes.

Each pannier has a 7-gallon (33-liter) capacity

Seat is electrically heated

Six-cylinder engine is just 22 in (56 cm) wide

Engine Spec

BMW made strenuous efforts to keep the bike's width to a minimum. The engine is not significantly wider than many four-cylinder units. This is achieved by having comparatively narrow cylinder bores, and keeping the space between the cylinders small.

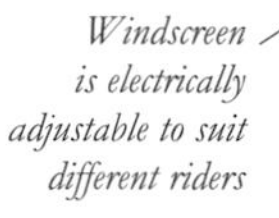

Windscreen is electrically adjustable to suit different riders

"Adaptive" headlight beam remains level when bike is loaded, or when braking or accelerating

Front fork moves on linkages controlled by a single shock absorber

SPECIFICATIONS
Model BMW K1600GT
Capacity 1649cc
Power Output 158 bhp @ 7750 rpm
Weight 703 lb (319 kg)
Top Speed 155 mph (249 km/h)
Country of Origin Germany

BÖHMERLAND

DESIGNED BY ALBIN LIEBISCH and built from 1925 until 1939, the long wheelbase Böhmerland was one of the strangest motorcycles ever made. Seating was provided for three people; a short wheelbase two-seater version was also made. The extraordinary frame design suggested a bridge rather than a motorcycle. The front fork springs operated in tension not compression, and the wheels were made of cast-alloy. Liebisch's long-stroke engine retained exposed valve gearing until production ended. Without the protection of a normal fuel tank, riders hoped that the engine did not eject its pushrods. This is a 1927 model.

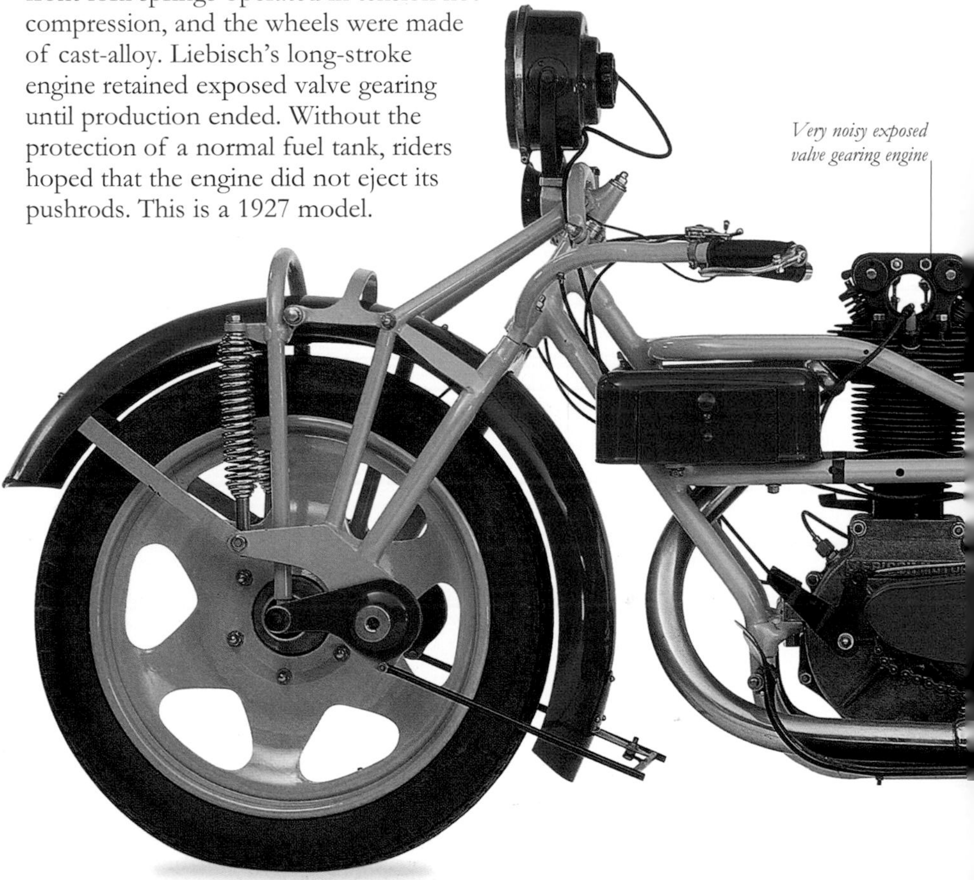

Very noisy exposed valve gearing engine

SPECIFICATIONS

MODEL Böhmerland
CAPACITY 603cc
POWER OUTPUT 16 bhp @ 3000 rpm
WEIGHT 500 lb (227 kg)
TOP SPEED 59 mph (95 km/h)
COUNTRY OF ORIGIN Czech Republic

NOVEL SUSPENSION
The unusual leading-link forks use springs in tension rather than the more conventional compressed position. The disc wheels are one of the earliest examples of the one-piece, cast-alloy variety.

The color scheme emphasizes the Böhmerland's eccentricity

Twin fuel tanks mounted beside the rear wheel

The carburetor is mounted on an extremely long inlet tube

BRIDGESTONE *Hurricane*

THE HURRICANE WAS introduced in 1968—taking advantage of interest in so-called "fun bikes"—and won immediate praise for both its performance and build. Dubbed a scrambler, it was really a dual-purpose machine that was more at home on the road than on the dirt. An unusual feature of the Hurricane was its gearbox: in one mode it had four speeds arranged in rotary sequence, so when in top gear one notch down was neutral, then another took you into bottom. This was not too popular with riders, but a conventional five-speed arrangement was available at the flick of a switch.

Steering shock

Twin leading-shoe drum brake

Road tires are standard

Bashplate protects engine when riding off-road

SPECIFICATIONS

MODEL Bridgestone Hurricane
CAPACITY 177cc
POWER OUTPUT 20 bhp @ 8000 rpm
WEIGHT 271 lb (123 kg)
TOP SPEED 78 mph (126 km/h)
COUNTRY OF ORIGIN Japan

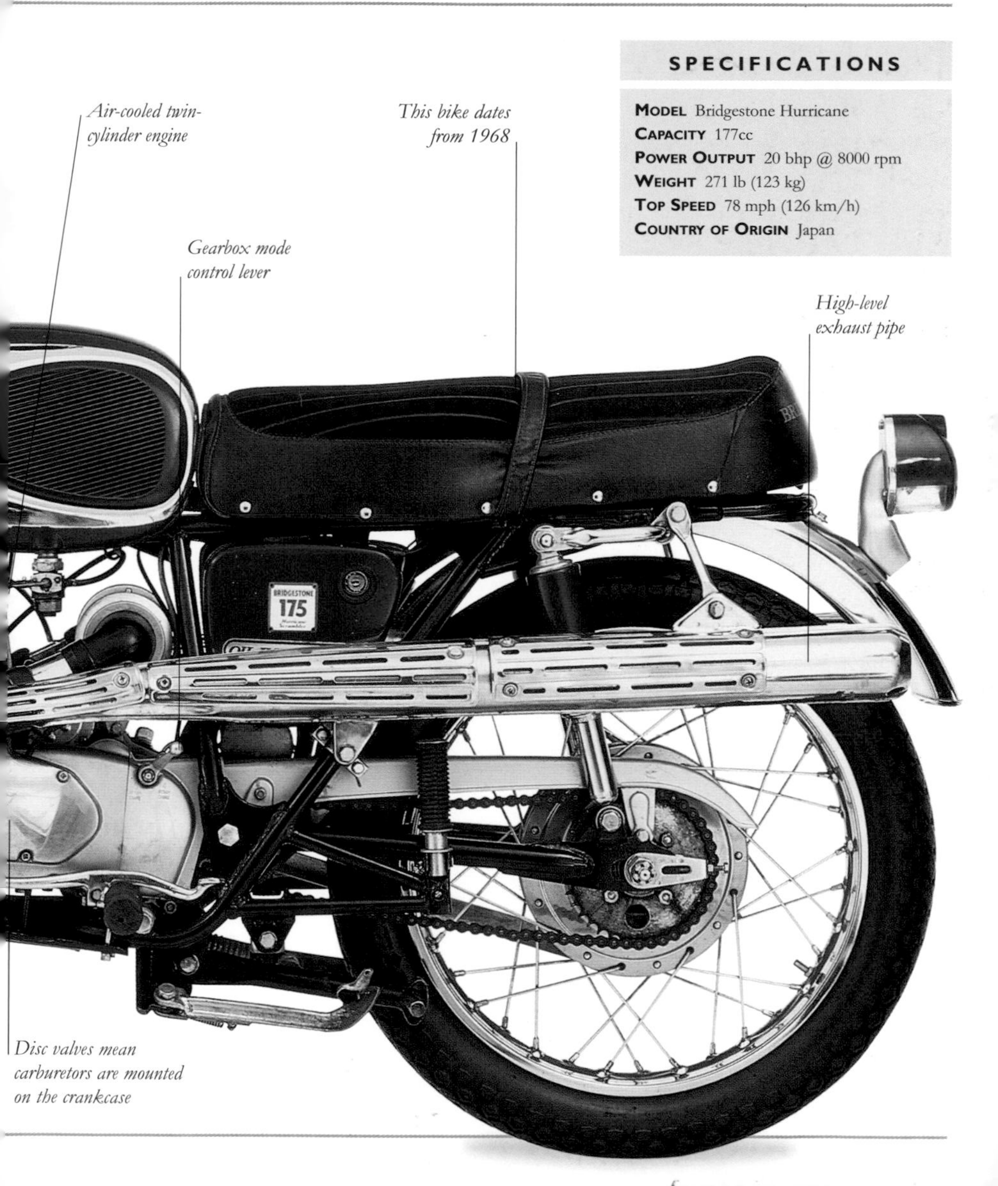

Air-cooled twin-cylinder engine

This bike dates from 1968

Gearbox mode control lever

High-level exhaust pipe

Disc valves mean carburetors are mounted on the crankcase

BRITTEN *V1000*

THE BRITTEN IS AN extraordinary bike built by an extraordinary man. John Britten ignored convention when he built this bike and came up with an elegant and inspired machine. The 60° V-twin engine is an integral part of the chassis to which the front and rear suspension are bolted. In fact, almost every part of the Britten performs at least two functions. It also looks great. The Britten performed well in International Battle of the Twins racing, and in the early Nineties, was practically unbeatable in this field, even when competing against factory Ducatis. Two great tragedies were that it was never able to race in any more worthy competition (only 11 bikes were built, ruling it out of Superbike racing), and that its creator, the amazing John Britten, died before the bike could show its real potential.

Radiator is concealed beneath the rider's seat

Three-spoke alloy wheel

Rear brake caliper

SPECIFICATIONS

MODEL Britten V1000
CAPACITY 985cc
POWER OUTPUT 165 bhp
WEIGHT 320 lb (145 kg)
TOP SPEED 189 mph (304 km/h)
COUNTRY OF ORIGIN New Zealand

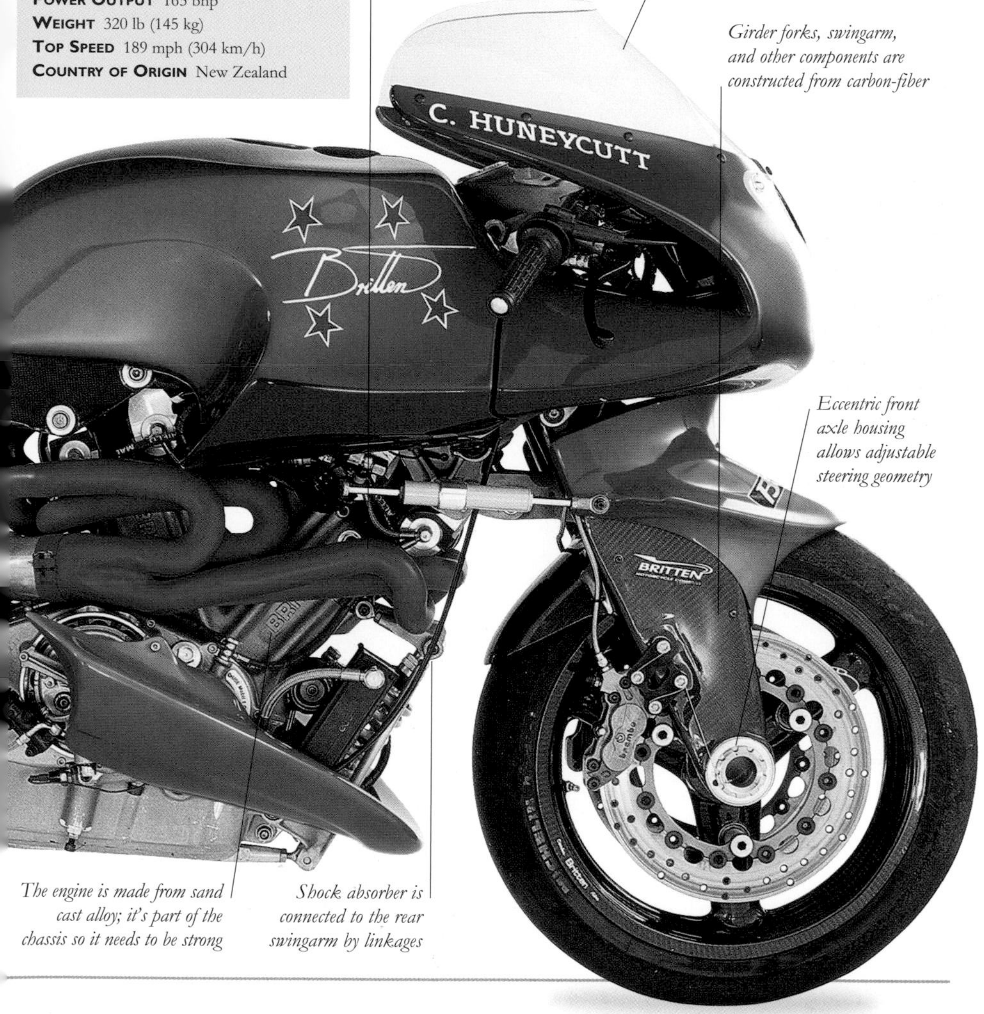
Intricate exhaust system is sculpted to achieve maximum performance; the extraordinary appearance is a bonus
Bikini fairing
Girder forks, swingarm, and other components are constructed from carbon-fiber
Eccentric front axle housing allows adjustable steering geometry
The engine is made from sand cast alloy; it's part of the chassis so it needs to be strong
Shock absorber is connected to the rear swingarm by linkages
C. HUNEYCUTT
Britten
BRITTEN

Brough *Superior Dream*

Master publicist George Brough repeatedly stunned visitors to Britain's annual motorcycle show with a series of innovative four-cylinder prototypes. His most remarkable creation came in 1938. The Brough Superior Dream was, in essence, two flat twins mounted one above the other. They shared a common crankcase, and both crankshafts were geared together. The gearbox bolted to the rear of the engine and drove the rear wheel by shaft. Only two or three prototypes were reputedly built before World War II ended the project, a low figure even by Brough's standards.

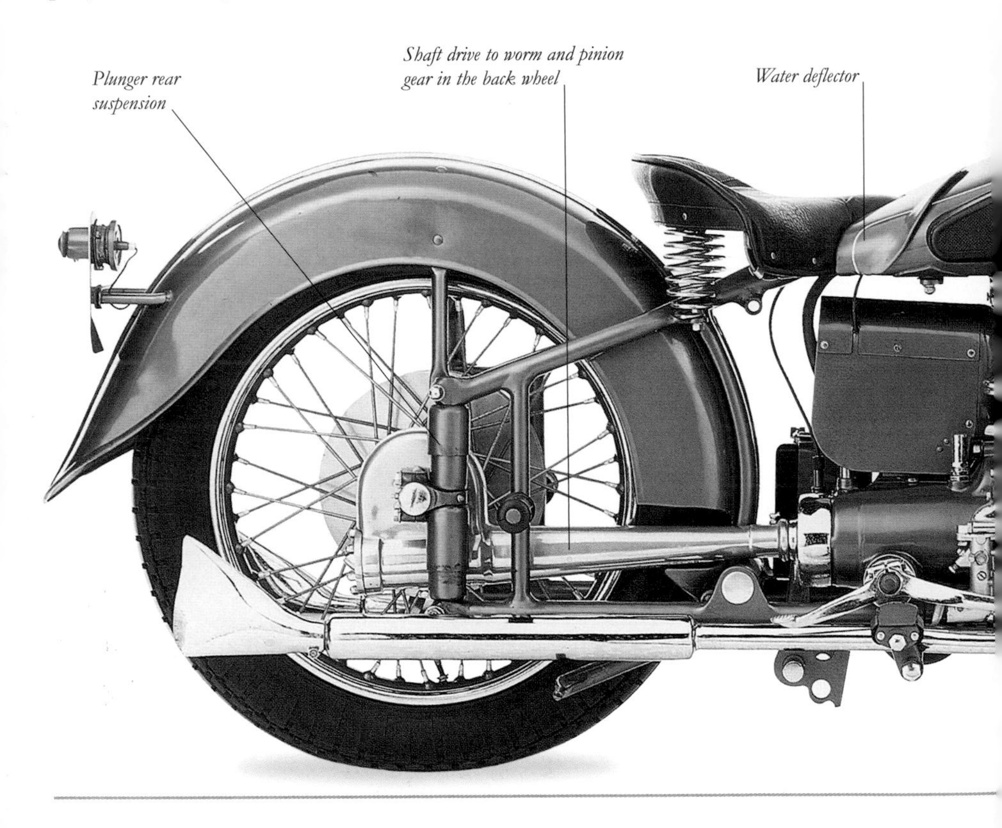

SPECIFICATIONS

MODEL Brough Superior Dream
CAPACITY 998cc
POWER OUTPUT Not known
WEIGHT Not known
TOP SPEED Not known
COUNTRY OF ORIGIN UK

TYPICAL FLAT-TWIN

The Dream inherited many of the pros and cons of the flat-twin layout. The cylinders were well positioned for air cooling and accident damage. The twin camshafts ran between the cylinder bores, operating the valves by pushrods fore and aft of the barrels.

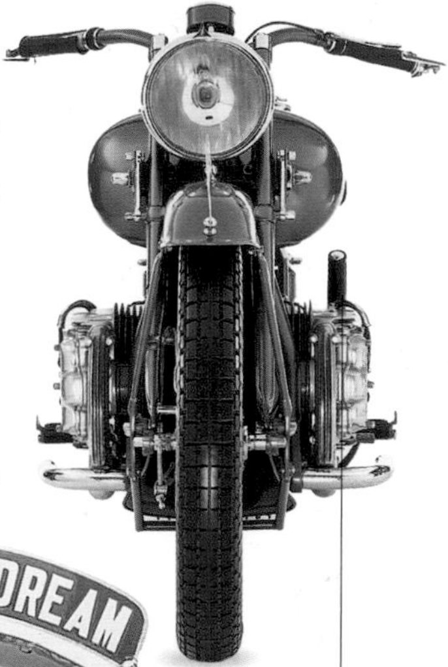

The engine was put into a duplex tubular cradle frame

Lever operates center stand

BSA *Sloper*

The sporty BSA Sloper machines launched in 1926 were so named to reflect the foward tilt of the engines in their frames. Six models were produced in all, in 350cc, 500cc, and 600cc engine capacities. The first model was called the S27 o.h.v. and was housed in a twin downtube frame. The 493cc single had wet-sump lubrication, a gear-driven magneto in front of the cylinder, and a single port head with two valves. By 1929 the Sloper came with a twin-exhaust port head and a 349cc model was available, but it was only marginally cheaper than the 500cc. This eye-catching and unusually quiet sports machine set new standards in motorcycle construction and style. The bike on this page is a 1930 model.

Specifications

Model BSA Sloper
Capacity 493cc
Power Output 20 bhp (est.)
Weight 337 lb (153 kg)
Top Speed 70–75 mph (113–120 km/h)
Country of Origin UK

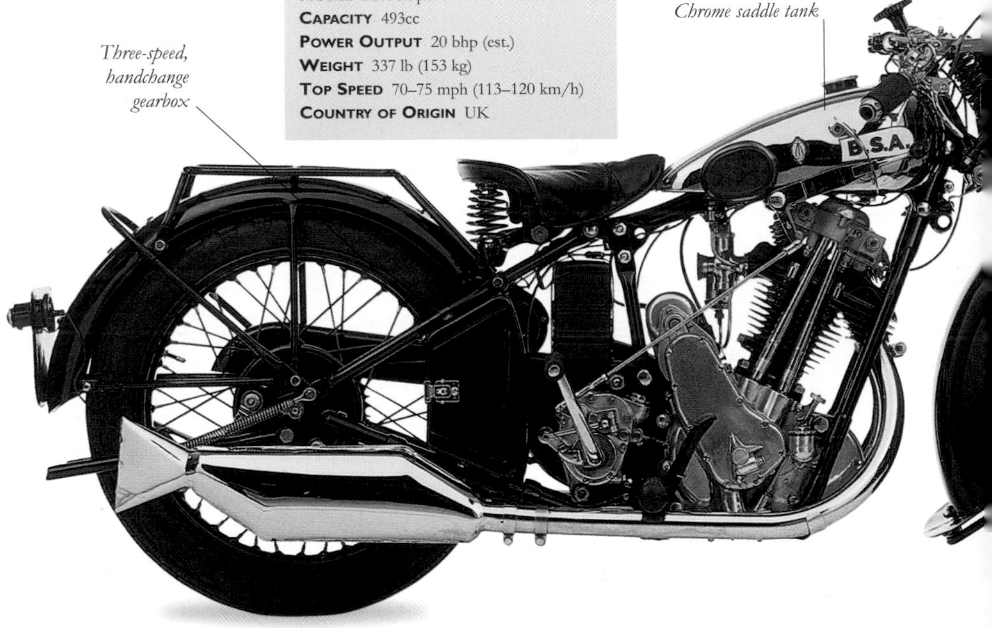

Three-speed, handchange gearbox

Chrome saddle tank

The Star Twin produced 4 bhp more than the standard A7

SPECIFICATIONS

Model BSA Star Twin
Capacity 495cc
Power Output 29 bhp @ 6000 rpm
Weight 375 lb (170 kg)
Top Speed 86 mph (138 km/h)
Country of Origin UK

BSA *Star Twin*

The launch of Triumph's innovative Speed Twin in 1937 *(see pp.408–09)* set British manufacturers in fierce competition with each other to produce similar twin-cylinder bikes. World War II prevented BSA from launching its entry until 1946. It was a 495cc o.h.v. twin that produced 25 bhp at 5800 rpm. The Star Twin is a sports model of the original A7. It only lasted until 1950, when the 650cc A10 took its place.

BSA *Gold Star DBD34*

Arguably the most evocative name in motorcycle history, the Gold Star took its name from a lapel pin awarded to riders who completed a race lap at Brooklands, in England, of over 161 km/h (100 mph). This was in the 1930s, and the first Gold Star was an o.h.v., all-alloy, 500cc single. It soon became known for its versatility, being equally competitive on and off road, and by 1960 it had become the chosen mount for racing riders. They demanded essential modifications, so clip-on handlebars and a swept-back exhaust pipe were used on the top-of-the-line Clubman model. Other classic Gold Star fixtures included an RRT2 close-ratio gearbox, an Amal 1½-in (38-mm) Grand Prix carburetor, a distinctive muffler, and a 7½-in (19-cm) front drum brake.

A swept-back exhaust pipe was attached to the top-of-the-line Clubman models

SPECIFICATIONS

Model BSA Gold Star DBD34
Capacity 499cc
Power Output 40 bhp @ 7000 rpm
Weight 308 lb (140 kg)
Top Speed 110 mph (177 km/h)
Country of Origin UK

Ride Quality

A ride on a typical Gold Star is characterized by tall gearing, which means awkward getaways and slipping the clutch at slow speeds. But once on the move, a speed of nearly 90 mph (145 km/h) in second should be possible.

Racing-style clip-on handlebars were an option chosen by the majority of buyers

BSA Gold Star pin

All-alloy engine was different from singles in BSA's line, which still used iron cylinder barrels

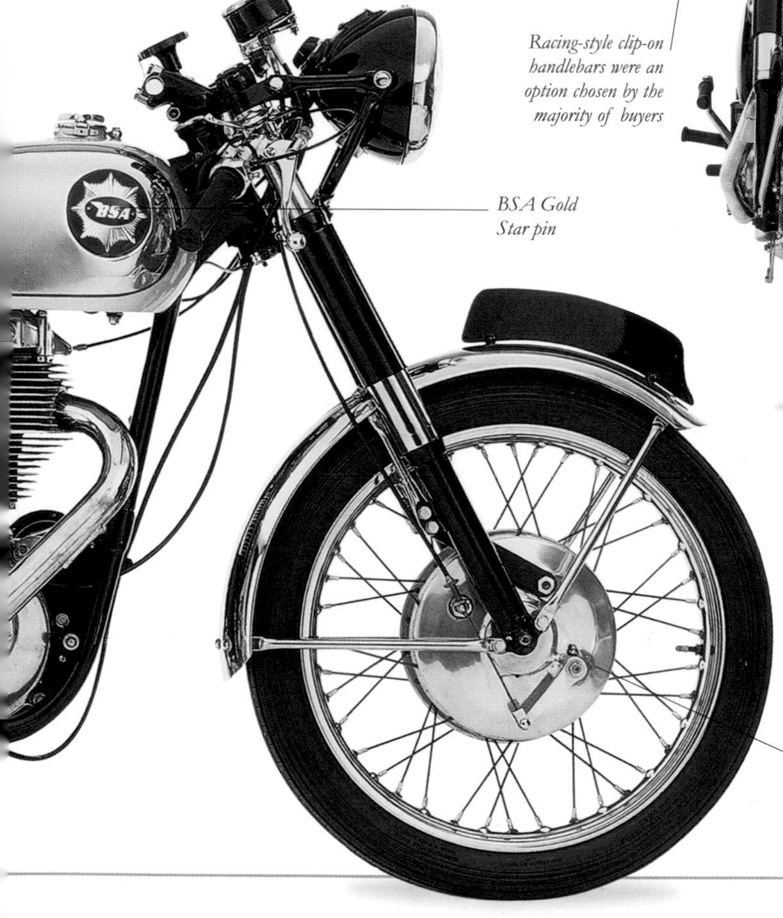

7½-in (19-cm) front drum brake

BSA *A65L*

UNIT CONSTRUCTION WAS adopted for BSA's 499cc (A50) and 654cc (A65) twins in 1962. A hump-backed seat, twin carburetors, and tach put the Lightning Clubman version of the A65 firmly in the sports class. Unfortunately for BSA, Triumph's 650cc Bonneville *(see p.411)* and Norton's 650SS *(see p.325)* had the performance edge, but the A65 was still a fine bike. Its smooth lines were reflected in the oval cases of the unit-construction engine and gearbox. Before production ceased, the Lightning was made into a 750cc version—the US-only A70L—by lengthening the stroke to 85 mm.

SPECIFICATIONS

MODEL BSA A65L
CAPACITY 654cc
POWER OUTPUT 53 bhp @ 7000 rpm
WEIGHT 421 lb (191 kg)
TOP SPEED 115 mph (185 km/h)
COUNTRY OF ORIGIN UK

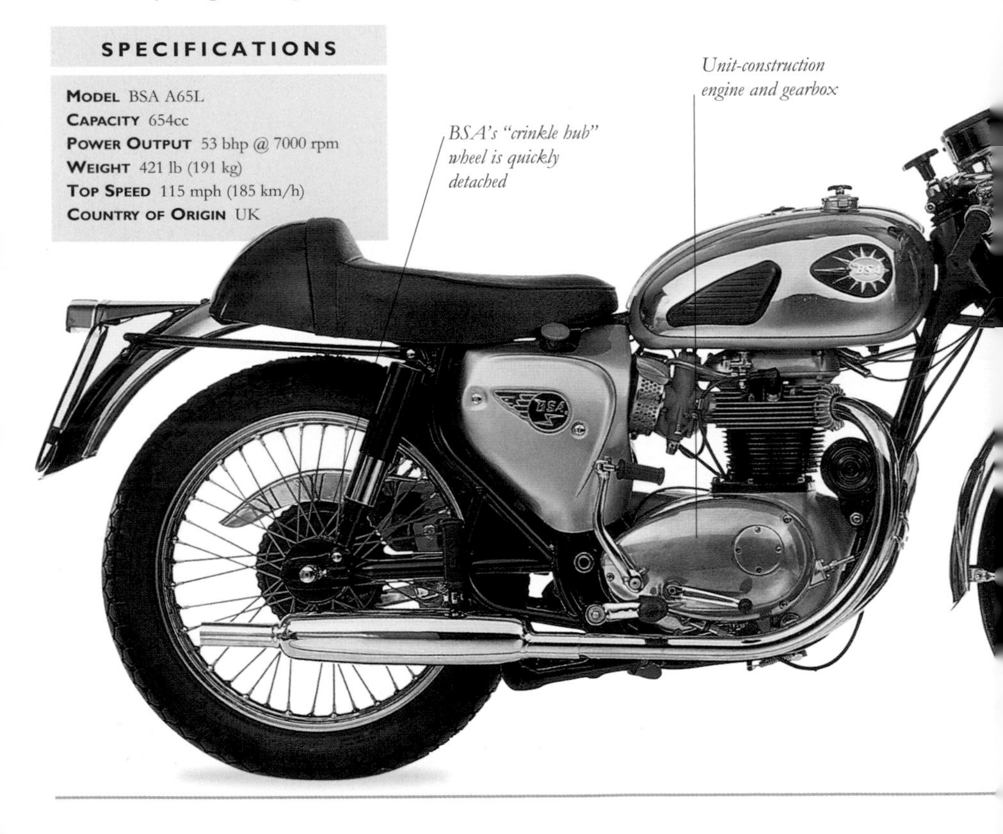

BSA's "crinkle hub" wheel is quickly detached

Unit-construction engine and gearbox

"Raygun" mufflers indicate an attempt at futuristic styling

Inclined cylinders distinguish BSA's 740cc triple from Triumph's version

Ground clearance on the Rocket 3 was poor

SPECIFICATIONS

Model BSA Rocket 3
Capacity 740cc
Power Output 60 bhp @ 7250 rpm
Weight 490 lb (222 kg)
Top Speed 122 mph (196 km/h)
Country of Origin UK

BSA *Rocket 3*

The three-cylindered Rocket 3—or A75—was launched in Britain in 1969. The BSA/Triumph group had to act quickly to catch up with Honda, which was known to be developing a four. The Rocket 3 and Triumph Trident *(see pp.414–15)* were similar, though BSA's engine design had its cylinder block tilted forward where the Triumph's was vertical. This bike is from 1970; BSA ceased production the following year.

BSA *Rocket 3 Racer*

Despite the short life of the Rocket 3 road bike, the racing version was a unqualified success, outpacing bikes such as Honda's 750 four on the track. Its high point came at Mallory Park's "Race of the Year" in 1971, when British rider John Cooper beat the world champion Giacomo Agostini, who was riding an MV. Good performances both in the United States and in Britain by riders like Mike Hailwood, Percy Tait, and Ray Pickrell bolstered the Rocket's reputation. Part of the Racer's success was due to the excellent handling given by the frames, which were specially built by Rob North for the Rocket 3 and Triumph Trident *(see pp.416–17)* racers.

SPECIFICATIONS

Model BSA Rocket 3 Racer
Capacity 740cc
Power Output 69 bhp @ 8500 rpm
Weight 380 lb (172 kg)
Top Speed 152 mph (245 km/h)
Country of Origin UK

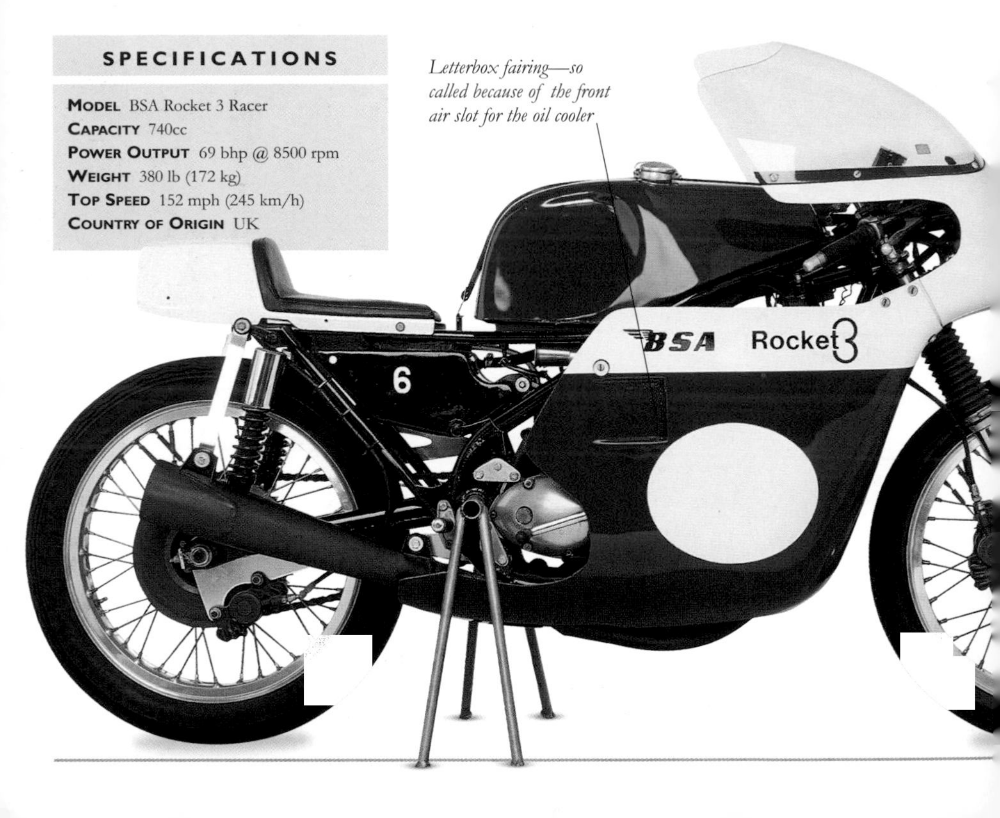

SPECIFICATIONS

MODEL Buell RR1000
CAPACITY 61cu. in. (998cc)
POWER OUTPUT 77 bhp @ 5600 rpm
WEIGHT 390 lb (177 kg)
TOP SPEED 135 mph (217 km/h) (est.)
COUNTRY OF ORIGIN US

Bulges reflect the position of the carburetors on the XR1000 engine

Aerodynamic tinted windshield

Italian Marzocchi forks

BUELL *RR1000*

THE FIRST RR1000 PROTOTYPE was built in 1984 by Eric Buell, a former Harley-Davidson employee, as a commission from the Vetter fairing company. Although Buell was still independent of Harley-Davidson at this point, the company would soon be incorporated into the Harley fold. The RR1000 used an engine from a Harley XR1000 *(see pp.172–73)* used in Buell's patented Uniplanar chassis, which restricted engine vibration by using a system of rods, joints, and rubber mountings. Only 50 RR1000s were built before the supply of XR1000 engines dried up. Buell also offered a 1200cc Sportster-engined version of the RR.

BUELL *X1 Lightning*

WHEN HARLEY BOUGHT ANOTHER chunk of Buell in the late 1990s—taking its stake in the sports bike company to 98 percent—the X1 Lightning followed soon afterward. Launched in 1998 for the 1999 model year, the Lightning was more polished and refined than earlier Buells, while retaining the oddball looks of the older machines. The ride was improved, the styling was cleaned up, and a new electronic fuel-injection system was added to the Sportster engine. But in refining the Lightning some of the raw charm of the earlier bikes was lost.

Alloy seat support subframe new for 1999

Lightweight carbon-fiber hugger and belt guard

17-in (43-cm) wheel with Dunlop Sportmax tire

SPECIFICATIONS

MODEL Buell X1 Lightning
CAPACITY 73cu. in. (1203cc)
POWER OUTPUT 101 bhp @ 6000 rpm
WEIGHT 200 kg (441 lb)
TOP SPEED 140 mph (225 km/h)
COUNTRY OF ORIGIN US

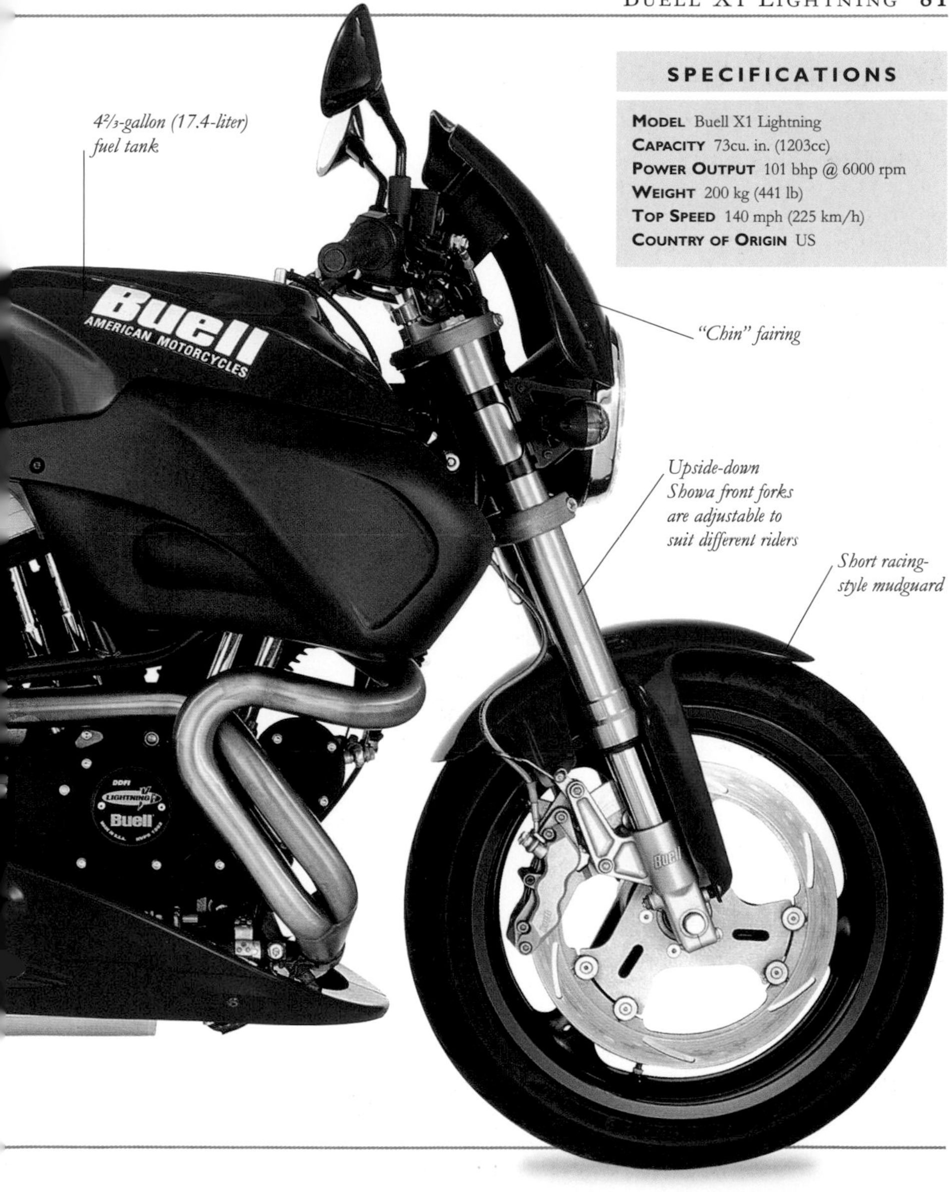
4⅔-gallon (17.4-liter) fuel tank
"Chin" fairing
Upside-down Showa front forks are adjustable to suit different riders
Short racing-style mudguard
Buell
AMERICAN MOTORCYCLES

BULTACO *Sherpa*

DEVELOPED BY SAMMY MILLER and introduced late in 1964, the Bultaco Sherpa was so successful that most other trials machines were immediately rendered obsolete. Never before had a combination of low-speed pulling power, hill-climbing ability, and instant throttle response, allied to docility when required, been available in one machine. Using many existing parts—basic engine, wheels, and forks—it was claimed that the prototype progressed from concept to finished product in just 12 days. The Sherpa shown here is Sammy Miller's development machine from 1964.

SPECIFICATIONS

MODEL Bultaco Sherpa
CAPACITY 244cc
POWER OUTPUT 18 bhp
WEIGHT 204 lb (93 kg)
TOP SPEED Not known
COUNTRY OF ORIGIN Spain

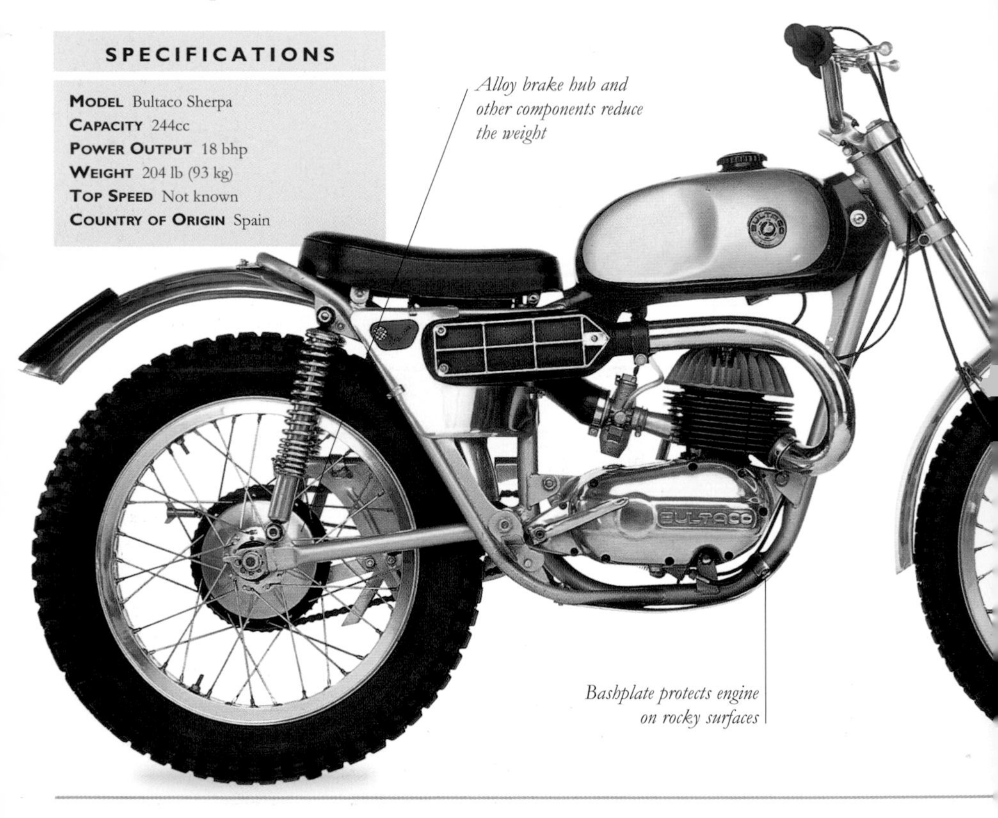

Enclosed final drive chain

Simple, piston-ported, two-stroke engine runs on gas/oil mixture

This bike dates from 1975

SPECIFICATIONS

MODEL Bultaco Metralla
CAPACITY 244cc
POWER OUTPUT 25 bhp
WEIGHT 271 lb (123 kg)
TOP SPEED 85 mph (137 km/h)
COUNTRY OF ORIGIN Spain

Air scoop for twin leading-shoe front drum brake

BULTACO *Metralla*

IN CONTRAST TO THE refined 250cc Japanese sports machines, the Spanish sports roadster was basically a motocross engine attached to race-bred cycle parts. Its success was demonstrated when Metrallas secured the first three places in the 250cc production class at the 1967 TT. The engine was devoid of performance aids such as reed or rotary valves. A six-speed box and expansion-chamber exhaust completed the specification.

Bultaco *Alpina*

When interest in off-road motorcycling rose in the early 1970s, Bultaco had a well-established reputation and a range of trials and motocross machines. The firm was in an excellent position to take advantage of the market with proven products. The Alpina probably appealed to the leisure rider who wanted to dabble in competition rather than the serious competitor. It incorporated the same five-speed 244cc engine as in the Sherpa *(see p.82)*, and though more of a trail bike than an enduro machine, it was still pretty good for either usage. The bike illustrated is a 1975 model.

Specifications

Model Bultaco Alpina
Capacity 244cc
Power Output 19 bhp
Weight 240 lb (109 kg)
Top Speed Not known
Country of Origin Spain

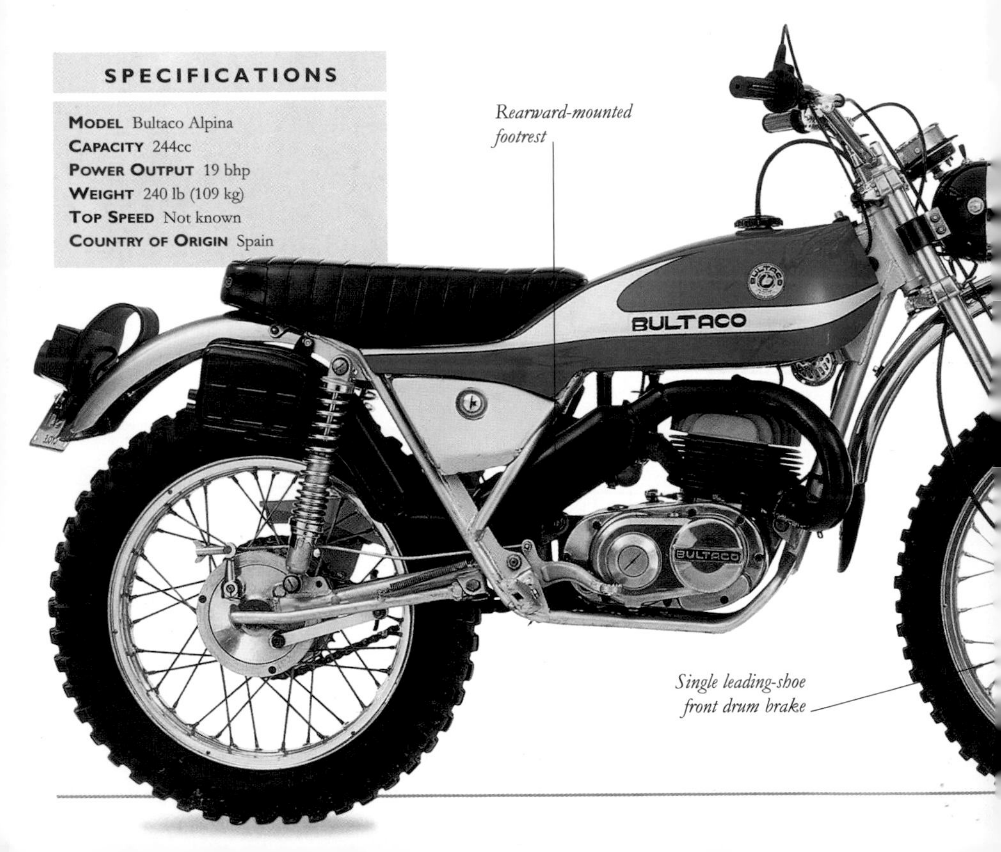

High-level alloy mudguard

Three-spoke wheel

SPECIFICATIONS

MODEL Cagiva Mito
CAPACITY 124cc
POWER OUTPUT 30 bhp @ 11,000 rpm
WEIGHT 275 lb (125 kg)
TOP SPEED 105 mph (169 km/h) (est.)
COUNTRY OF ORIGIN Italy

CAGIVA *Mito*

FIFTEEN YEARS AFTER Cagiva's SST, the 125 market had moved on. Buyers demanded looks and performance as delivered by bikes like the 1994 Cagiva Mito. In return, motorcycle manufacturers demanded that buyers have fat wallets. The price for "just" a 125 was high. The comprehensive specification, stunning performance, and superlative handling almost justified it. The Mito was styled to look like the 916 Ducati *(see pp.110–11)* and had a twin-spar alloy frame.

CAGIVA *V-Raptor*

THE DUCATI MONSTER *(see p.108)* created a new class when it first appeared in 1992. Eight years later the Cagiva Raptor arrived looking for a piece of the same market. Both bikes were designed by the same man, Miguel Galluzzi, thereby explaining the similarity in appearance. A 90° V-twin engine and tubular trellis frame are essential to the concept of the machine. Cagiva couldn't use the Ducati engine so they used Suzuki engines, which were much more powerful than the two-valve unit used in the original Monster.

Sculpted 4-gallon (18-liter) fuel tank

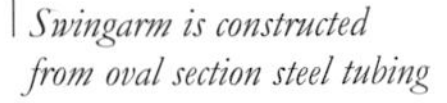

Swingarm is constructed from oval section steel tubing

SPECIFICATIONS

MODEL Cagiva V-Raptor
CAPACITY 996cc
POWER OUTPUT 98 bhp
WEIGHT 423 lb (192 kg)
TOP SPEED 138 mph (222 km/h)
COUNTRY OF ORIGIN Italy

Art deco–style triangular tachometer

Style of the tachometer as well as ribbing on passenger seat footrest scuff plates are unique to Raptors

Radiator

Aerodynamically styled exhaust

MODEL DIFFERENCES

The V-Raptor has different handlebars and mirrors from the basic Raptor model. It also has an extraordinary headlight fairing arrangement featuring indicators mounted in sidepods and plastic spars over the steering head.

Brembo brake caliper

Suzuki TL1000 engine uses fuel injection. A 650 version is also available

CLEVELAND *Tornado*

AFTER PRODUCING PROBABLY the most successful two-stroke to be built in the United States, Cleveland built its first four in 1925, based on a design by L.E. Fowler. The new machine was not a success and was replaced the following season by a completely new four designed by Everitt DeLong. Original capacity was 45cu. in. (737cc), but by the time this improved Tornado model appeared in 1929, capacity had been increased to 61cu. in. (1000cc). It meant that the bike could output approximately 31 bhp and reach a speed of 100 mph (161 km/h). The machine used i.o.e. valve gearing—which achieved greater power by incorporating alloy pistons and bigger valves—and a three-speed handchange gearbox with chain final drive.

SPECIFICATIONS

MODEL Cleveland Tornado
CAPACITY 61cu. in. (1000cc)
POWER OUTPUT 31 bhp (est.)
WEIGHT 540 lb (245 kg)
TOP SPEED 100 mph (161 km/h)
COUNTRY OF ORIGIN US

The main stand pivots at the end of the rigid frame

Large, wet-sump engine with horizontally split crankcase

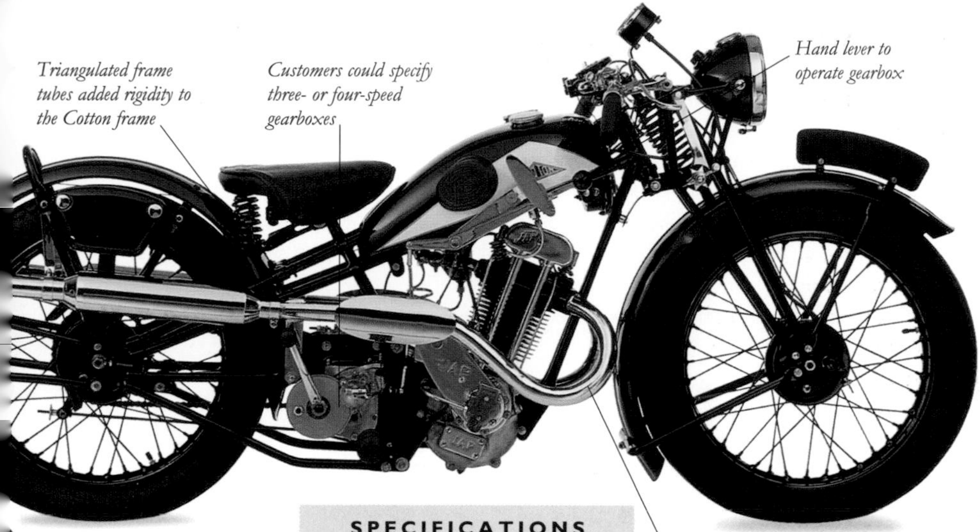

Triangulated frame tubes added rigidity to the Cotton frame

Customers could specify three- or four-speed gearboxes

Hand lever to operate gearbox

SPECIFICATIONS
MODEL Cotton 500
CAPACITY 490cc
POWER OUTPUT Not known
WEIGHT 255 lb (116 kg)
TOP SPEED 75 mph (121 km/h) (est.)
COUNTRY OF ORIGIN UK

Leading-link forks

Exhaust pipe exits engine and rises up at the back

COTTON *500*

THE DISTINCTIVE FEATURE of Cotton machines was the triangulated frame developed and patented by Frank Willoughby Cotton. The steering head was connected to the rear axle by four straight tubes. This structure made the frame very rigid, thus endowing Cotton machines with excellent handling. Equipped with Blackburne o.h.v. engines, the bikes achieved considerable success in competition in the 1920s, including winning the 1923 Junior TT. This machine dates from 1937.

CUSHMAN *Eagle*

AN EXTENSION OF THE CRAZE for conventional scooters that swept America in the 1950s was a fashion for small-wheeled motorcycles that were a little bit different from the average offerings of the time. These combined the small wheels and the industrial engines of scooters with styling inspired (loosely) by big Indians and Harley-Davidsons. The Eagle was Cushman's offering in this class, in which the other key players were the Mustang and the Powell A-V-8. The first Eagle appeared in 1949, and the line continued until the firm's production stopped in 1965. The machine pictured here is a 1958 model.

SPECIFICATIONS

MODEL Cushman Eagle
CAPACITY 19.44cu. in. (319cc)
POWER OUTPUT 8 bhp
WEIGHT Not known
TOP SPEED Not known
COUNTRY OF ORIGIN US

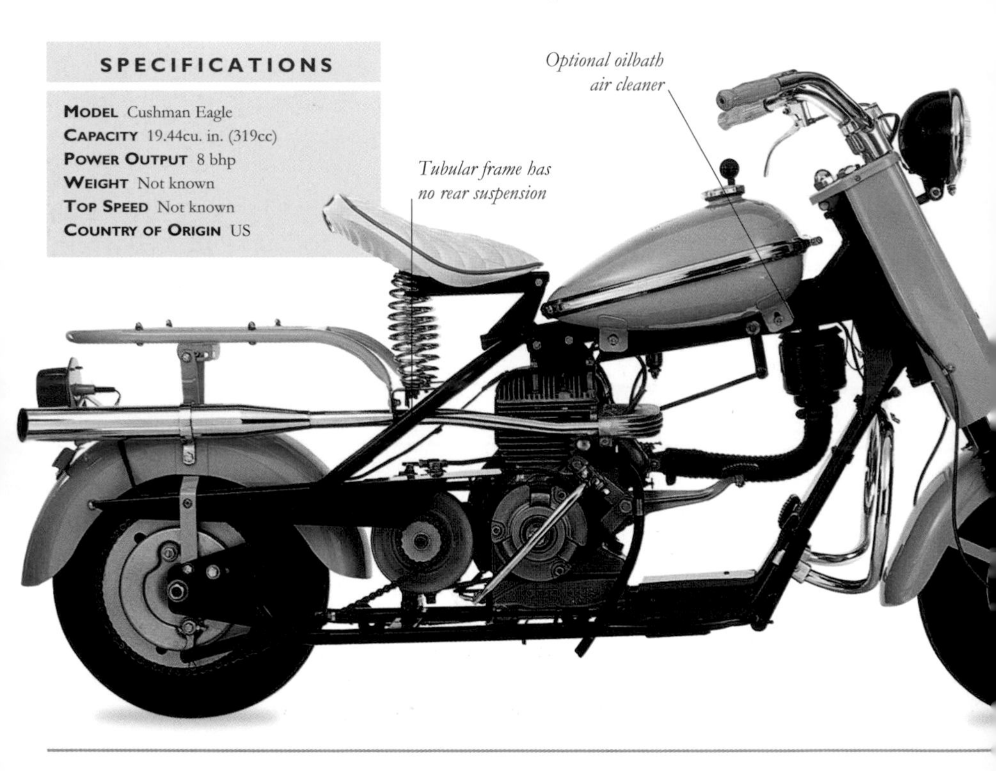

SPECIFICATIONS

Model CZ 380 Motocrosser
Capacity 380cc
Power Output 42 bhp @ 6800 rpm
Weight Not known
Top Speed 75 mph (121 km/h)
Country of Origin Czech Republic

Braced handlebars

Light alloy wheel rim

Knobby motocross tire

Radially finned cylinder head

Under-engine expansion chamber exhaust pipe

CZ *380 Motocrosser*

CZ WAS NATIONALIZED after the war but continued to build its two-stroke machines as well as o.h.c. road racers. The CZ was typical of period machines in the early Seventies—a light and powerful air-cooled two-stroke engine in a tubular chassis with twin-shock rear suspension and drum brakes, and an exhaust that ran under the engine. Many of these features were obsolete by the end of the decade as different manufacturers, both Japanese and European, tried to gain a competitive edge. Unfortunately, while other names rose to prominence CZ slipped into the background. The company's heyday was the Sixties and early Seventies.

DAIMLER *Einspur*

THE GERMAN WORD *EINSPURIG* means "single track," an inaccurate description of this early machine, as it was equipped with spring-loaded outrigger wheels to keep it upright; these were necessary because the saddle was so high above the engine that the rider's feet could not reach the ground. The top of the engine projected up between the frame members, and above it was mounted a curved saddle more suited to the back of a horse. On the first machine, the handlebars were attached to a tiller, but this was later replaced by the link arrangement shown here. The engine had an automatic inlet valve fed from a surface carburetor.

SPECIFICATIONS

MODEL Daimler Einspur
CAPACITY 264cc
POWER OUTPUT 0.5 bhp @ 700 rpm
WEIGHT 198 lb (90 kg)
TOP SPEED 7½ mph (12 km/h)
COUNTRY OF ORIGIN Germany

Ignition is by hot tube, heated by an enclosed burner attached to the cylinder head

Engine has a crankshaft of two flywheels joined by a crankpin and is enclosed within a cast-aluminum crankcase

Crude clutch is provided by a movable jockey wheel bearing against the belt to vary its tension

This 1929 model has no rear suspension: a sprung saddle is attached to the rigid frame

Radiator for water-cooling system

Separate three-speed gearbox

The original Einspur was constructed in 1885 simply as a mobile testing ground for the gas engine that the German engineer and inventor Gottlieb Daimler was developing.

SPECIFICATIONS

Model DKW ZSW500
Capacity 494cc
Power Output 14 bhp @ 4000 rpm
Weight 330 lb (150 kg)
Top Speed 65 mph (105 km/h)
Country of Origin Germany

DKW *ZSW500*

By the late 1920s, DKW was the biggest motorcycle manufacturer in the world. It offered a wide range of two-stroke machines from lightweight singles to large water-cooled twins. Like many manufacturers during the 1920s, it even experimented with scooter design. The ZSW500, though, with its flat tank design and separate gearbox, appealed to buyers suspicious of trendy innovations of the time, such as pressed-steel frames. The water-cooled engine coped well when, as was common, the bike was used with a sidecar.

DKW *SS250*

ALTHOUGH THIS 1939 PRODUCTION RACER was inspired by specially built factory bikes, the SS250 lacked the sophistication of those machines. To reduce production costs the commercially available SS250 was equipped with a horizontal displacer piston to replace cylindrical rotary or reed valves. Factory bikes also tried various engine designs, but the SS250 used only a split single. The combination of the limited braking effect available from the SS250's two-stroke engine and its substantial weight resulted in DKW paying particular attention to brake design. Wide and well-constructed alloy wheel hubs housed similarly generous brake linings. The SS250 was confined to national racing until 1951, and as a result, few have survived.

The engine is mounted in a duplex frame with swingarm rear suspension linked to plungers

SPECIFICATIONS

Model DKW SS250
Capacity 243cc
Power Output 21 bhp @ 5000 rpm
Weight 310 lb (141 kg)
Top Speed 90 mph (145 km/h)
Country of Origin Germany

Restricted Power

The extra charge drawn into the SS250's engine by the displacer piston did not produce the intended increase in power output. Peak revolutions were limited due to the long path of combustion gases through the engine.

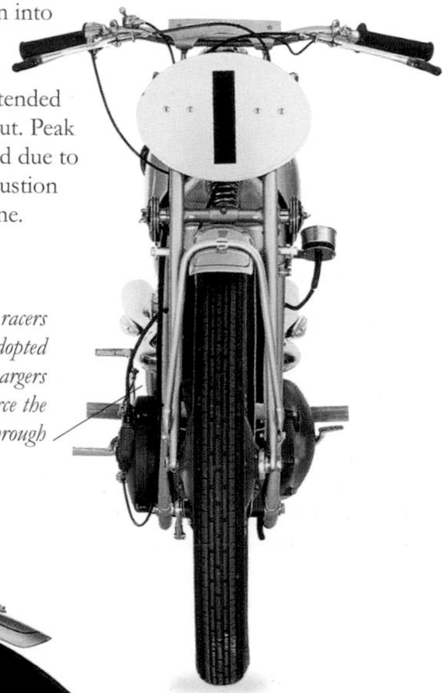

A large-capacity fuel (and oil) tank was essential on a machine that achieved as little as 15 mpg (5 km/l)

Factory racers eventually adopted rotary superchargers to force the mixture through

As it was only sold in racing specification, the SS250 was not equipped with any accessories

Horizontal displacer cylinder increases the swept volume of the crankcase

Specially designed, wide, alloy brake drums

DOT *Supersports*

THE DOT COMPANY was founded in 1902 by pioneer motorcycle racer Harry Reed. Like many British manufacturers, Dot relied on engines it purchased throughout its 60-year history. During the 1920s JAP and Blackburne were its major suppliers of four-stroke engines; however, this Supersports model of 1923 has an o.h.v. Bradshaw engine with oil cooling. This feature was nicknamed "the Oilboiler." The otherwise orthodox Bradshaw-engined model remained in the Dot catalog for five years. Founder and owner of Dot, Harry Reed, won a solo TT race in 1908 for his company.

SPECIFICATIONS

MODEL Dot Supersports
CAPACITY 349cc
POWER OUTPUT 2.7 bhp
WEIGHT Not known
TOP SPEED Not known
COUNTRY OF ORIGIN UK

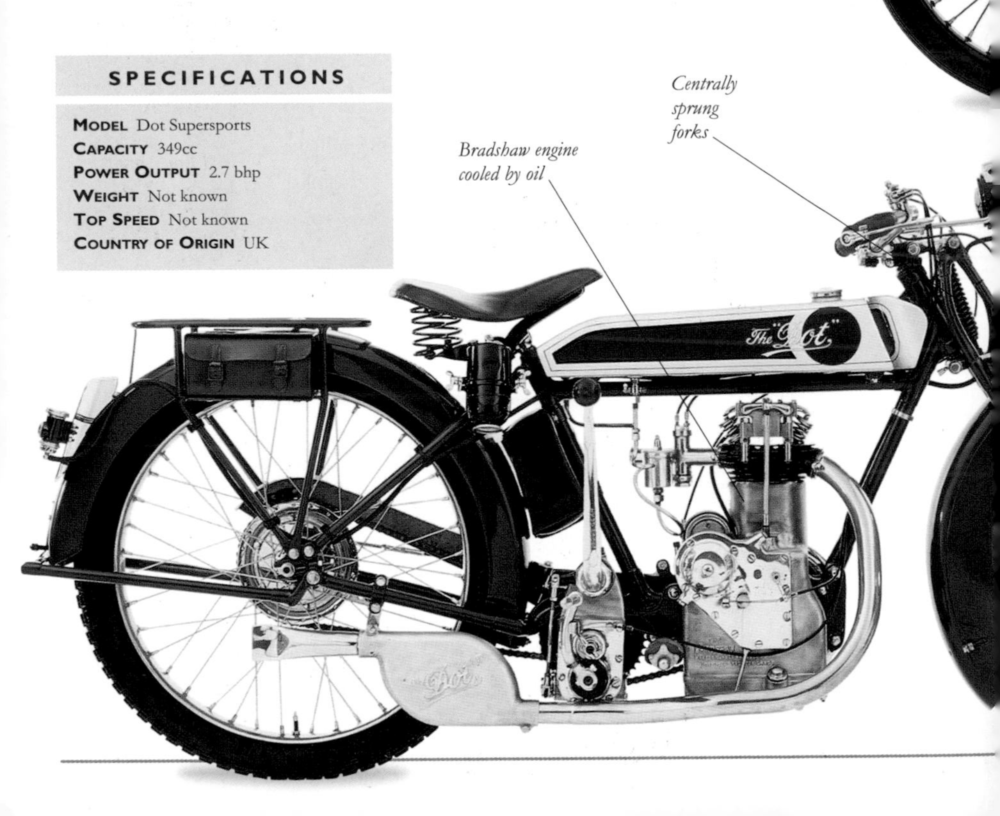

DT5 has no clutch because rolling starts were used and the three-speed gearbox remained in top gear throughout

The standard engine produced 27 bhp, but a 32 bhp engine was available at an extra cost

SPECIFICATIONS

MODEL Douglas DT5
CAPACITY 494cc
POWER OUTPUT 27 bhp
WEIGHT Not known
TOP SPEED Not known
COUNTRY OF ORIGIN UK

DOUGLAS *DT5*

SPEEDWAY RACING, or "dirt track" as it was then known (hence "DT"), came to England from Australia in 1928, although its roots were probably in the US. Top Australian riders used machines based on the Douglas RA, and soon Douglas produced a specialized speedway bike—the DT5. Long and low, it was ideal for the tracks and riding styles of the day, though the shorter Rudge- and JAP-engined machines had rendered the Douglas obsolete by 1931.

DOUGLAS *80 Plus*

UNLIKE ALL PREVIOUS DOUGLAS twins, with the exception of the short-lived Endeavour, the postwar machines had their engines mounted transversely in the frame but did not then use the more logical shaft drive. The frame was equipped with leading-link front forks and pivoted-fork rear suspension using torsion bars in the lower frame tubes, actuated via bell cranks. Two sports variants were announced for 1950: the 90 Plus was built almost to racing standards, while engines that did not meet the 28 bhp 90 Plus threshold were put into sports roadsters and designated "80 Plus." The bike pictured dates from 1953.

SPECIFICATIONS

MODEL Douglas 80 Plus
CAPACITY 348cc
POWER OUTPUT 25 bhp @ 7000 rpm
WEIGHT 393 lb (178 kg)
TOP SPEED 85 mph (137 km/h)
COUNTRY OF ORIGIN UK

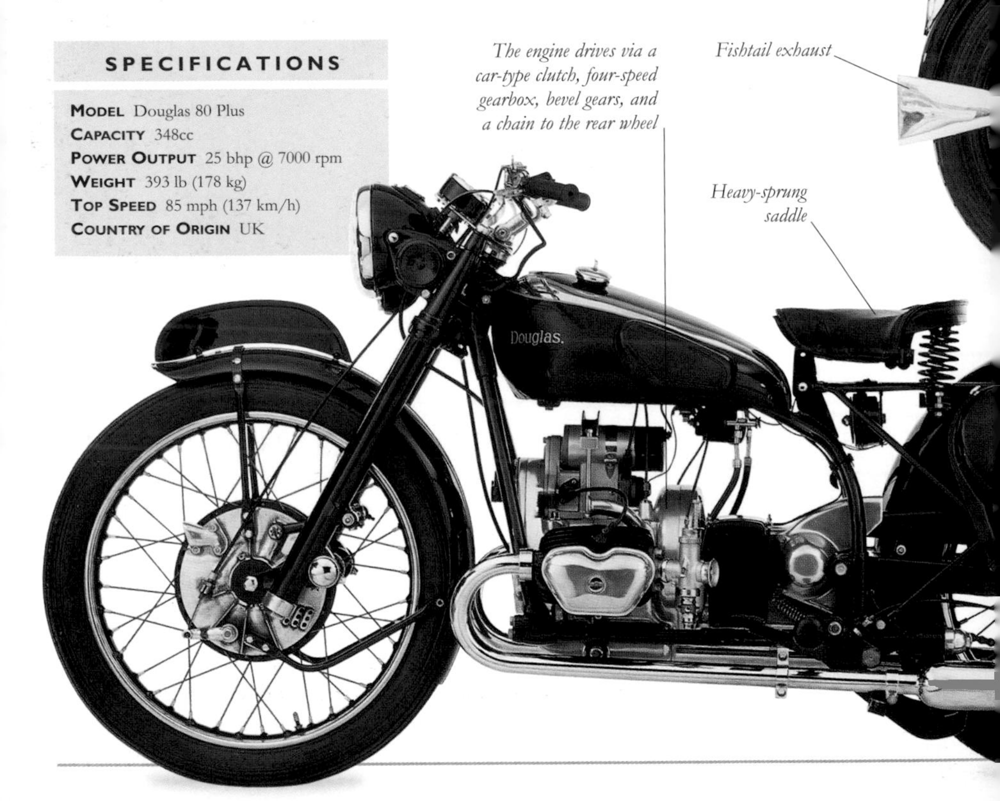

The engine drives via a car-type clutch, four-speed gearbox, bevel gears, and a chain to the rear wheel

Fishtail exhaust

Heavy-sprung saddle

SPECIFICATIONS

MODEL Dresch
CAPACITY 495cc
POWER OUTPUT 18 bhp
WEIGHT 310 lb (141 kg)
TOP SPEED 75 mph (121 km/h)
COUNTRY OF ORIGIN France

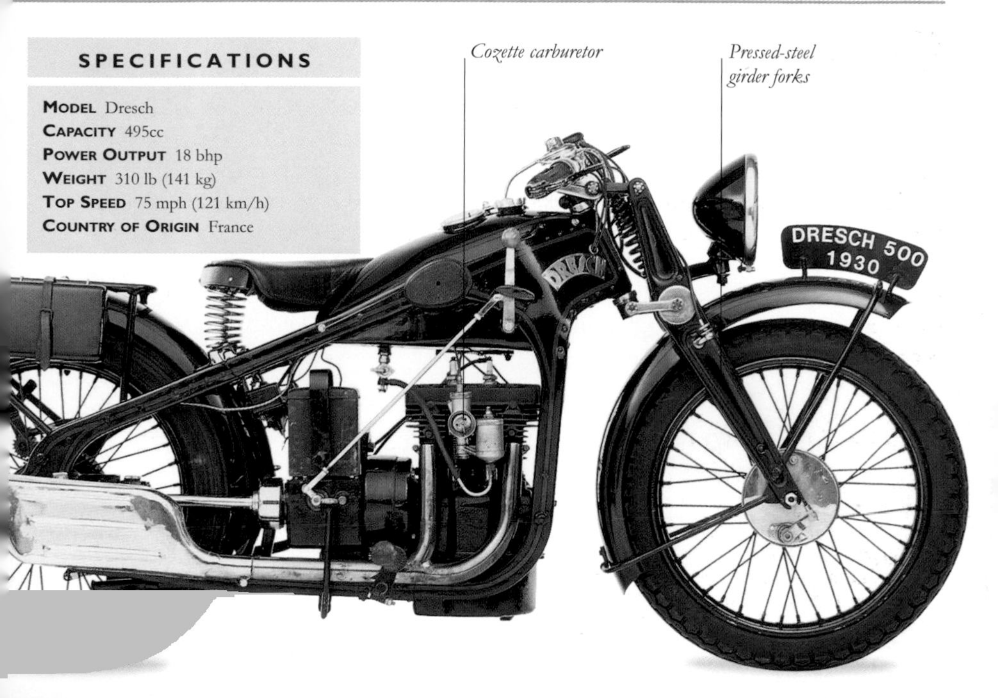

DRESCH

LAUNCHED AT THE Paris salon in 1930, the 500 Dresch twin had an impressive specification. The in-line, twin-cylinder, side-valve engine was mounted in a pressed-steel frame. There was a three-speed hand-operated gearbox and shaft final drive. The instruments—a speedometer, ammeter, and eight-day clock—were set into the fuel tank. Electric lighting and sprung handlebars came as standard. In France it was cheaper than most comparable machines. An o.h.v. version of the bike was also produced.

DUCATI *Mach 1*

INTRODUCED IN 1964, the Mach 1 was the fastest 250 road bike available at that time. The new model was based on the earlier Diana model (called the Daytona in the UK) but was equipped with a five-speed gear cluster, high compression piston, a big carburetor, and a number of other tuning aids. With its high performance and capable handling, the Mach 1 was a popular machine for competition use, and it greatly enhanced Ducati's growing reputation. The model lasted until 1967, when revised "wide-case" engines were introduced. The example on this page dates from 1964. A cheaper and less heavily tuned model was also available.

SPECIFICATIONS

MODEL Ducati Mach 1
CAPACITY 249cc
POWER OUTPUT 26 bhp @ 8500 rpm
WEIGHT 256 lb (116 kg)
TOP SPEED 106 mph (171 km/h)
COUNTRY OF ORIGIN Italy

Sports-style clip-on handlebars

Dell'Orto SS1 carburetor with open bell mouth

Leading-axle telescopic forks

SPECIFICATIONS

Model Ducati 750 Sport
Capacity 748cc
Power Output 55 bhp @ 7900 rpm
Weight 482 lb (219 kg)
Top Speed 115 mph (185 km/h)
Country of Origin Italy

One-piece single-seat and tail unit; a dual seat was an option on late model 750 Sports

Alloy wheel rim

Ducati *750 Sport*

The V-twin was a logical way for Ducati to enter the big bike class. Two of the company's o.h.c. singles were attached to a common crankshaft to produce a large-capacity machine that relied on existing Ducati technology. The 90° V-twin offered exceptional engine balance at the expense of size; the length of the engine is responsible for the 60-in (152-cm) wheelbase of the 750 Sport. Production versions of the 750 appeared in 1971. The Sport followed in 1972 and a Desmodromic machine in '73. The 90° V-twin became a Ducati trademark.

Ducati *350 Desmo*

Desmodromic valve gearing had been a feature of Ducati racers since the later 1950s. With this system the valve is closed by the camshaft rather than by a spring. Its accuracy allows high revs to be used without the risk of valve bounce. It became available on Ducati road bikes in 1971, when 250, 350, and 450cc versions of the single were produced with Desmo cylinder heads. The singles had a long production run but grew to be increasingly outdated. Production of these singles finally ceased in 1974. However, Desmodromic valvegear is still a remarkable feature of Ducati machines.

Five-speed gearbox has a right-foot shift

SPECIFICATIONS

MODEL Ducati 350 Desmo
CAPACITY 340cc
POWER OUTPUT 38 bhp @ 7500 rpm
WEIGHT 282 lb (128 kg)
TOP SPEED 105 mph (169 km/h)
COUNTRY OF ORIGIN Italy

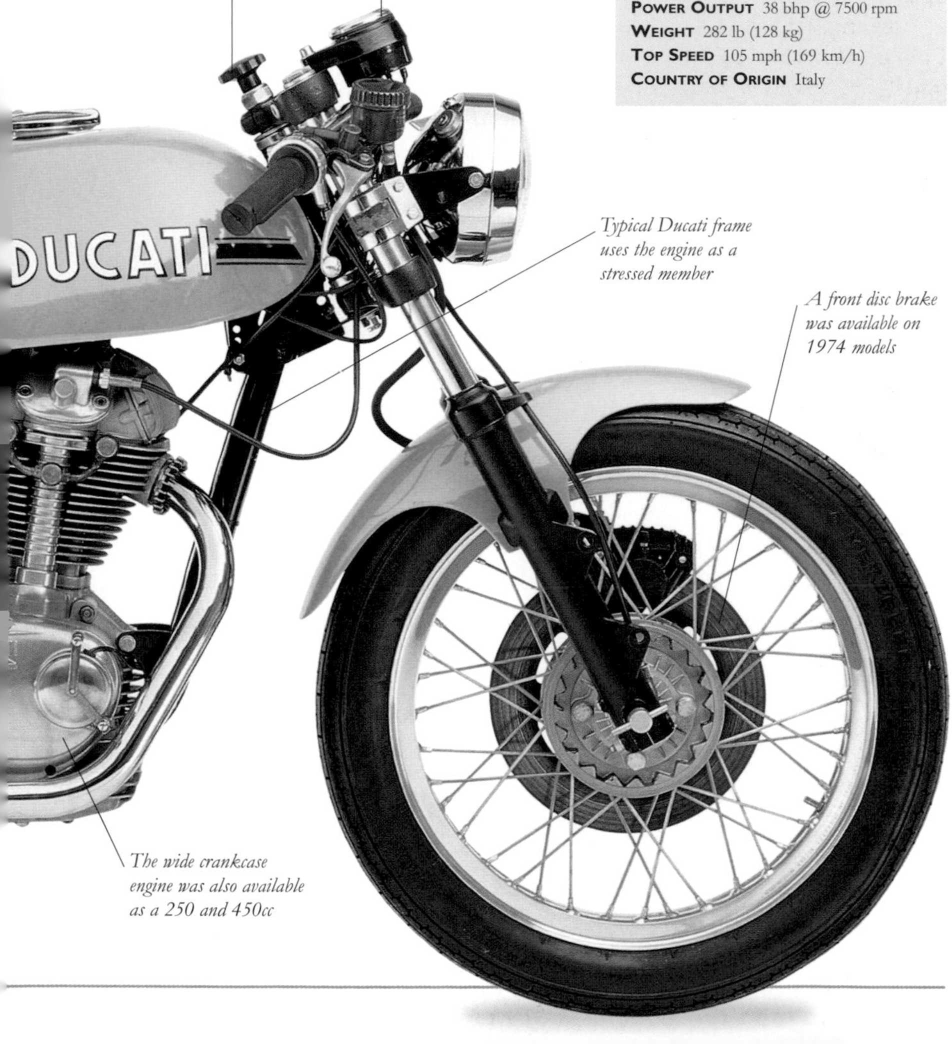

Steering damper

Top speed was 105 mph (169 km/h)

Typical Ducati frame uses the engine as a stressed member

A front disc brake was available on 1974 models

The wide crankcase engine was also available as a 250 and 450cc

DUCATI *MHR 1000*

THE MIKE HAILWOOD REPLICA was built to commemorate Hailwood's victory in the 1978 Formula One TT. Introduced in 1980 and based on the 900SS, the body and paintwork were revised to echo Hailwood's bike. This was one of Ducati's biggest sellers in the early 1980s, but it was gradually overshadowed by smaller machines with belt-driven camshafts. The MHR was axed in 1986, after Cagiva took over Ducati. However, with its enormous stability and deceptive performance, it was destined to become something of a classic. In the late Seventies, before the Japanese manufacturers got their chassis right, the bevel gear Ducati V-twin was the ultimate performance machine. By the time the MHR 1000 appeared, though, it was rather long in the tooth.

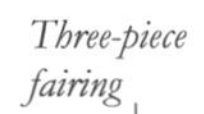

Three-piece fairing

UPDATED AND IMPROVED

The MHR "Mille" was introduced in 1985, its capacity was increased to 973cc, and the engine's bottom end was stronger with improved lubrication. This finally ended the machine's crankshaft reliability problems.

Sculpted fuel tank

New cast wheels used on the Mille

SPECIFICATIONS

MODEL Ducati MHR 1000
CAPACITY 973cc
POWER OUTPUT 83 bhp
WEIGHT Not known
TOP SPEED 137 mph (221 km/h)
COUNTRY OF ORIGIN Italy

Ducati *851*

The 851 was the most complex Ducati road bike ever built when introduced in 1987. It retained Desmodromic valve operation and the 90° V-twin engine layout, but also incorporated four camshafts, eight valves, water cooling, and fuel injection—all firsts for a road Ducati and introduced to meet increasingly strict emissions and noise legislation. These features were also brought in to meet World Superbike race rules, and the 851 was the basis for Ducati's successful 888 Superbike racer.

SPECIFICATIONS

MODEL Ducati 851
CAPACITY 851cc
POWER OUTPUT 100 bhp @ 9250 rpm
WEIGHT 396 lb (180 kg)
TOP SPEED 150 mph (241 km/h) (est.)
COUNTRY OF ORIGIN Italy

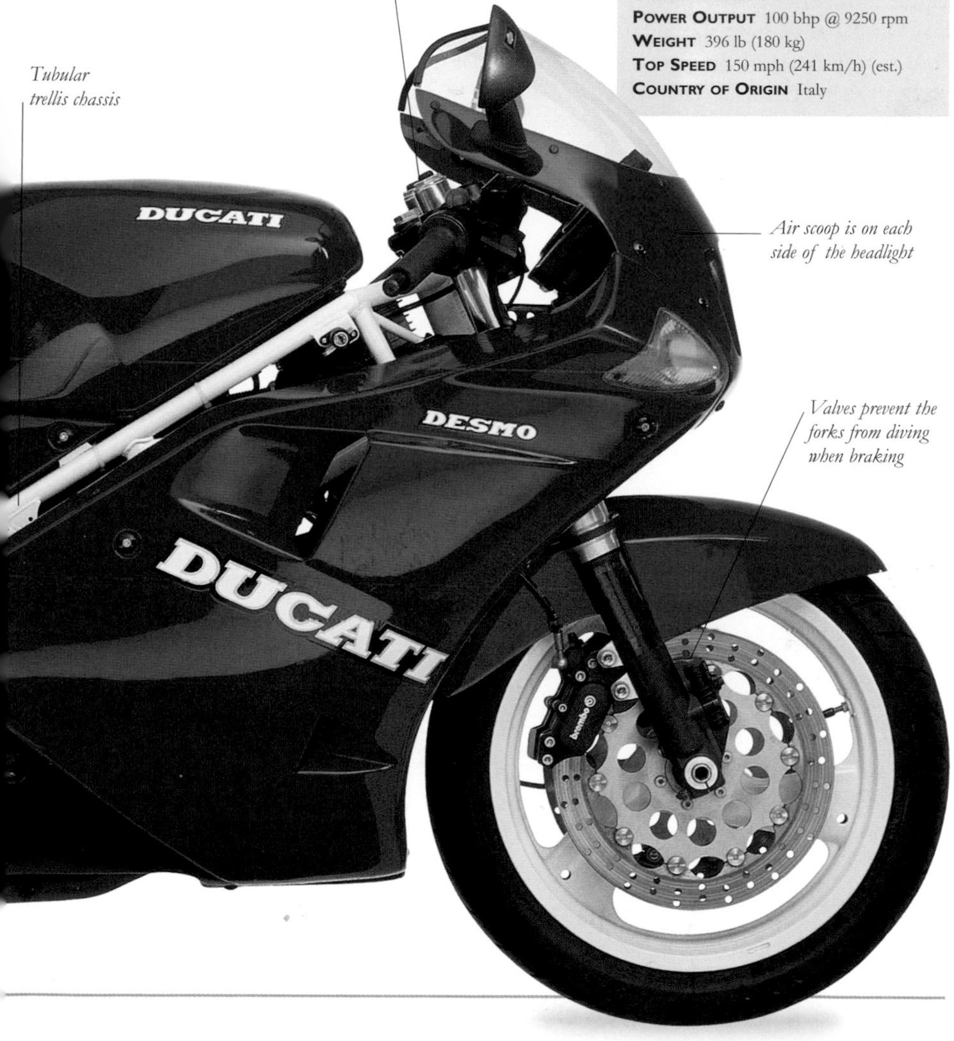
Suspension damping adjustment controls
Tubular trellis chassis
Air scoop is on each side of the headlight
Valves prevent the forks from diving when braking
DUCATI
DESMO
DUCATI
brembo

Ducati *M900 Monster*

Recognizing a market trend toward unfaired bikes, Ducati introduced the Monster in 1994. Unlike much of the opposition in this category, Ducati did not compromise the technology in the new model. The tubular trellis frame had rising-rate rear suspension and was derived from Ducati's Superbike machinery. The Monster's brakes and suspension components were all premium quality. The engine was a four-valve V-twin taken from the 900SS. The combination of torquey engine and lightweight chassis produced an exceptional fun bike.

SPECIFICATIONS

Model Ducati M900 Monster
Capacity 904cc
Power Output 73 bhp @ 7000 rpm
Weight 408 lb (185 kg)
Top Speed 119 mph (192 km/h)
Country of Origin Italy

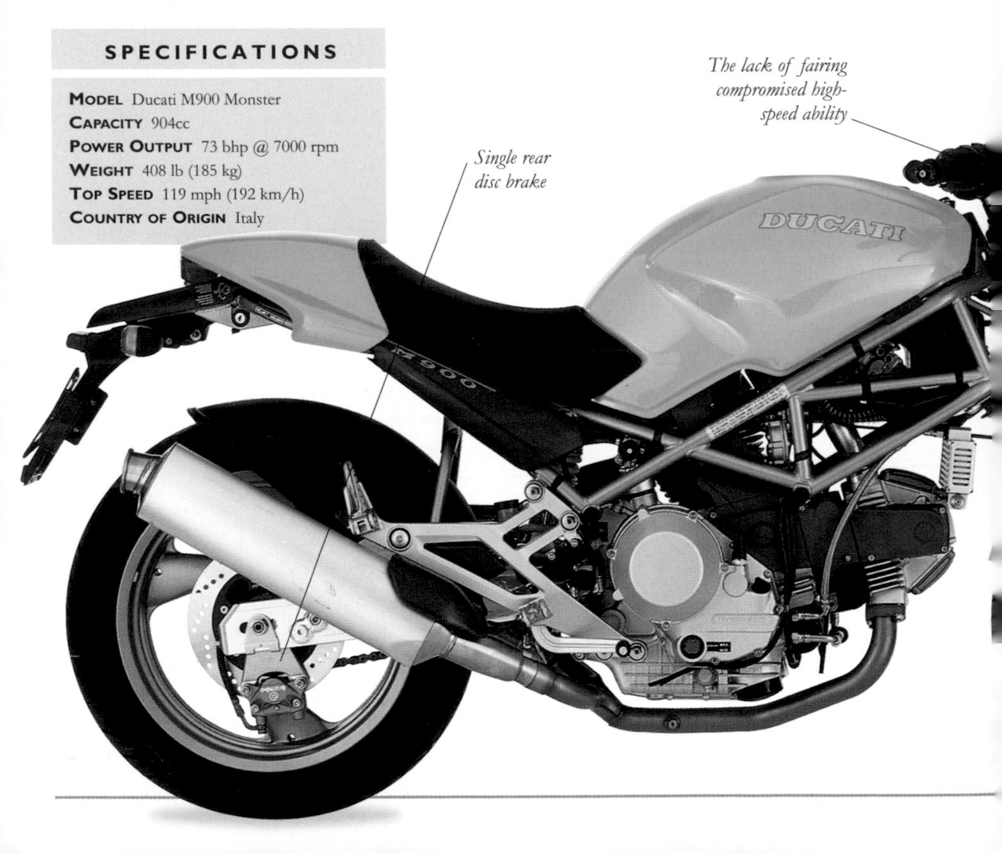

Horizontally mounted shock absorber

SPECIFICATIONS

MODEL Ducati Supermono
CAPACITY 550cc
POWER OUTPUT 65 bhp
WEIGHT 279 lb (127 kg)
TOP SPEED 150 mph (241 km/h)
COUNTRY OF ORIGIN Italy

Instrument console does not include a tachometer

DUCATI *Supermono*

SINGLE CYLINDER FOUR-STROKES have inherent disadvantages as racing engines, but the rules for single-cylinder racing are pleasingly free. So when Ducati wanted to build a racing single in the early 1990s it came up with some slick solutions to get around the problems. Vibration is eliminated because a second conrod works a clapper weight that balances the movement of the real piston—the engine thinks it's a V-twin. And fuel injection gets around the carburation difficulties that arise when a high performance single tries to suck through a carburetor.

Ducati *916SPS*

The Ducati 916 is one of the best looking and most successful bikes ever built. Hailed as a design classic, it has won six World Superbike championships, and it remains an object of desire for ordinary motorcyclists. The technology for the 916 was developed in the mid-Eighties by Ducati engineer Massimo Bordi, who took Ducati's trademark 90° V-twin and added water-cooling, four-valve cylinder heads, and fuel injection. The chassis is also a Ducati trademark. The tubular steel trellis, utilizing the engine as a stressed member, has proved the ideal complement to the V-twin engine.

SPECIFICATIONS

MODEL Ducati 916SPS
CAPACITY 996cc
POWER OUTPUT 120 bhp
WEIGHT 430 lb (195 kg)
TOP SPEED 165 mph (266 km/h)
COUNTRY OF ORIGIN Italy

TOP BIKE

The 916SPS model was the top-of-the-line in 1997. The 916's styling was the work of Massimo Tamburini and, like all true design classics, it still looks fresh several years after it first appeared.

DUCATI *1098S*

INTRODUCED IN 2007, the 1098 replaced the 999 model which, though excellent, had never captured the hearts of Ducati enthusiasts because of its looks. The new bike featured horizontal headlights, an underseat exhaust, and a single-sided swingarm that harkened back to the earlier and much loved 916 range. The capacity of the V-twin engine was increased to 1098cc to match the performance of the 1000cc four-cylinder opposition. The bike produced 160 bhp in standard form and three versions were made, with the S having better quality suspension components than the base model, and the R version intended for racing. The 1098's racing pedigree was established with World Superbike Championship wins in 2008 by Troy Bayliss, and in 2011 by Carlos Checa.

Seat is mounted on detachable rear subframe

Termignoni exhaust system from the parts catalog

Marchesini alloy wheel carries 190/55 section 43-cm (17-in) rear tire

10-in (245-mm) rear disc brake

SPECIFICATIONS

MODEL Ducati 1098S
CAPACITY 1098cc
POWER OUTPUT 160 bhp @ 9750 rpm
WEIGHT 381 lb (173 kg)
TOP SPEED 180 mph (290 km/h)
COUNTRY OF ORIGIN Italy

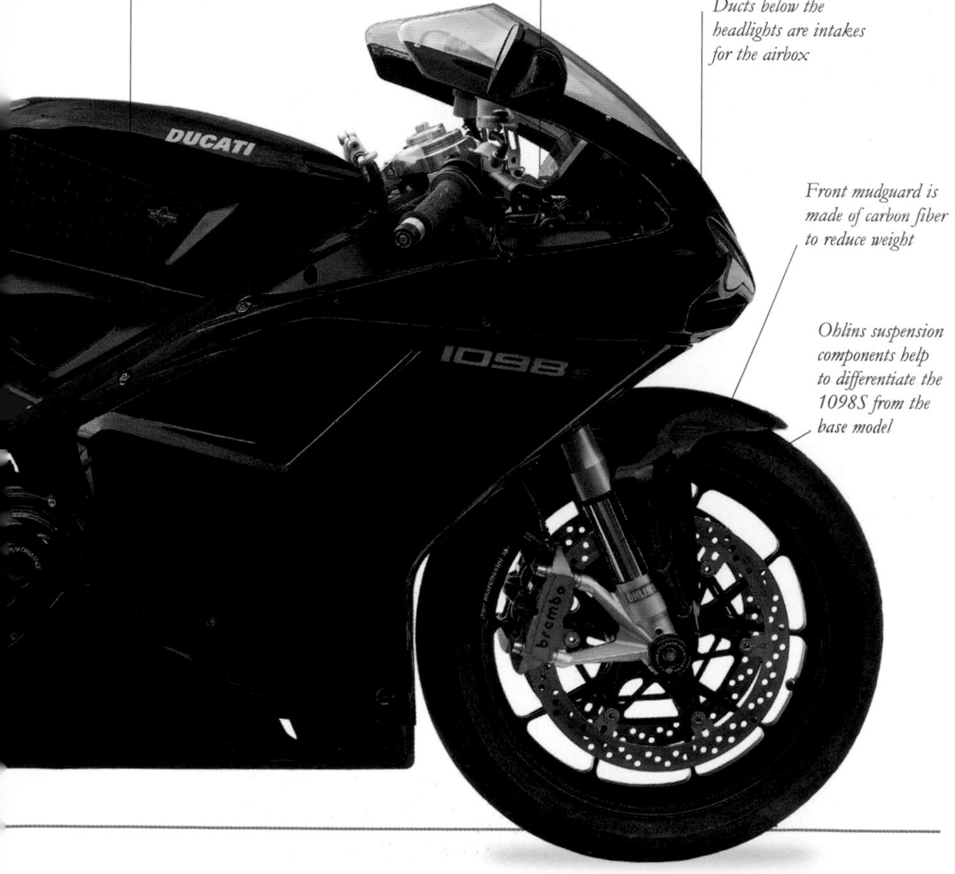
Dash panel includes lap timing facility for track use
Fuel tank has 4-gallon (15.5-liter) capacity, giving a range of near 120 miles (193 km) on the road
Ducts below the headlights are intakes for the airbox
Front mudguard is made of carbon fiber to reduce weight
Ohlins suspension components help to differentiate the 1098S from the base model
DUCATI
1098S
brembo
ÖHLINS

Eso *Speedway*

When speedway racing first became popular, the Douglas was preeminent; but as tracks and riding styles changed there was a swing toward shorter machines such as the Rudge, which provided faster—if less spectacular—racing. After the introduction of the speedway JAP engine in 1931, machines changed little until the Eso started to be seen outside its native Czechoslovakia in the late 1950s. Eso had built speedway machines—with their typically light but strong construction—since 1949, and the company continued on its own until 1962 when Jawa took over. Though still building speedway bikes, Jawa no longer uses the Eso name. The bike shown here dates from 1966.

Specifications

Model Eso Speedway
Capacity 497cc
Power Output 50 bhp @ 8000 rpm
Weight 182 lb (83 kg)
Top Speed Not known
Country of Origin Czech Republic

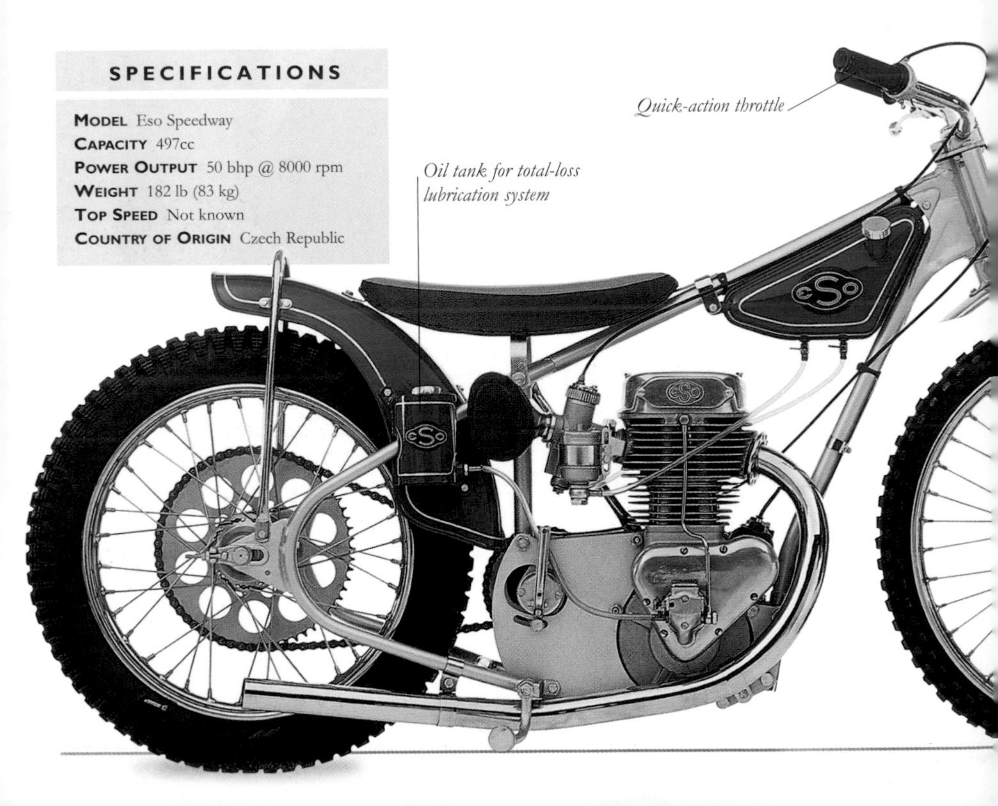

Two rear brakes are used

Detachable frame rail allows the engine to be removed

SPECIFICATIONS

MODEL Excelsior 20R
CAPACITY 61cu. in. (1000cc)
POWER OUTPUT 20 bhp (est.)
WEIGHT 500 lb (227 kg) (est.)
TOP SPEED 100 mph (161 km/h)
COUNTRY OF ORIGIN US

Spindly telescopic forks

A puncture caused the tire to leave Clincher rims, which were abandoned by the mid-1920s

EXCELSIOR *20R*

EXCELSIOR FOLLOWED the example of other American manufacturers and introduced a V-twin model in 1910. It had cylinders spaced at 45° and had a mechanical i.o.e. valve layout. Capacity of the first machines was 50cu. in. (819cc), but in 1912 it grew to 61cu. in. (1000cc). Excellent publicity for the revised model was created when a 61cu. in. Excelsior became the first bike to officially break the 100-mph (161-km/h) barrier. The machine shown here is a three-speed version from 1920. The 61cu. in. model was dropped when the Super X was introduced.

EXCELSIOR *Super X*

INTRODUCED IN 1925, the Super X was the first of a new class of American 45cu. in. (738cc) machines. It was quickly followed into the marketplace by Indian and Harley-Davidson forty-fives. The new model featured a neat unit-construction engine/gearbox. Primary drive to the three-speed gearbox was by helical gear, and the engine was mounted in a duplex cradle frame with leading-link forks. The Super X had exceptional performance and earned an enviable reputation; it is Excelsior's most famous model. The design was restyled for 1929, receiving the "Streamline" look. The bike shown dates from 1930, the year before it was dropped.

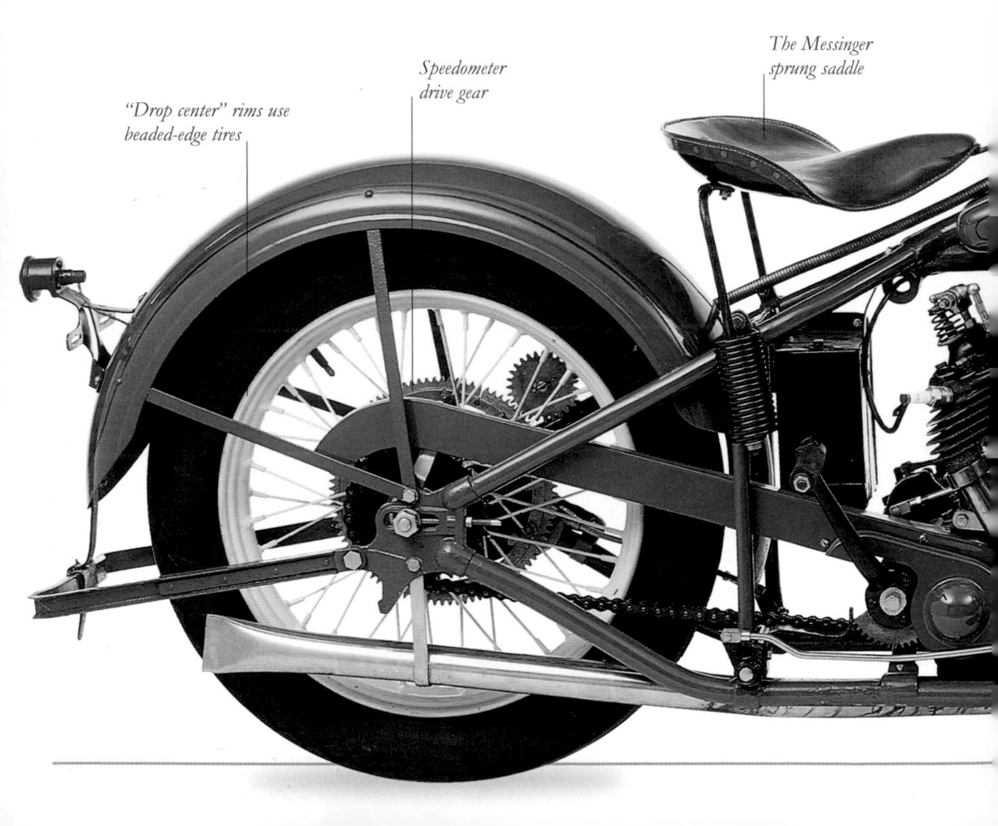

SPECIFICATIONS

MODEL Excelsior Super X
CAPACITY 45.5cu. in. (746cc)
POWER OUTPUT 20 bhp (est.)
WEIGHT 450 lb (204 kg) (est.)
TOP SPEED 65 mph (105 km/h)
COUNTRY OF ORIGIN US

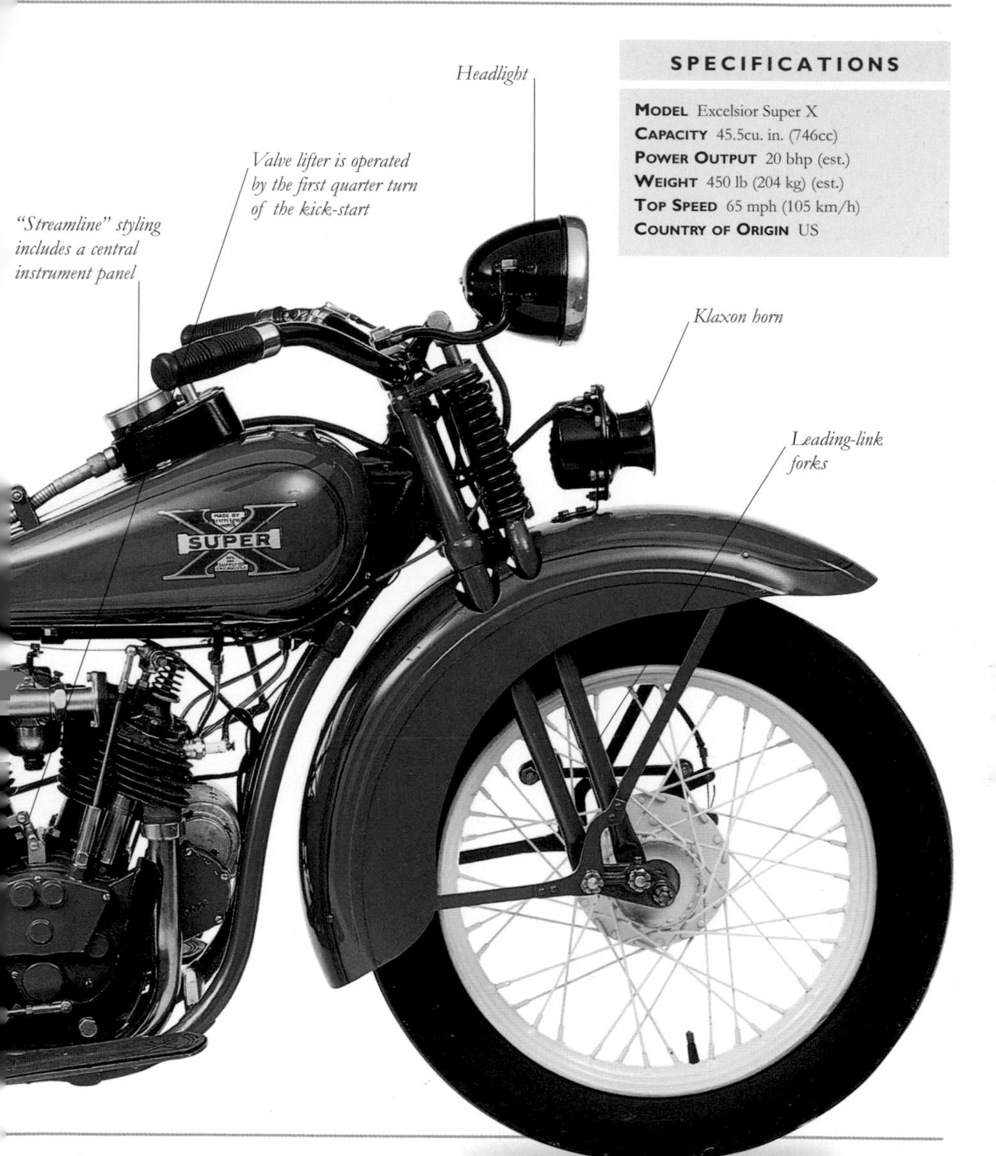

EXCELSIOR *Manxman*

BAYLISS, THOMAS & CO., maker of Excelsior bicycles, is credited with having been the first manufacturer to produce and sell a motorcycle in Britain: the firm marketed a Minerva-engined machine in 1896. Thirty-nine years later, the company introduced the Manxman, a sports machine offered in road or racing trim that superseded the mechanically complex Mechanical Marvel. A more conventional design than its predecessor, the Manxman had a shaft-driven overhead camshaft, and the bronze cylinder head contained just two valves. Initially produced in 250cc and 350cc forms, a 500cc followed in 1936.

SPECIFICATIONS

Model Excelsior Manxman
Capacity 349cc
Power Output 23 bhp @ 6000 rpm
Weight 335 lb (152 kg)
Top Speed 85 mph (137 km/h)
Country of Origin UK

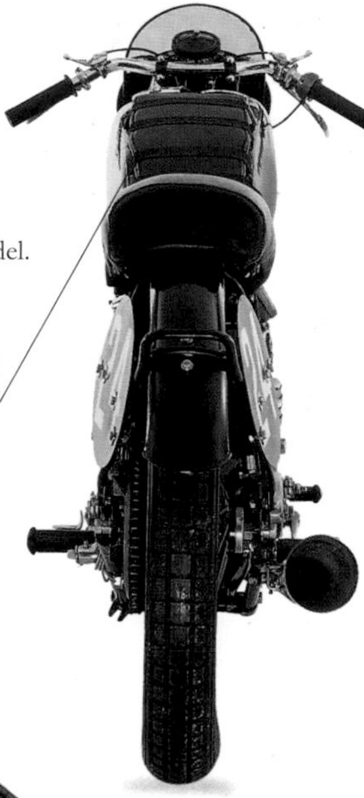

Limited Success

The Manxman was raced successfully by the factory team and by private riders, but despite deriving its name from Britain's Isle of Man, it never won a TT race there. This bike is a 1936 FR12 model.

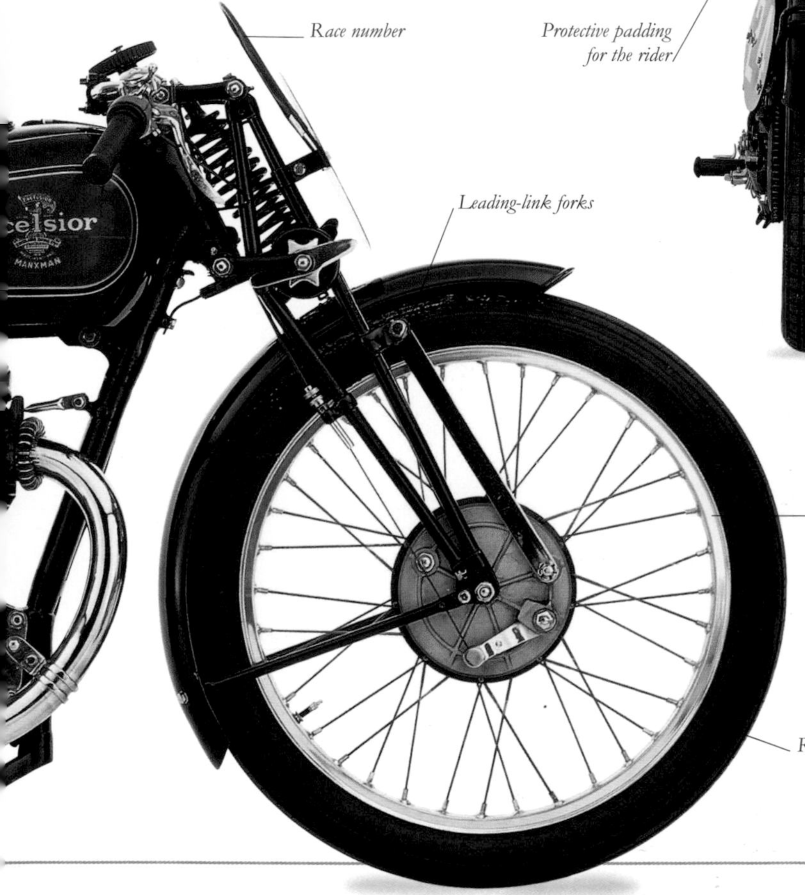

Race number

Protective padding for the rider

Leading-link forks

21-in (53-cm) front wheel

Road-race tire

FANTIC *Chopper*

ONE OF THE WACKIEST BIKES built in the 1970s, the Fantic Chopper was an ideal mount for those aspiring to the wild world of the movie *Easy Rider*. But where the actors enjoyed the effortless throb of a large-capacity Harley-Davidson V-twin, Fantic riders had to endure the manic wail of a small-capacity two-stroke engine. And it must have been embarrassing riding the bike at full speed and then being overtaken by scooters. Surprisingly, the extended forks and weird riding position did not make the Fantic dangerous, even if they did make it uncomfortable. Perhaps fortunately, the Chopper was comparatively short-lived and in the 1980s Fantic concentrated instead on building an excellent series of off-road bikes. This is a 1977 model.

SPECIFICATIONS

MODEL Fantic Chopper
CAPACITY 123cc
POWER OUTPUT 13.2 bhp @ 6600 rpm
WEIGHT 260 lb (118 kg)
TOP SPEED 65 mph (105 km/h)
COUNTRY OF ORIGIN Italy

Fuel tank is sculpted to allow the rider to adopt an aerodynamic crouch

Camshaft drive housing

Sporty dual seat

Alloy brake hubs

Steel mudguard

SPECIFICATIONS

Model FB Mondial
Capacity 173cc
Power Output 10 bhp @ 6700 rpm
Weight 264 lb (120 kg)
Top Speed 68 mph (110 km/h)
Country of Origin Italy

FB *Mondial*

Fratelli Boselli Mondial chose to advertise its products through racing and won the 125cc World Championship in 1949, 1950, 1951, and 1957. While the racing machines were exotic o.h.c. and d.o.h.c. devices, most production models were orthodox o.h.v. and two-stroke lightweights. Like other Italian manufacturers, FB Mondial equipped its sports models with low handlebars and often chose red for its color schemes. The machine shown here is a 1956 sports model that combines an o.h.c. engine with the cycle parts of its more ordinary stablemates. Italian lightweight machines often featured a specification far superior to the products of other countries and this model is typical. High-quality alloy engine castings are used for the engine and the full-width brake hubs.

FN *Four*

INTRODUCED IN 1904, the FN was the first successful four-cylinder motorcycle. Designed by Paul Kelecom, the original design had shaft drive, magneto ignition, splash lubrication, and the benefits of the vibration-free, four-cylinder engine. It was one of the first serious attempts at an integrated design for a motorcycle rather than the usual bicycle-with-engine-style machines of the period. The capacity of the original machine was 362cc, but by the time the 1907 model (shown here) was produced, it had increased. There were also other detail improvements.

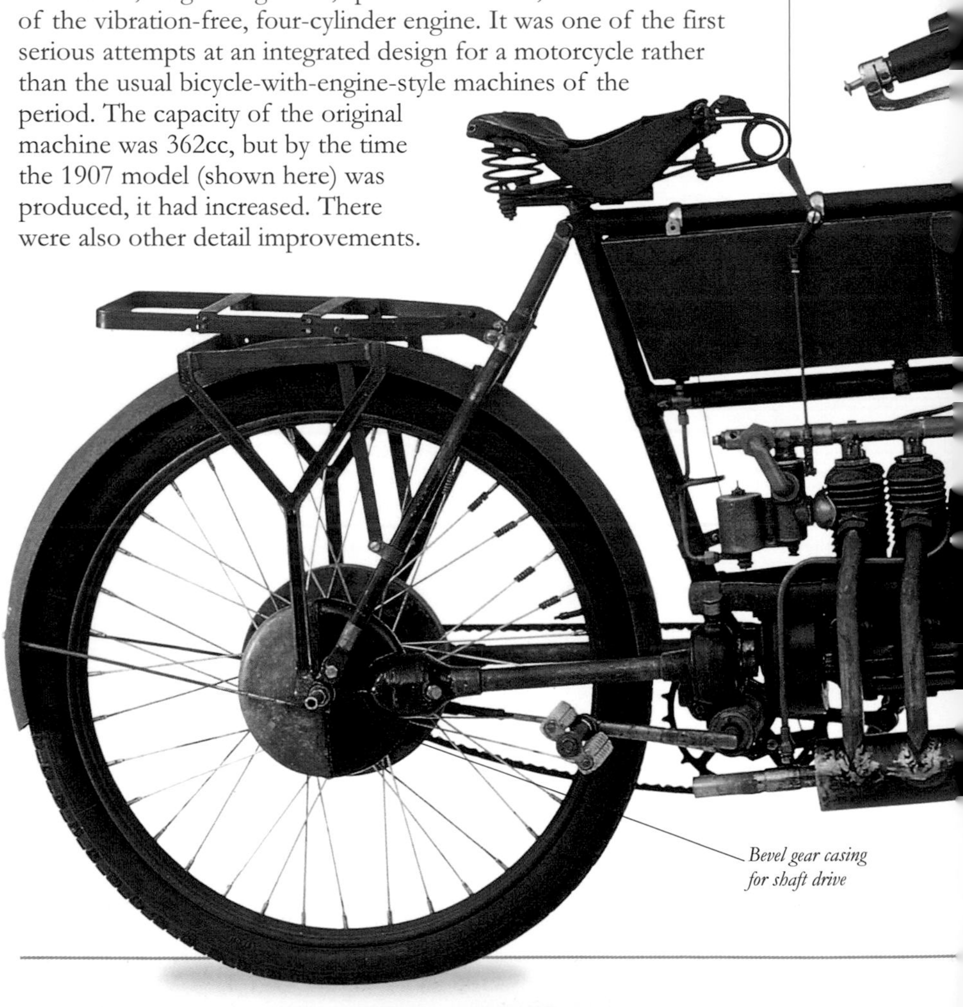

Throttle

Bevel gear casing for shaft drive

SPECIFICATIONS

Model FN Four
Capacity 410cc
Power Output Not known
Weight Not known
Top Speed 42 mph (68 km/h) (est.)
Country of Origin Belgium

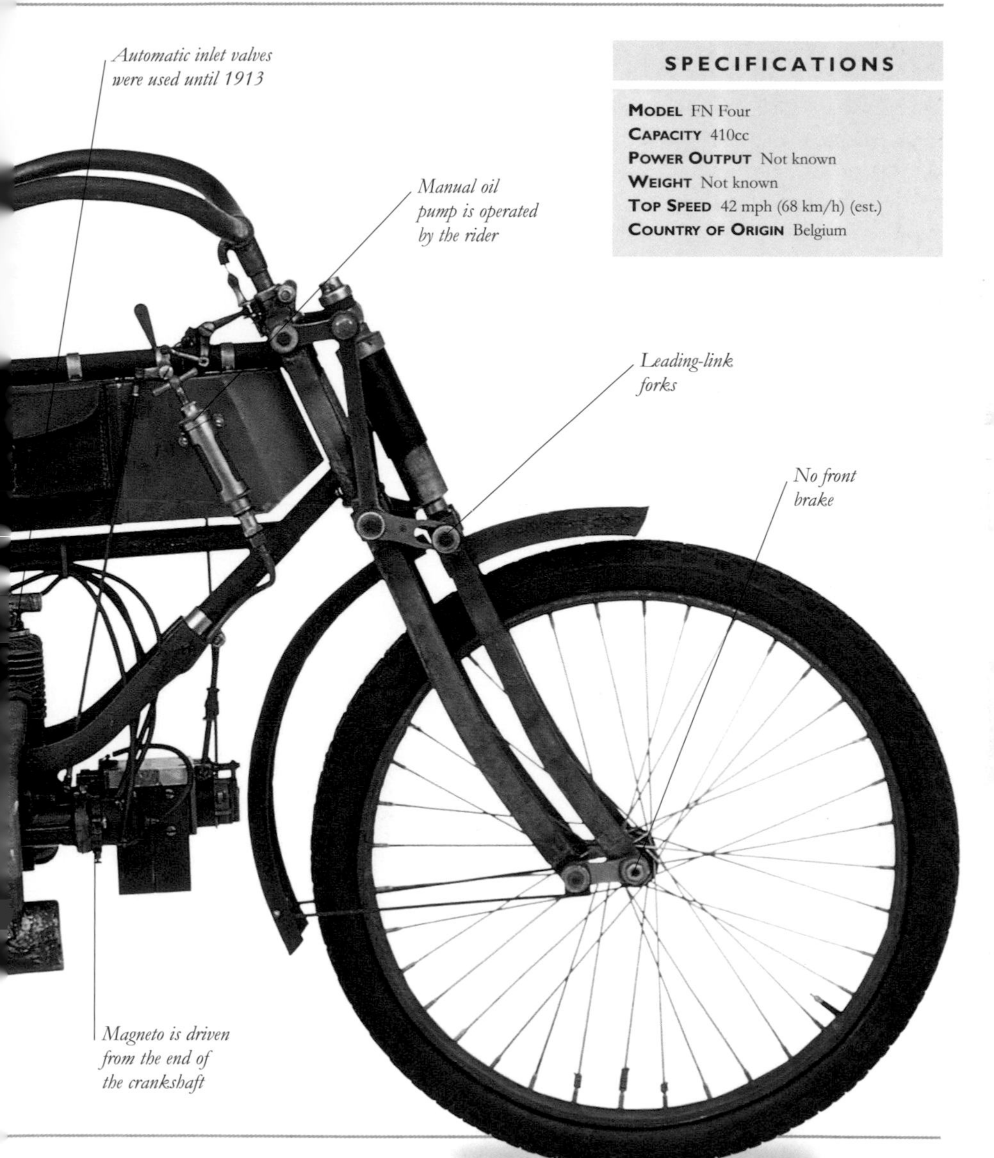

GILERA *Saturno San Remo*

GILERA'S MOST FAMOUS RACERS were the World Championship–winning four-cylinder machines produced following Gilera's takeover of the Rondine company in 1937. For use at lesser events and for customers, Gilera produced the single-cylinder San Remo machine, which was based on the Saturno road bike *(see p.126)*. It had a victorious debut at the 1947 Ospedaletti Grand Prix near San Remo, hence the bike's name, but never seriously challenged the multicylinder or overhead-camshaft machines of the opposition. It retained the o.h.v. unit-construction of the road machines, but clever modification gave it more power. This is the 1949 model.

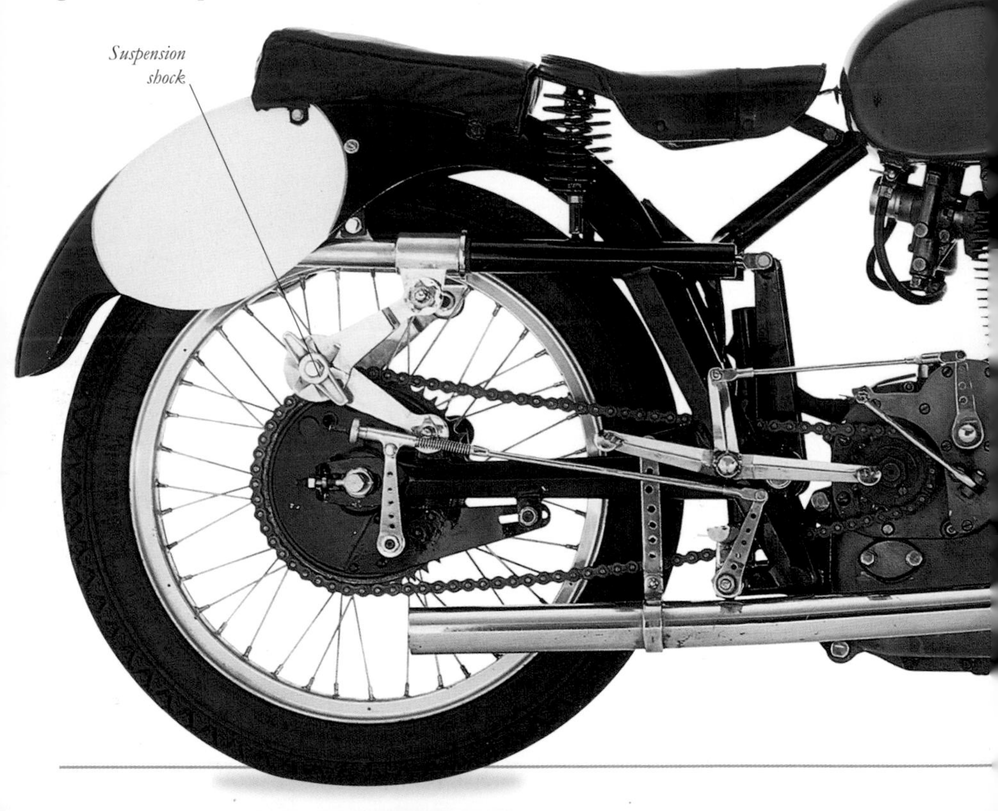

SPECIFICATIONS

MODEL Gilera Saturno San Remo
CAPACITY 498cc
POWER OUTPUT 38 bhp @ 6000 rpm
WEIGHT 282 lb (128 kg)
TOP SPEED 115 mph (185 km/h)
COUNTRY OF ORIGIN Italy

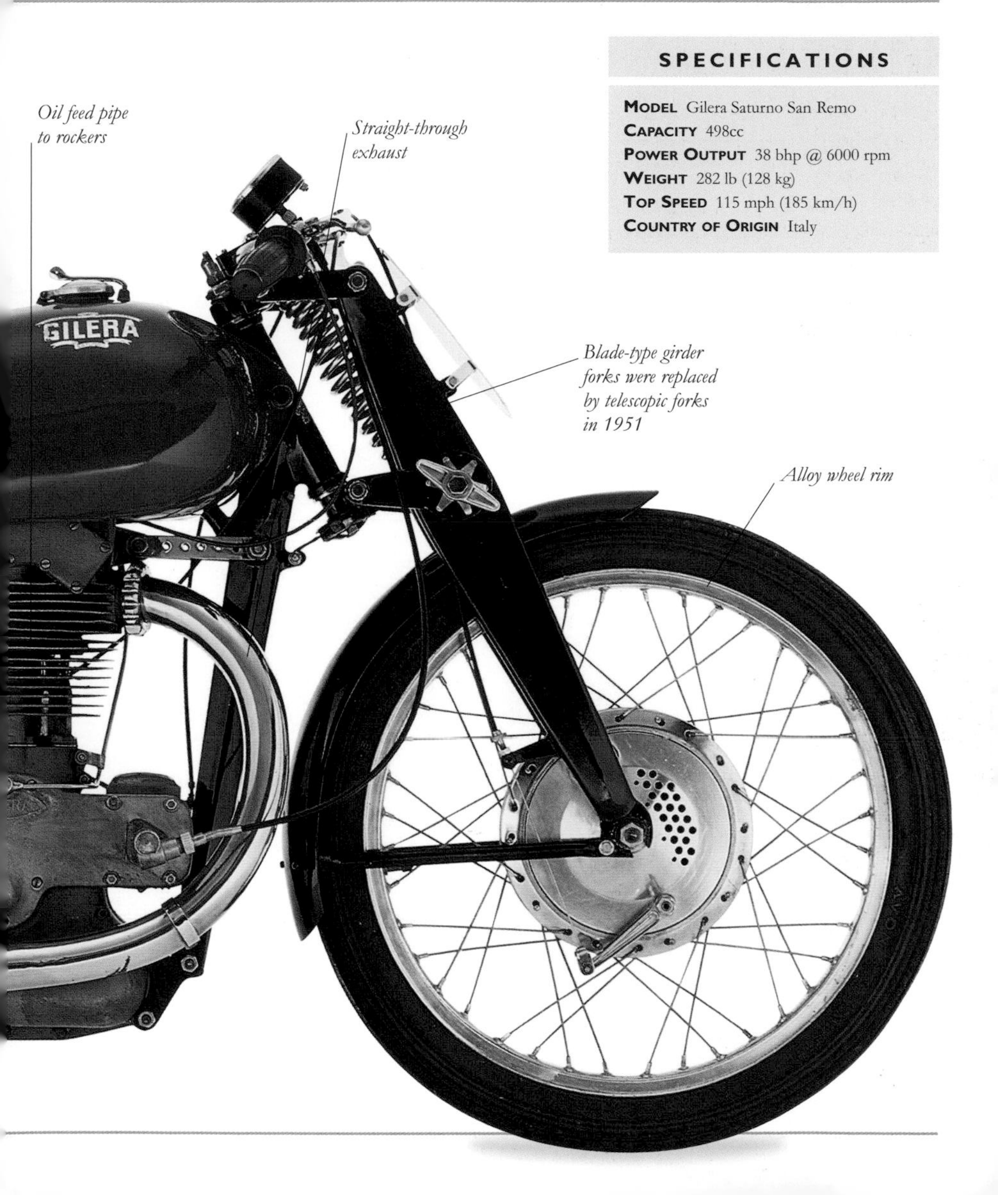

GILERA *Saturno*

BUILT FROM 1940, THE SATURNO was a logical development of previous Gilera designs. The 3.3 x 3.5-in (84 x 90-mm) engine had overhead valves that were closed by hairpin valve springs. Iron was used for the head and barrel, but the engine cases, which also housed the four-speed box, were alloy. Primary drive was by gear. Front forks were centrally sprung girders, and rear suspension was provided by Gilera's horizontal-spring system. The machine shown is a 1951 model. Later, uprated "Sports" versions with alloy cylinder heads and telescopic forks were produced. Though the bike had few novel features to speak of, it was attractive and popular; it stayed in the Gilera line right through until 1959.

SPECIFICATIONS

MODEL Gilera Saturno
CAPACITY 498cc
POWER OUTPUT 22 bhp
WEIGHT 386 lb (175 kg)
TOP SPEED 85 mph (180 km/h)
COUNTRY OF ORIGIN Italy

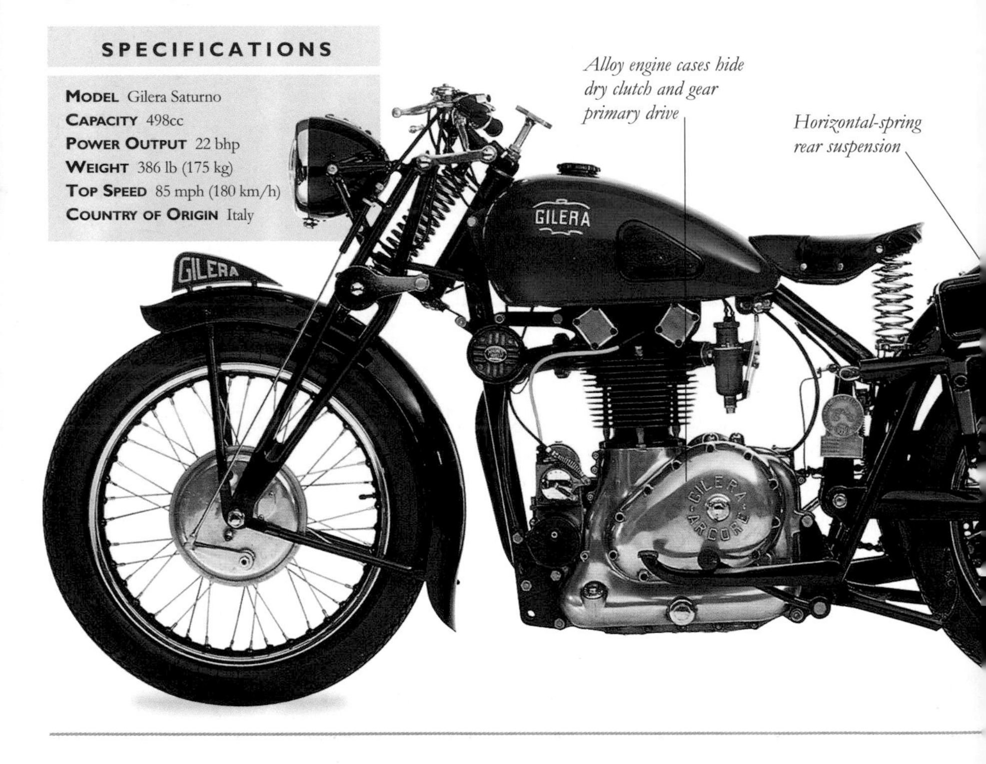

Racing-style saddle

Instrumentation includes a tachometer

SPECIFICATIONS

MODEL Gilera Speciale Strada
CAPACITY 124cc
POWER OUTPUT 10.5 bhp @ 8500 rpm
WEIGHT 205 lb (93 kg)
TOP SPEED 70 mph (113 km/h) (est.)
COUNTRY OF ORIGIN Italy

17-in (43-cm) wheels feature alloy rims and brake hubs

GILERA *Speciale Strada*

GILERA INTRODUCED A NEW 125 in 1959, and production continued in various forms until 1970. The neat, wet-sump engine unit incorporates a four-speed gearbox with gear primary drive. A single camshaft is gear-driven from the crankshaft and operates the valves via pushrods and conventional tappets with screw and locknut adjustment. Ignition is by battery and coil. A 150cc version of the engine was also produced. The model shown here dates from 1966.

GILERA *Nuovo Saturno*

ORIGINALLY INTENDED AS A limited edition model for the Japanese market, the Nuovo Saturno appeared in 1989 and was also sold in Europe. It combined a tubular-steel trellis frame with an excellent four-valve, twin-cam, water-cooled, single-cylinder engine, which was taken from Gilera's existing trail bike. Its light weight and slim profile helped provide excellent handling and also (for a 500 single) performance. Red paint and race styling gave it the desired look. The Gilera logo gave it a name that had authentic history; Gilera had won several world championships in the 1950s. Sadly, it was also very expensive and never sold in significant numbers. A 350 version was also sold in some markets.

SPECIFICATIONS

MODEL Gilera Nuovo Saturno
CAPACITY 492cc
POWER OUTPUT 37 bhp @ 7500 rpm
WEIGHT 309 lb (140 kg)
TOP SPEED 105 mph (169 km/h) (est.)
COUNTRY OF ORIGIN Italy

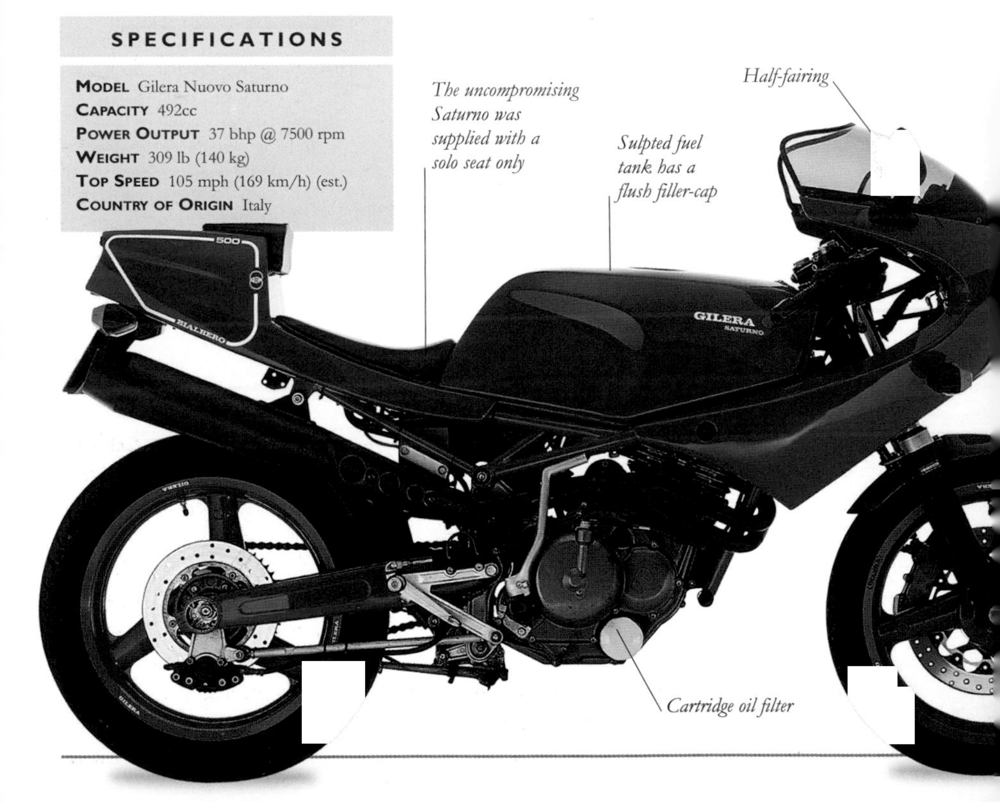

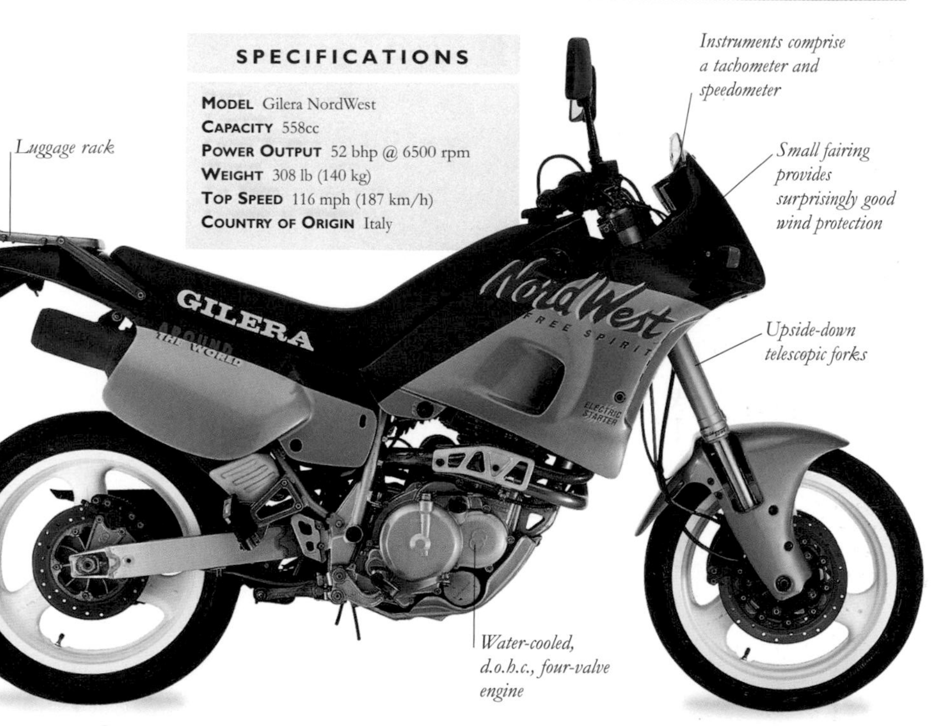

SPECIFICATIONS

MODEL Gilera NordWest
CAPACITY 558cc
POWER OUTPUT 52 bhp @ 6500 rpm
WEIGHT 308 lb (140 kg)
TOP SPEED 116 mph (187 km/h)
COUNTRY OF ORIGIN Italy

GILERA *NordWest*

GILERA TOOK THE BIG trail bike to its logical conclusion with the NordWest. As the bikes don't work off-road, why compromise on-road performance with dual purpose tires, brakes, wheels, and suspension? This model was based on the RC600 trail bike but the trail suspension, brakes, and wheels were replaced with a pure road setup. Sticky tires and sensational brakes, combined with a rev-happy engine, made the NordWest a stunning road bike. As long as the road was twisting it could stay ahead of genuine sports machines. The model disappeared with the Gilera brand in 1993. The name was revived by parent company Piaggio and used on scooters. They also won the 250cc World Championship in 2008.

GNOME & RHÔNE

THE FIRST MOTORCYCLES THAT WERE BUILT by aircraft-engine maker Gnome & Rhône in 1919 were copies of the British ABC flat-twin made under license. Gnome & Rhône subsequently made conventional single-cylinder machines before returning to the flat-twin in 1930. The o.h.v. Type X was introduced in 1935. With a 750cc engine, shaft drive, and a pressed-steel frame and forks, the new machine was one of the largest and most prestigious produced in France at that time. It was also ideal for pulling sidecars, and many were put to that use. The model shown here dates from 1939, shortly before the war stopped production. After the war Gnome & Rhône built 125–200cc two-strokes before getting out of motorcycle manufacturing in 1959.

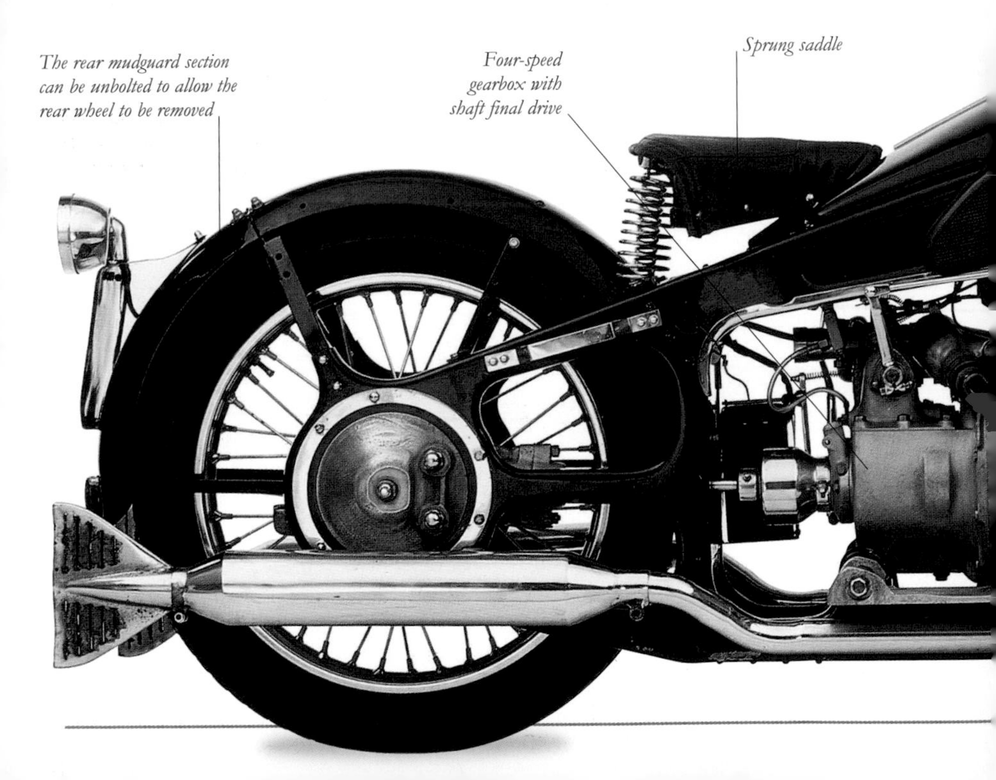

SPECIFICATIONS

MODEL Gnome & Rhône
CAPACITY 724cc
POWER OUTPUT 30 bhp @ 5500 rpm
WEIGHT Not known
TOP SPEED 90 mph (145 km/h) (est.)
COUNTRY OF ORIGIN France

CONFIGURATION
The flat-twin engine layout allows the cylinders to be perfectly positioned for air-cooling but also makes them vulnerable to accident damage if the bike is dropped.

Girder forks and pressed-steel frame

Alloy footboards are visible under the cylinders

Valanced mudguard

Large alloy brake drum

GREEVES *20T*

THE FIRST Greeves prototype motorcycles appeared in 1951; the company also made "invalid" cars. The first production models, launched in 1953, had novel features such as a distinctive cast-alloy beam in place of a downtube and headstock, and a rubber-in-torsion suspension system. These models included road, scramblers, and trials machines powered by the 197cc Villiers 8E single-cylinder two-stroke engine. This 20T trials model dates from 1955.

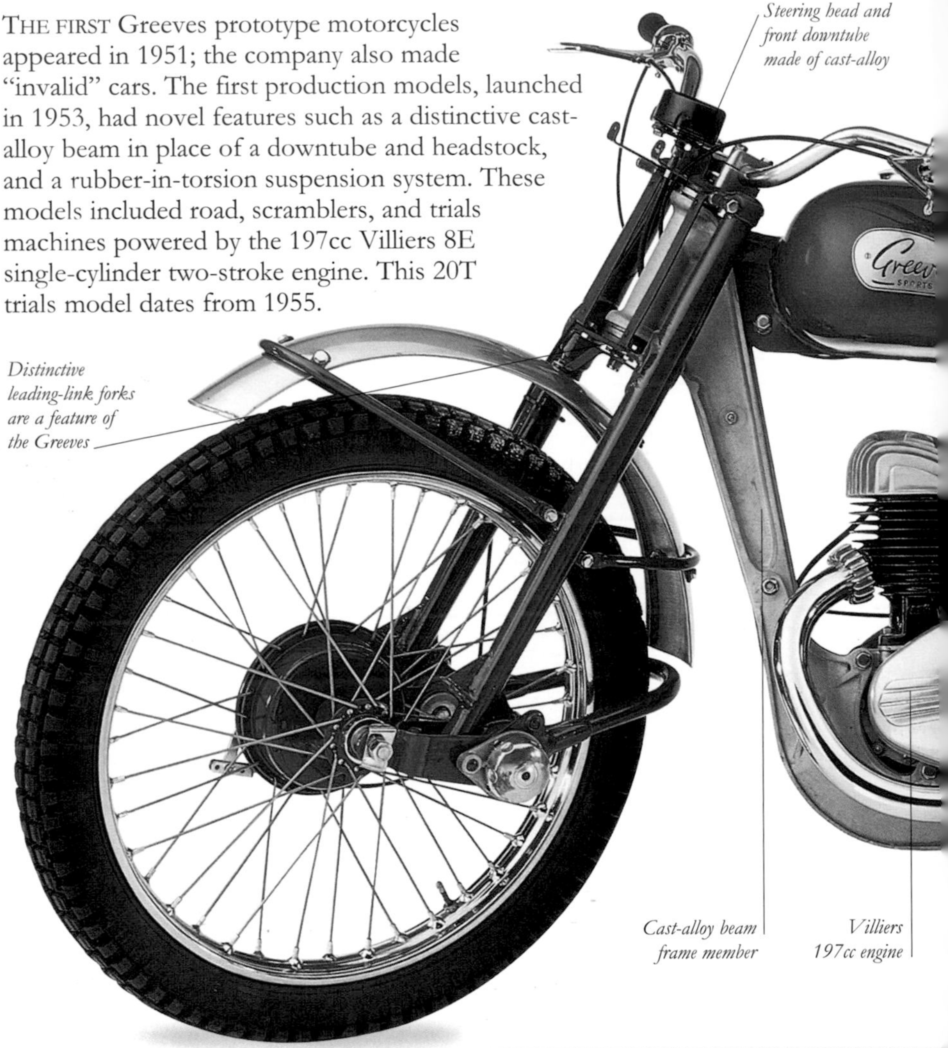
Steering head and front downtube made of cast-alloy
Distinctive leading-link forks are a feature of the Greeves
Cast-alloy beam frame member
Villiers 197cc engine
Greeves
SPORTS

SPECIFICATIONS

MODEL Greeves 20T
CAPACITY 197cc
POWER OUTPUT 8.5 bhp @ 4000 rpm
WEIGHT 228 lb (103 kg)
TOP SPEED 50 mph (80 km/h) (est.)
COUNTRY OF ORIGIN UK

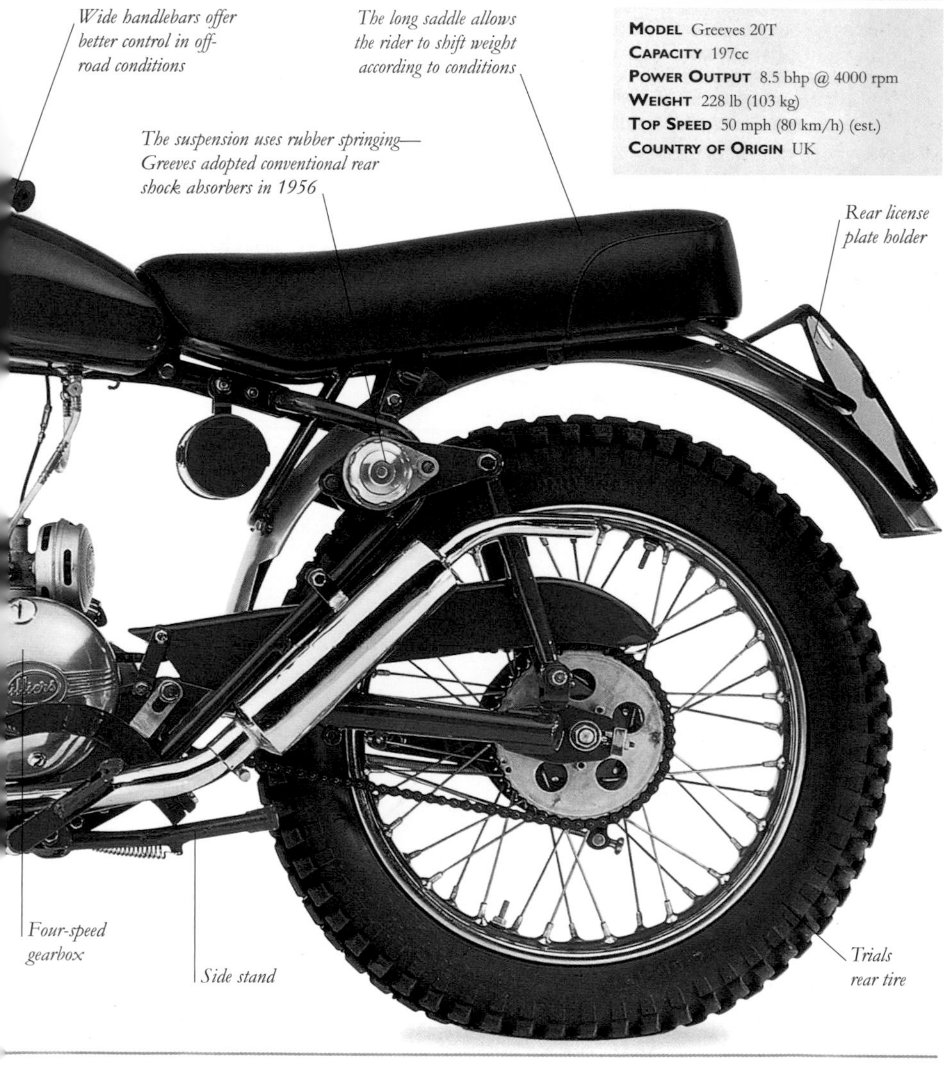

Wide handlebars offer better control in off-road conditions

The long saddle allows the rider to shift weight according to conditions

The suspension uses rubber springing—Greeves adopted conventional rear shock absorbers in 1956

Rear license plate holder

Four-speed gearbox

Side stand

Trials rear tire

GRITZNER *Monza Super Sport*

THE MONZA SUPER SPORT was introduced by the long-established Mars motorcycle company in 1957. When Mars folded in 1958, Gritzner took over its production. Unlike many lightweights of the period, the Monza Super Sport was built as a true motorcycle: it did not have pedals, and the three-speed gearbox was foot-operated rather than the twistgrip type used on other machines. Sadly, the Monza's good looks and features did not prevent its failure—Gritzner followed Mars and ceased operating in 1962. This is a 1960 model.

Flat, narrow, sports style handlebars

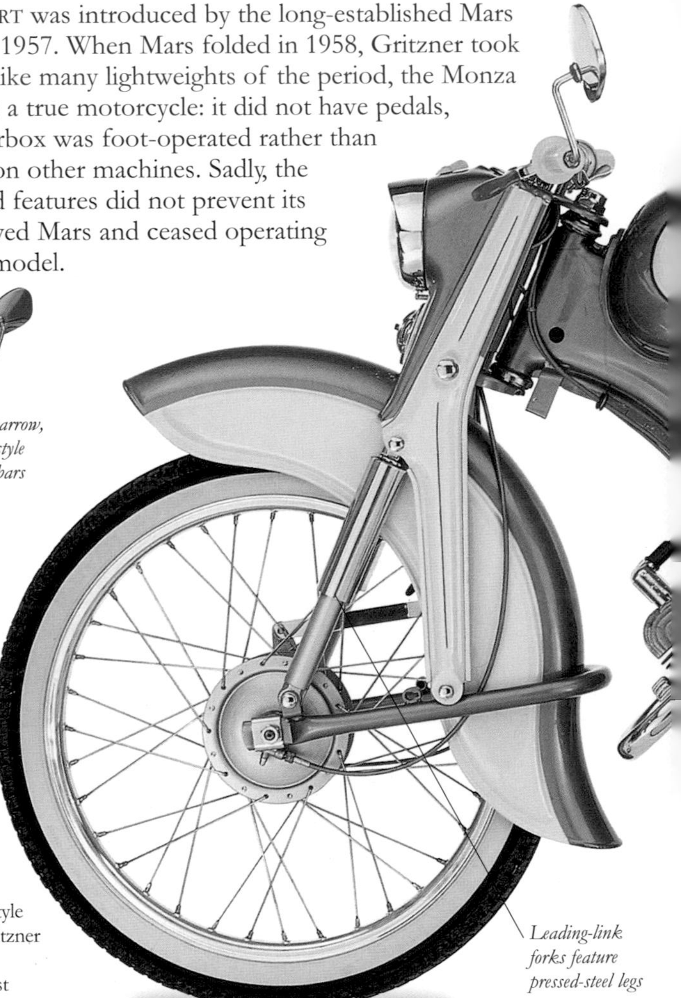

FOOTREST STYLING

The conventional, motorcycle-style footrest arrangement of the Gritzner means that it doesn't have the bicycle-style pedals fitted to most machines of this capacity.

Leading-link forks feature pressed-steel legs

SPECIFICATIONS

MODEL Gritzner Monza Super Sport
CAPACITY 50cc
POWER OUTPUT 3 bhp @ 6000 rpm
WEIGHT 145 lb (70 kg)
TOP SPEED 37 mph (60 km/h)
COUNTRY OF ORIGIN Germany

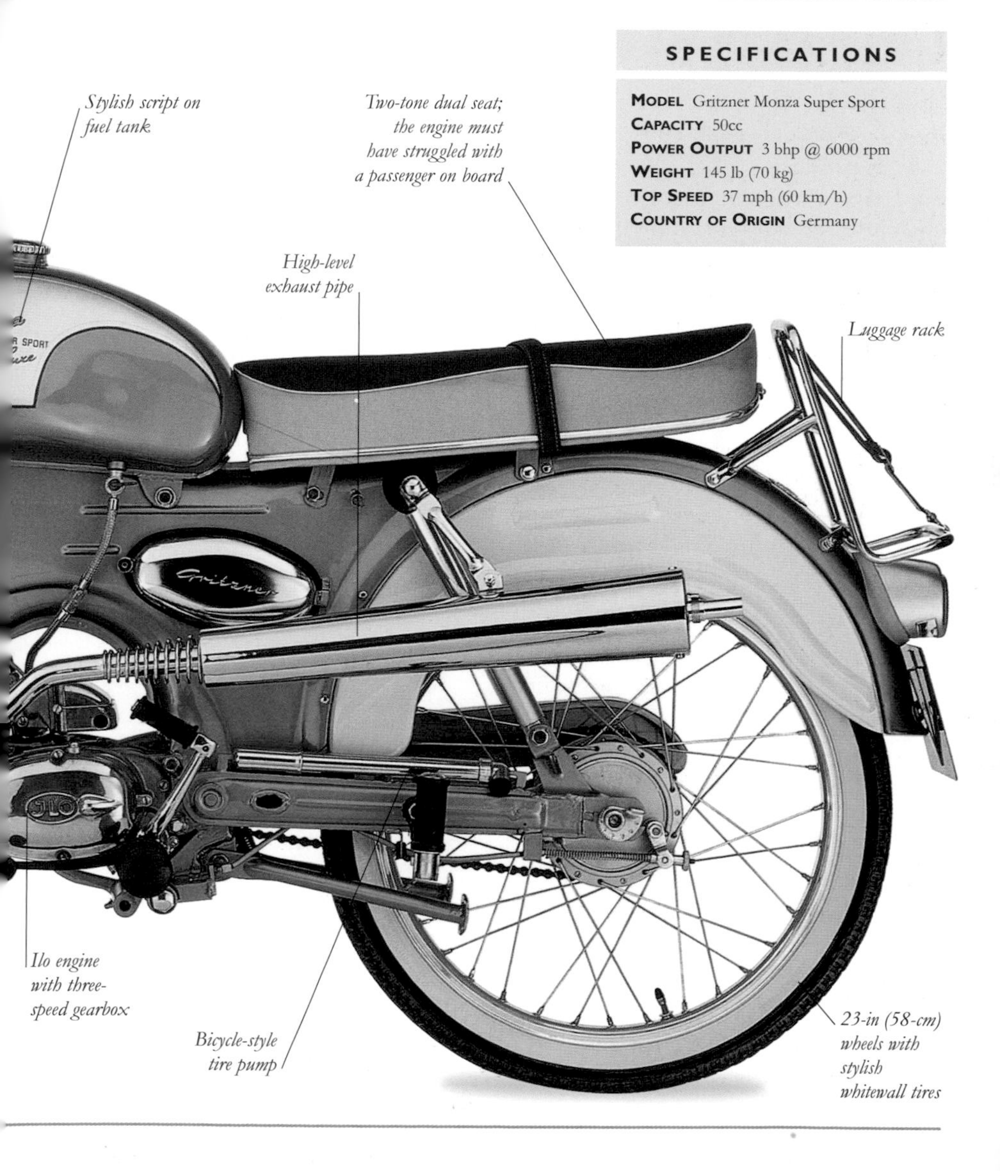

Stylish script on fuel tank

Two-tone dual seat; the engine must have struggled with a passenger on board

High-level exhaust pipe

Luggage rack

Ilo engine with three-speed gearbox

Bicycle-style tire pump

23-in (58-cm) wheels with stylish whitewall tires

HARLEY-DAVIDSON *KT Board Racer*

HARLEY-DAVIDSON DID NOT officially participate in racing until 1914, when it decided to exploit the potential benefits of publicity and development that could be derived from racing success. Board-track racing was reaching new levels of popularity, with promoters able to attract huge paying crowds to the meets, so Harley's decision to enter into competition made a lot of sense. And the move paid off almost immediately, as the Harley race team began to achieve significant results in 1915 on bikes such as this KT. In September 1915, an F-head Harley set a 100-mile (161-km) record of 89.11 mph (143.46 km/h) on a board track in Chicago.

Pedaling forward starts the bike, pedaling backward operates the rear brake—just like on a simple bicycle

SPECIFICATIONS

MODEL Harley-Davidson KT Board Racer
CAPACITY 61cu. in. (1000cc)
POWER OUTPUT 15 bhp
WEIGHT 325 lb (147 kg)
TOP SPEED 80 mph (130 km/h)
COUNTRY OF ORIGIN US

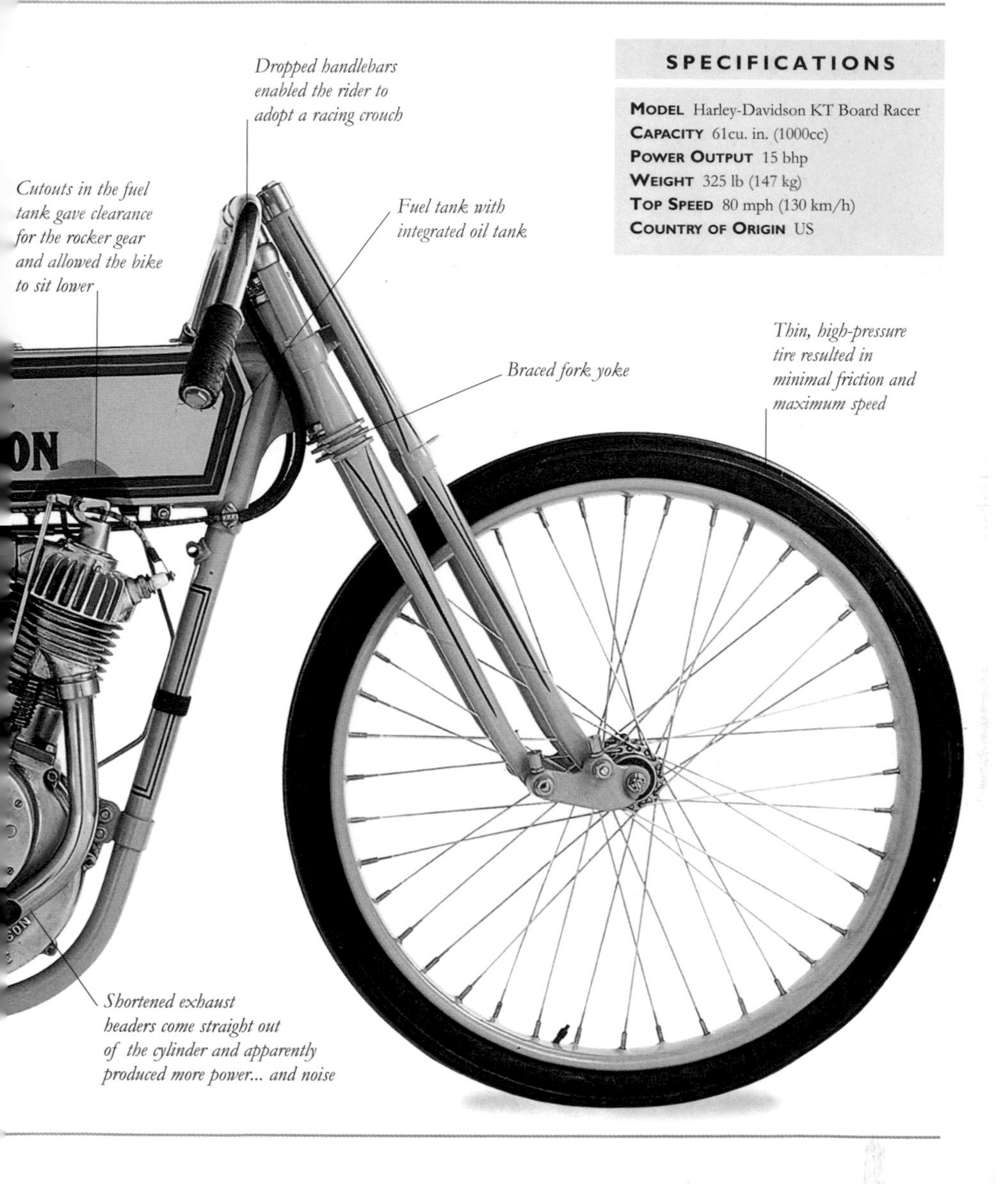

HARLEY-DAVIDSON *Eight-Valve Racer*

HAVING COMMITTED itself to bike racing in 1914, Harley soon began to take the sport seriously. Special eight-valve racing twins were introduced in 1916; these were built in limited numbers until 1927 for the exclusive use of the factory's own race team. Four versions of the machine were produced over an 11-year period, giving serious credibility to Harley as a race-bike manufacturer. The team secured many victories on the eight-valve racers and earned itself the nickname "The Wrecking Crew." The sight and sound of these, quite literally, fire-breathing machines must have been incredible as they reached speeds of around 120 mph (193 km/h) on the steeply banked wooden tracks where they were used.

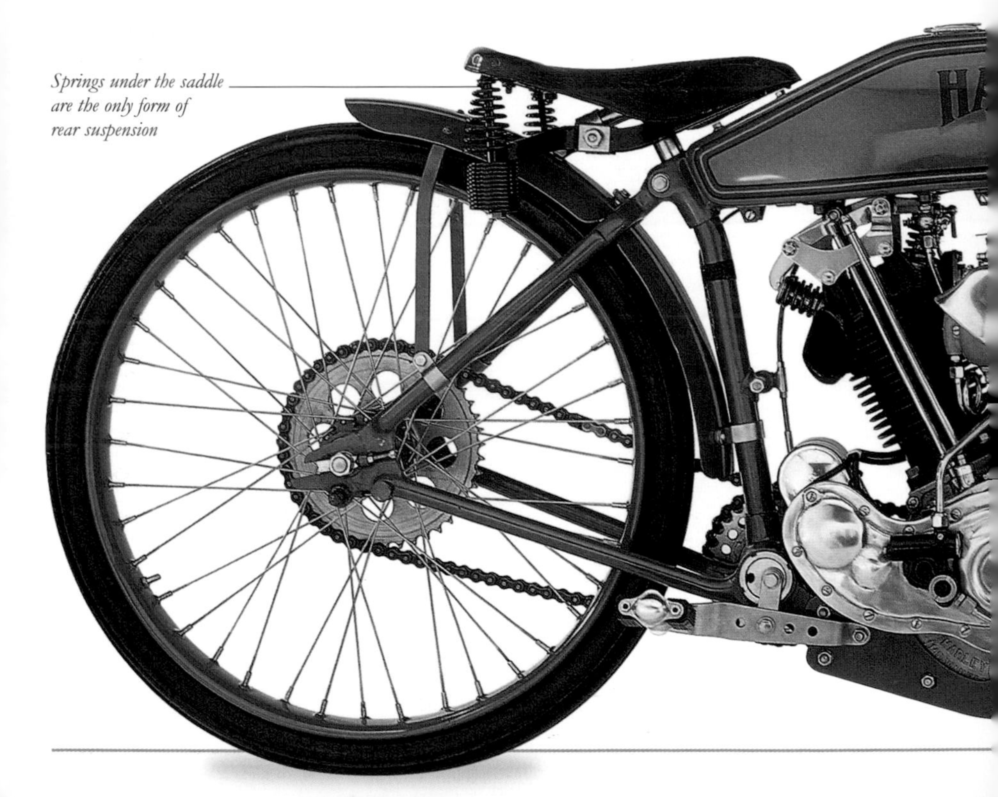

Springs under the saddle are the only form of rear suspension

SPECIFICATIONS

MODEL Harley-Davidson Eight-Valve Racer
CAPACITY 61cu. in. (1000cc)
POWER OUTPUT 15 bhp
WEIGHT 692 lb (314 kg)
TOP SPEED 120 mph (193 km/h)
COUNTRY OF ORIGIN US

BRAKE-FREE
The eight-valve racer had no gearbox or brakes. Riders slowed their machines using a combination of the throttle, the engine-kill button, and old-fashioned boot leather.

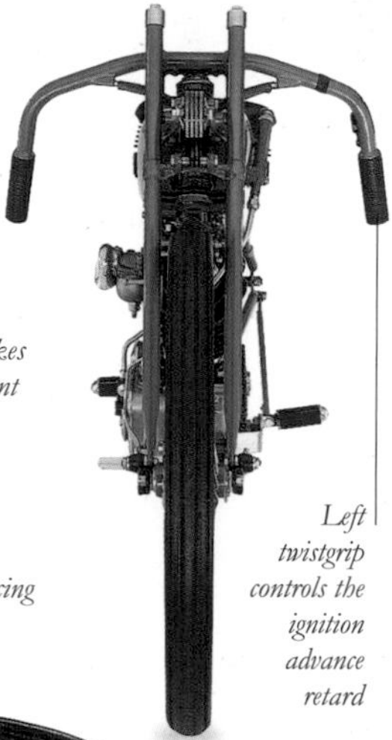

Exposed valve gear

Friction shock makes suspension movement more controllable

Spoked racing wheel

Left twistgrip controls the ignition advance retard

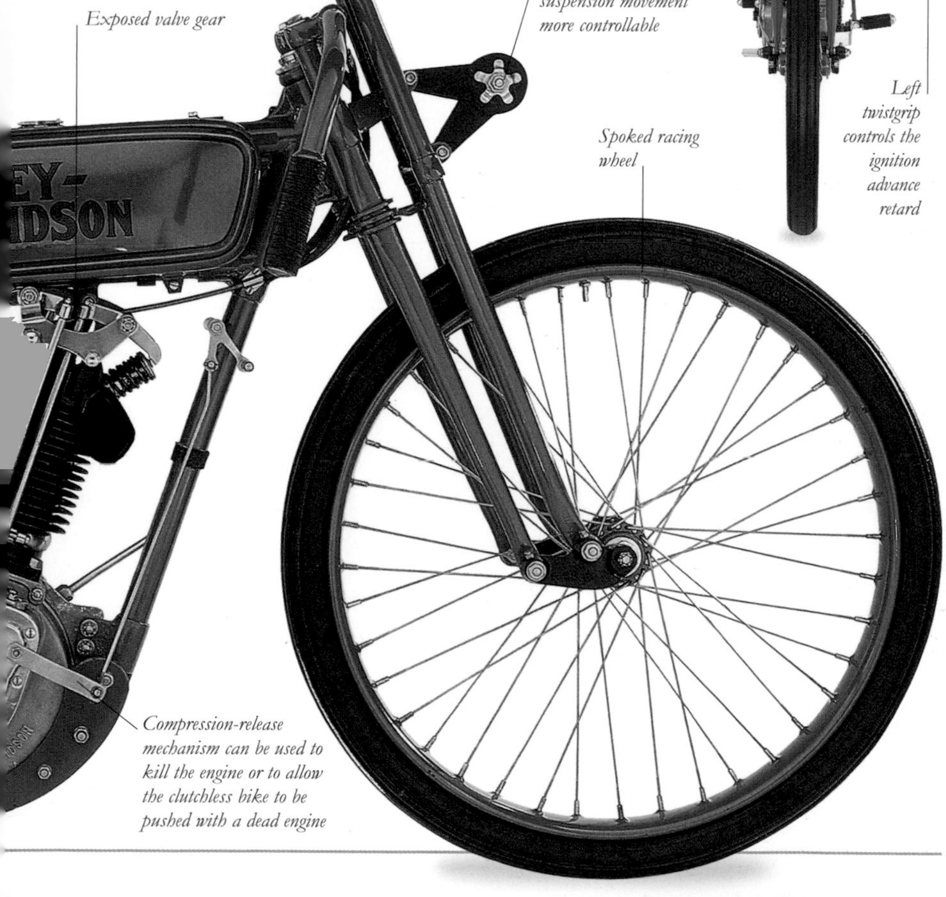

Compression-release mechanism can be used to kill the engine or to allow the clutchless bike to be pushed with a dead engine

HARLEY-DAVIDSON *Model S Racer*

A NEW 350CC RACING CLASS was created soon after Harley unveiled its "Peashooter" racer in the summer of 1925. The bike was based on its new 21cu. in. (346cc) o.h.v. single-cylinder economy road bike. To make it competitive for dirt-track racing, the bike had a shortened frame and simple telescopic forks that were triangulated for greater strength. The legendary Joe Petrali was among several riders who achieved success on Peashooters as Harleys swept the boards in the new class. Petrali was one of the best riders in the history of American bike racing, and in 1935 he won all 13 rounds of the US dirt-track championships on a team Peashooter. As well as success at home, the bikes were also raced in Britain and Australia.

SPECIFICATIONS

MODEL Harley-Davidson Model S Racer
CAPACITY 21cu. in. (346cc)
POWER OUTPUT 12 bhp
WEIGHT 109 kg (240 lb)
TOP SPEED 70 mph (113 km/h) (est.)
COUNTRY OF ORIGIN US

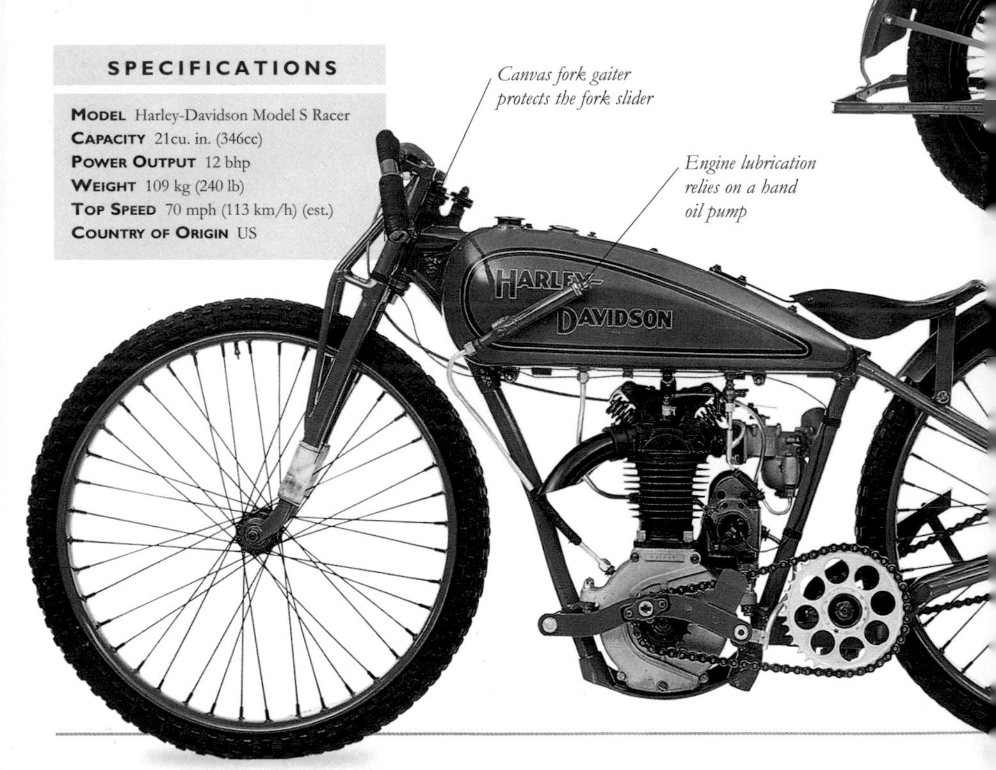

SPECIFICATIONS

MODEL Harley-Davidson JD28
CAPACITY 74 cu. in. (1213cc)
POWER OUTPUT 9.5 bhp @ 4500 rpm
WEIGHT 365 lb (166 kg)
TOP SPEED 75 mph (121 km/h)
COUNTRY OF ORIGIN US

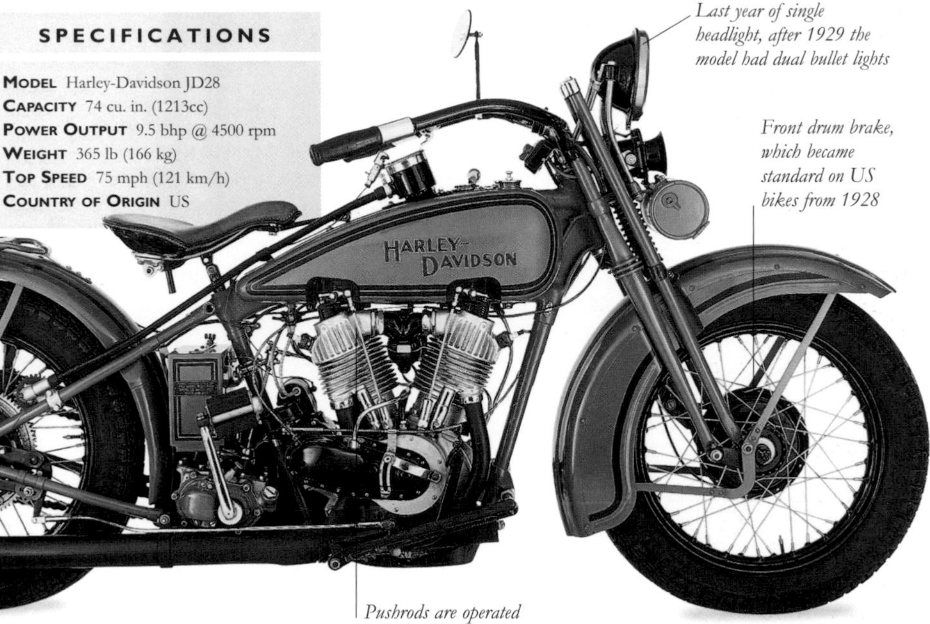

Last year of single headlight, after 1929 the model had dual bullet lights

Front drum brake, which became standard on US bikes from 1928

Pushrods are operated by a single camshaft

HARLEY-DAVIDSON *JD28*

HARLEY RARELY RUSHED change, and it always knew the value of cubic inches. The history of the 45° F-head V-twin goes back to the original prototype twin of 1907, and by 1928 it was approaching its expiration date. In 1922 Harley created the JD model by increasing capacity from 61 to 74 cubic inches (1213cc). The result was a high-performance machine capable of outrunning almost any other vehicle on the road in the 1920s. No wonder these models were popular with police departments. Harley dropped the big F-head twins and replaced them with side-valve machines in 1930.

HARLEY-DAVIDSON *Hill Climber*

THE INGREDIENTS OF A hill-climb bike appear simple, even if the reality is a lot more complicated. The essential element is power, and in the case of this machine a methanol-burning eight-valve engine was enough in 1930 to make it a competitive bike. A long wheelbase and weight at the front to prevent the bike from tipping over backward are both essential, as is grip, which is why this bike's rear tire is wrapped in chains. These crude facts belie the level of expertise involved in handling these machines, and once again it was Joe Petrali who took the honors. Between 1932 and 1938, he won six national hill-climbing titles on a Harley.

Large-diameter rear sprocket to give lower gearing and better climbing ability

Gearbox only provides one gear and neutral

SPECIFICATIONS

Model Harley-Davidson Hill Climber
Capacity 74cu. in. (1213cc)
Power Output Not known
Weight 350 lb (147 kg) (est.)
Top Speed Determined by chosen gearing
Country of Origin US

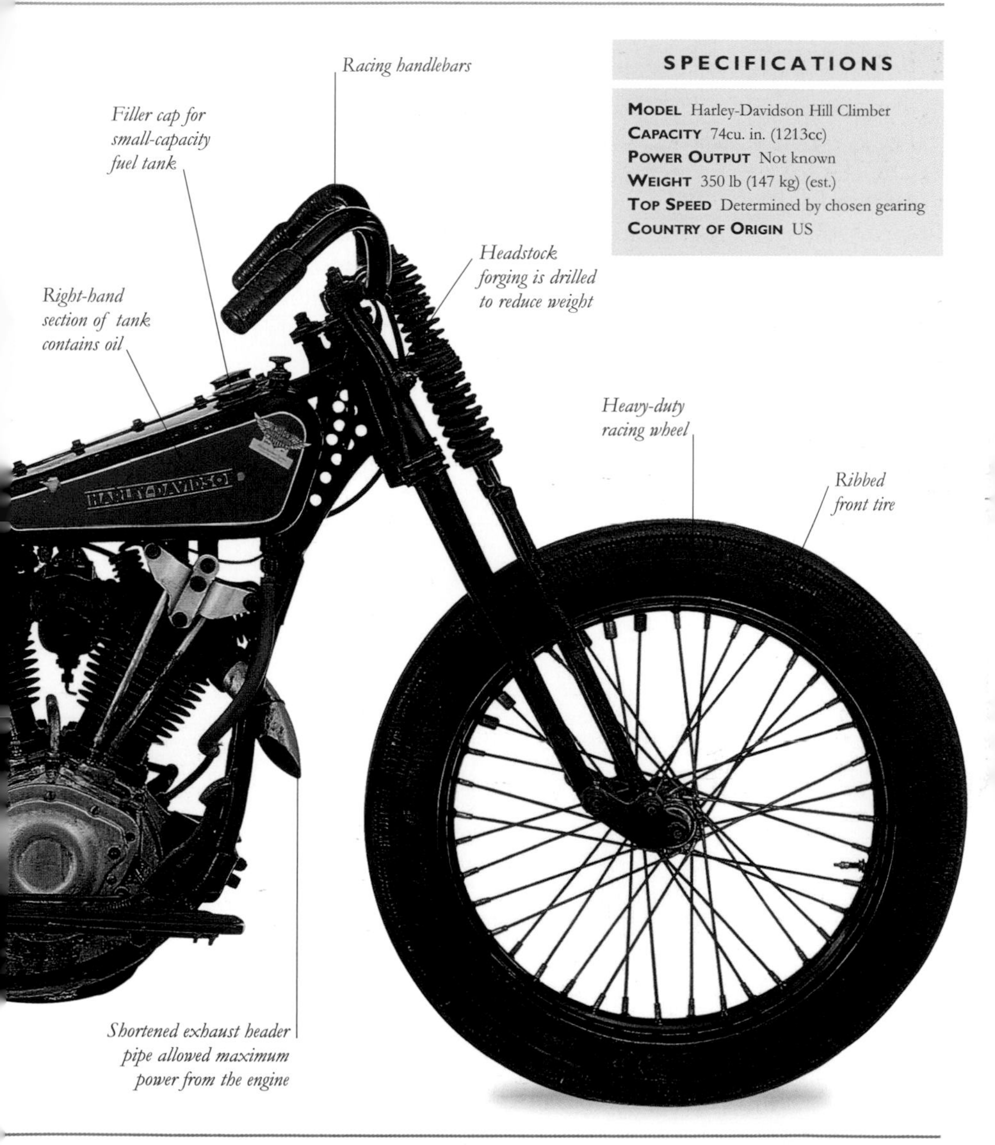

HARLEY-DAVIDSON *VLE*

HARLEY INTRODUCED THE V-series in 1930, 14 years after rival Indian had made its first side-valve big twins; but the bike suffered a number of teething problems. The first two months' production had to be recalled so that frames, flywheels, engine cases, valves, springs, and kick-start mechanisms could be changed. The work was carried out for owners free of charge, but it was a costly exercise that did little for the reputation of the V-series. After the shaky start, the side-valves evolved into rugged, fast, dependable bikes, and a VLE—the high-compression model in the series—even went on to establish the American production bike speed record in 1933 of 104 mph (167 km/h).

SPECIFICATIONS

MODEL Harley-Davidson VLE
CAPACITY 74cu. in. (1213cc)
POWER OUTPUT 22 bhp
WEIGHT 390 lb (177 kg)
TOP SPEED 65 mph (105 km/h)
COUNTRY OF ORIGIN US

SPECIFICATIONS

MODEL Harley-Davidson RL
CAPACITY 45cu. in. (738cc)
POWER OUTPUT 22 bhp (est.)
WEIGHT 390 lb (177 kg) (est.)
TOP SPEED 65 mph (105 km/h)
COUNTRY OF ORIGIN US

Diamond graphic tank detail was used in 1934 and '35

Horn replaced toolbox in 1935

Two-tone teak red-and-black paint finish

HARLEY-DAVIDSON *RL*

THE ORIGINAL D-series Harley 45s were introduced in 1929, produced by Harley in response to the success of the popular Indian Scout. The series was not considered a success, and it was replaced, in 1932, by the new R-series. The critical change was the new frame, which now featured a curved front downtube and allowed the use of a conventional horizontal generator in front of the engine. These 45s were available in four versions: the basic R model, the high-compression RL, the RLD, and the sidecar RS.

HARLEY-DAVIDSON *61EL Knucklehead*

SOME PEOPLE CONSIDER the 61 "Knucklehead" to be the bike that put Indian out of business; others claim it was the bike that saved Harley-Davidson. Either way, this was Harley's first real production overhead-valve twin and, introduced in 1936, it was a groundbreaking machine. The crucial new feature on the bike was its all-new overhead-valve Knucklehead engine that, for the first time on a Harley, also had a recirculating lubrication system. But the 61 wasn't just about improved technology—it was also one of the best-looking bikes that Harley ever built, and elements of its design can be seen in the cruisers of today. The teardrop fuel tank, curved mudguards, and elegant detailing gave the bike a tight, purposeful, and modern look.

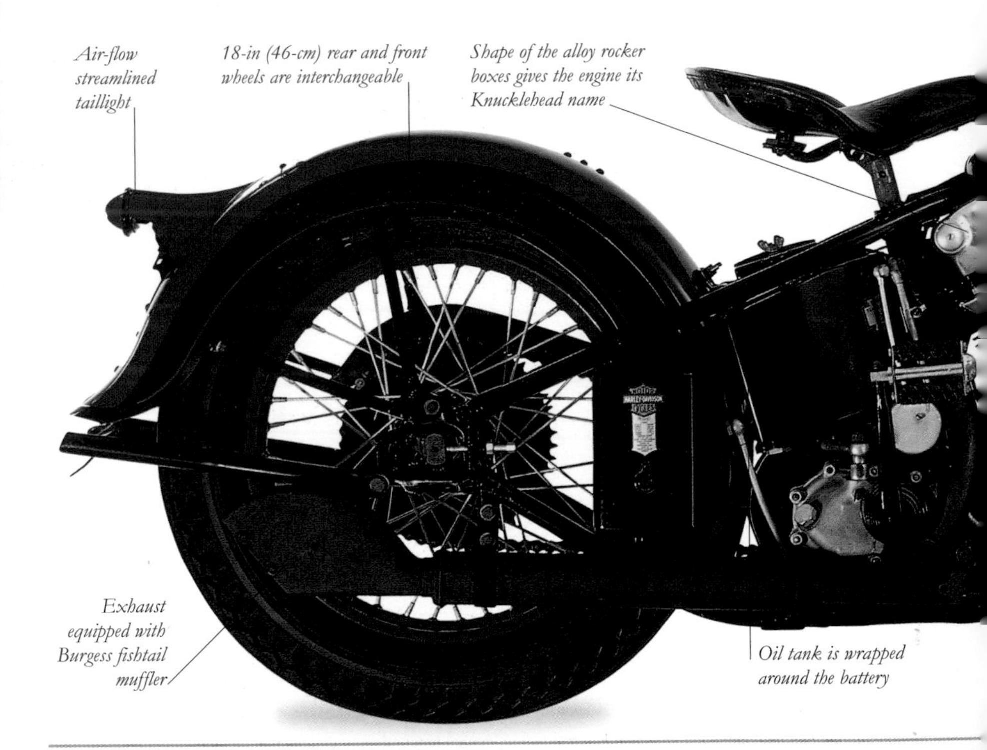

SPECIFICATIONS

MODEL Harley-Davidson 61EL Knucklehead
CAPACITY 61cu. in. (1000cc)
POWER OUTPUT 40 bhp @ 4800 rpm
WEIGHT 515 lb (234 kg)
TOP SPEED 100 mph (161 km/h)
COUNTRY OF ORIGIN US

GEARBOX INNOVATION
The 61 carried Harley-Davidson's first four-speed, constant mesh gearbox, yet another feature that in technological terms pushed Harley ahead of its rival Indian.

Control cables run inside the handlebars

Teardrop fuel tank

Friction suspension shock

Crash bar was a standard fixture

Diagonal air intake only used on 1936 models

HARLEY-DAVIDSON *WLD*

HARLEY-DAVIDSON HAD originally followed Indian when the latter produced its first 45cu. in. (738cc) side-valve machine in 1927. Initially, the Indian 45s were the most highly regarded; but by the time Harley introduced its W-series in 1937, it was the Milwaukee-built bikes that enjoyed the better reputation. Replacing the R-series, the three models in the original lineup were the basic W, this high-compression WLD, and the competition WLDR. The main difference over the Rs was in the new styling, which mimicked the classy 61 Knucklehead (*see pp.146–47*) introduced the previous year. Just like their big brother, the 45s now had teardrop tanks with an integrated instrument panel and curved mudguards, creating a quality line that further established Harley as the market leader.

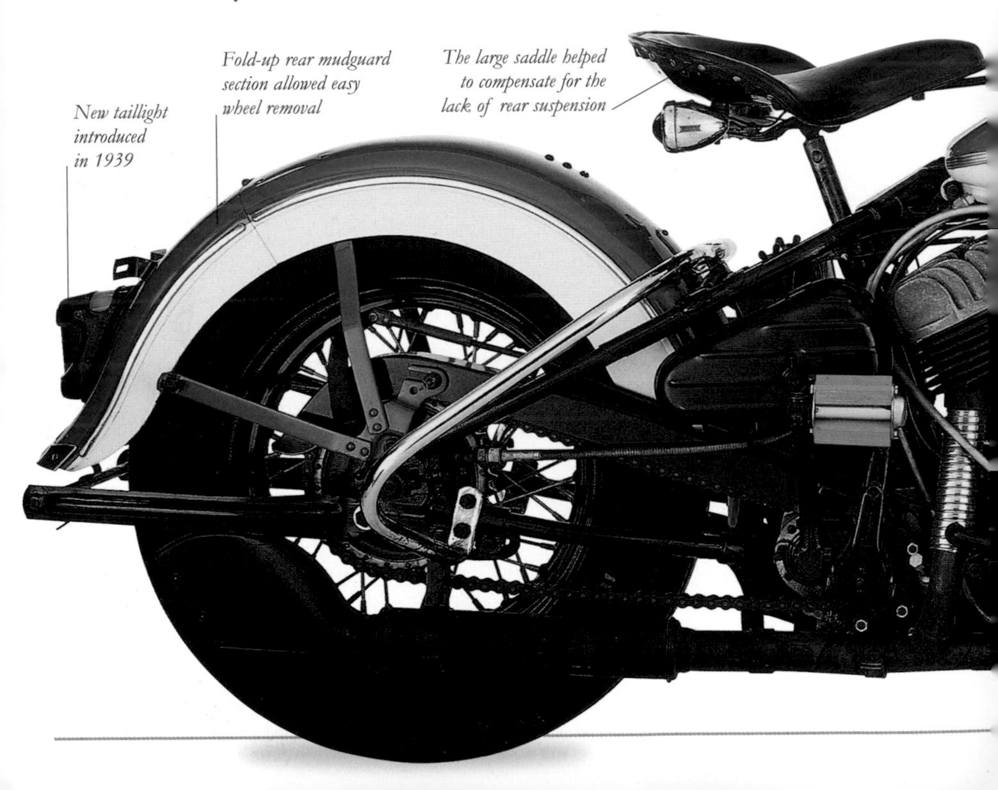

New taillight introduced in 1939

Fold-up rear mudguard section allowed easy wheel removal

The large saddle helped to compensate for the lack of rear suspension

SPECIFICATIONS

MODEL Harley-Davidson WLD
CAPACITY 45cu. in. (738cc)
POWER OUTPUT 25 bhp (est.)
WEIGHT 692 lb (314 kg)
TOP SPEED 96 mph (155 km/h)
COUNTRY OF ORIGIN US

COMPACT BIKES
The Ws were Harley's smallest machines of the period and matched the 45s put out by Indian. In terms of quality control, Harley had been ahead for some time.

New recirculating lubrication system meant that the oil was contained on the left side of the fuel tank

HARLEY-DAVIDSON *U Navy*

THE U MODEL WAS INTRODUCED in 1937 as a replacement for the V-series 74 and 80cu. in. (1213 and 1312cc) twins. The redesigned engine, which had a recirculating lubrication system, was placed into a chassis taken from the 61E Knucklehead *(see pp.146–47)* that had been introduced the previous year. The styling of other components, including the fuel tank and running gear, was also based on the sublime Knucklehead. Side-valve fans got an improved engine with a four-speed gearbox in a much more modern-looking package, and these large-capacity side-valve machines proved to be especially useful for sidecar work. With the outbreak of World War II, Harley began supplying large numbers of machines to the Allied war effort, mainly 45cu. in. (738cc) bikes but also some 74cu. in. U models. This example was used by the US Navy in Guam.

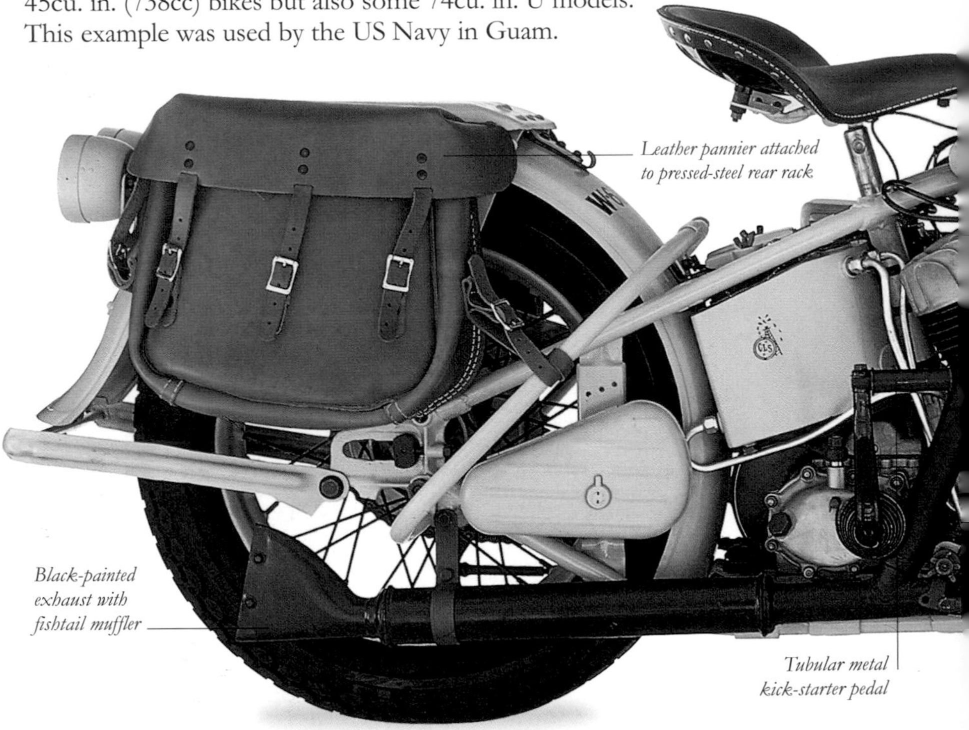

Leather pannier attached to pressed-steel rear rack

Black-painted exhaust with fishtail muffler

Tubular metal kick-starter pedal

SPECIFICATIONS

Model Harley-Davidson U Navy
Capacity 74cu. in. (1213cc)
Power Output 22 bhp
Weight 390 lb (177 kg)
Top Speed 75 mph (120 km/h)
Country of Origin US

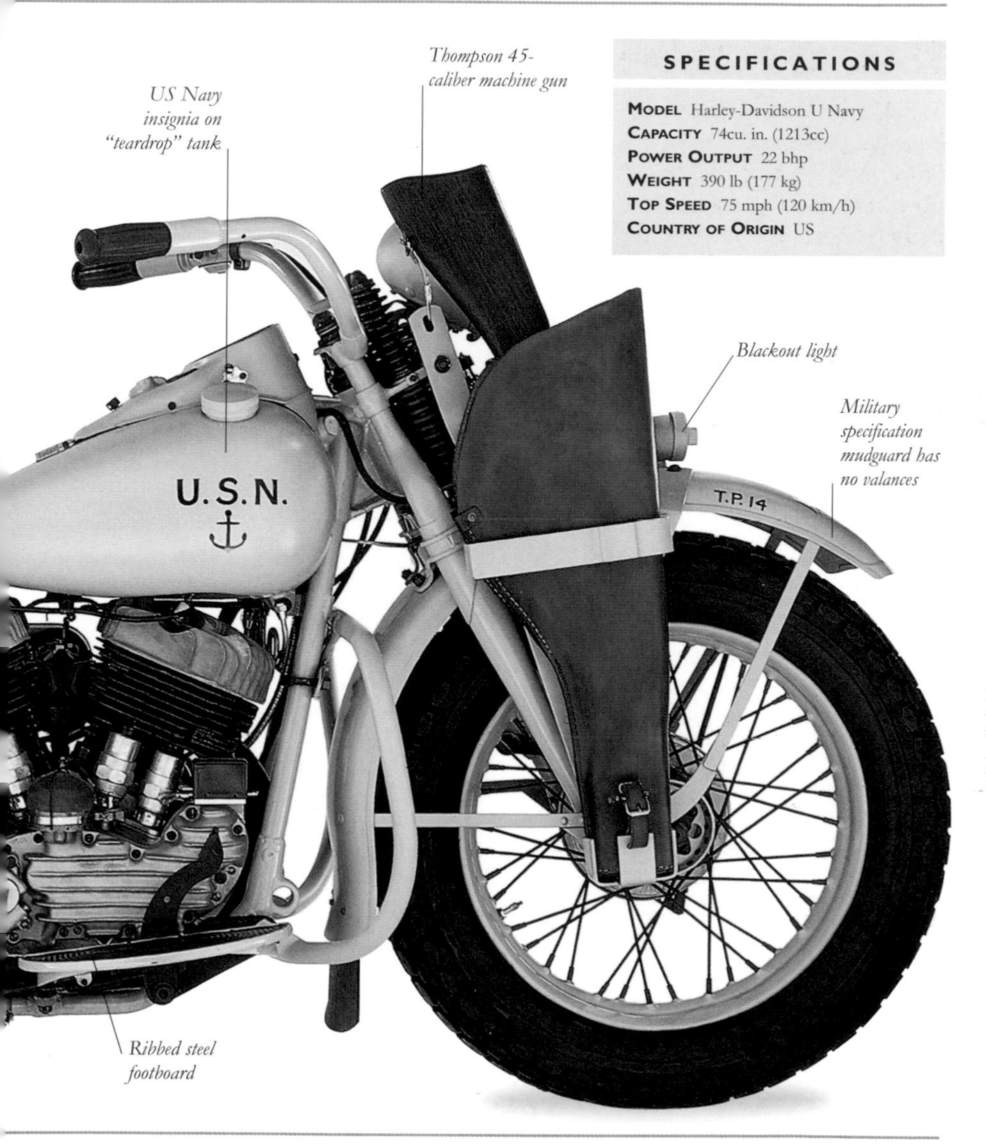

HARLEY-DAVIDSON *WR Racer*

IN 1934 THE RULES OF US racing were changed to encourage the participation of amateur riders on cheaper, production-based motorcycles. Though influenced by the fact that Harley-Davidson and Indian's 45cu. in. (738cc) twins were comparatively cheap and popular at the time, the change meant that Harley had to put out some new models to meet the challenge of the class. In 1937 Harley offered the souped-up WLDR, but the real response came in 1941 when the WR (flat-track) and WRTT (TT) models were introduced. These pure racing machines were supplied without any extraneous equipment. More importantly, the engine was much more powerful than the basic W models *(see pp.148–49)* on which the bike was based.

Different sprockets allowed gearing changes

Folding footrests were attached to the WR; WRTT models came with footboards, brakes, and a different frame

SPECIFICATIONS

MODEL Harley-Davidson WR Racer
CAPACITY 45cu. in. (738cc)
POWER OUTPUT 38 bhp
WEIGHT 300 lb (136 kg)
TOP SPEED 110 mph (177 km/h) (est.)
COUNTRY OF ORIGIN US

HARLEY-DAVIDSON *74FL Hydra-Glide*

HARLEY HAD BEEN keeping its riders comfortable using the springer leading-link fork (introduced 1907) and the sprung seatpost (introduced 1912) for years. By 1949, though, the merits of hydraulically damped telescopic forks were obvious, and from that year they were used on the 61 and 74cu. in. (1000 and 1213cc) twins; hence the name Hydra-Glide. Another development had taken place the previous year with the arrival of the Panhead engine to replace the Knucklehead. The new unit sought to address the Knucklehead's oil leak problem—the Panhead moniker referred to the large rocker covers that enclosed the valvegear and kept the oil inside the engine.

SPECIFICATIONS

MODEL Harley-Davidson 74FL Hydra-Glide
CAPACITY 74cu. in. (1213cc)
POWER OUTPUT 55 bhp @ 4800 rpm
WEIGHT 598 lb (271 kg)
TOP SPEED 102 mph (164 km/h)
COUNTRY OF ORIGIN US

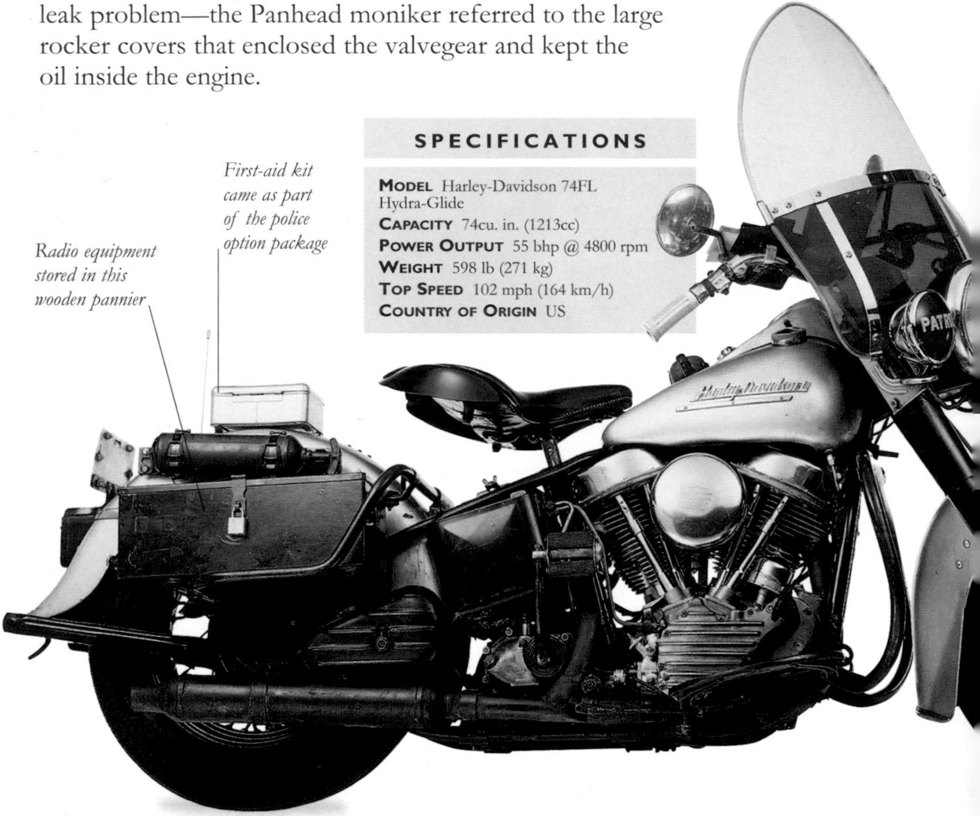

First-aid kit came as part of the police option package

Radio equipment stored in this wooden pannier

SPECIFICATIONS
MODEL Harley-Davidson Model K
CAPACITY 45cu. in. (738cc)
POWER OUTPUT 30 bhp (est.)
WEIGHT 300 lb (181 kg)
TOP SPEED 85 mph (136 km/h) (est.)
COUNTRY OF ORIGIN US

HARLEY-DAVIDSON *Model K*

IN THE EARLY 1950s, American motorcyclists wanted more from their motorcycles. They were buying faster, better handling, and better looking British imports rather than Harley's traditional 45cu. in. (738cc) V-twins. In 1952 Harley struck back with the K model, its first significant new machine since the Knucklehead *(see pp.146–47)*. While the bike incorporated several novel features on a 45cu. in. side-valve block, it still could not match the performance of the smaller-capacity British machines. Increasing capacity to 54cu. in. (883cc) on the KH in 1954 helped, but there was only so much that could be done with the side-valve layout.

HARLEY-DAVIDSON *XL Sportster*

HARLEY FINALLY ATTACHED overhead cylinder heads to its smaller V-twin in 1957 to create the Sportster, a model that was to become one of the longest surviving production motorcycles in the world. The Sportster combined the good looks of the earlier K *(see p.155)* and KH models with enough power to match the performance of contemporary imported bikes. Although capacity remained at 54cu. in. (883cc), the same as for the KH, the larger bore and shorter stroke resulted in increased horsepower. It meant that US buyers could now invest in a domestic product without suffering the indignity of being blown away by their Triumph- and BSA-mounted friends. As with the bigger twins, the factory offered a variety of accessories.

Windshield was a Harley-Davidson optional extra

SPECIFICATIONS

MODEL Harley-Davidson XL Sportster
CAPACITY 54cu. in. (883cc)
POWER OUTPUT 32 bhp @ 4200 rpm
WEIGHT 463 lb (210 kg)
TOP SPEED 92 mph (148 km/h) (est.)
COUNTRY OF ORIGIN US

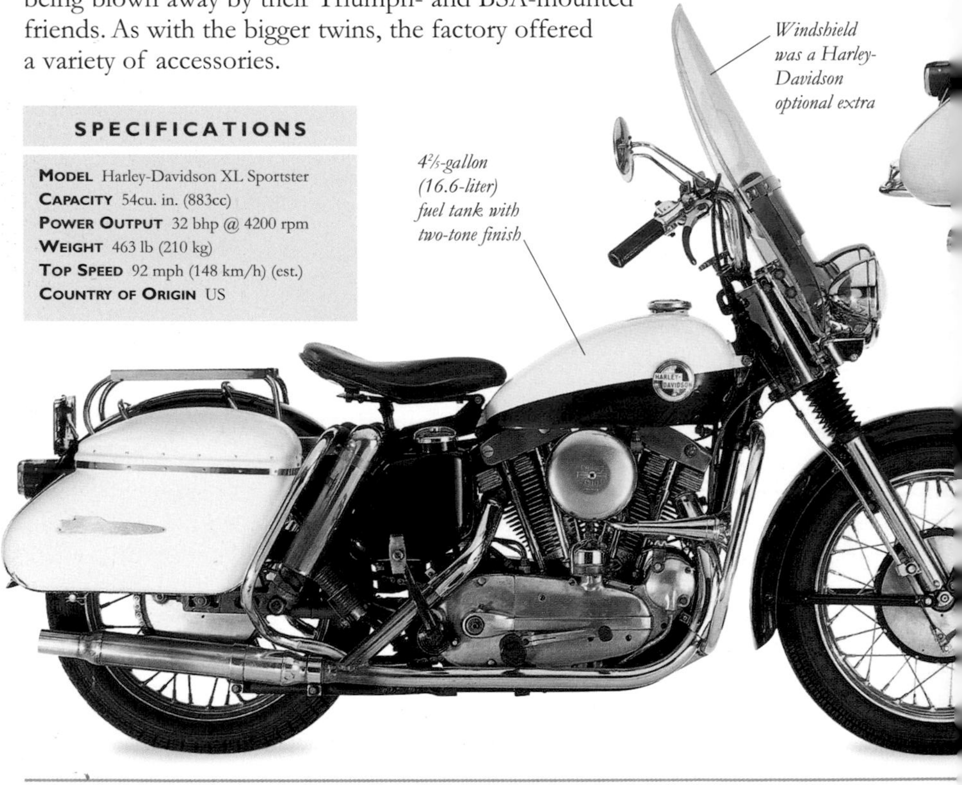

$4\frac{4}{5}$-gallon (16.6-liter) fuel tank with two-tone finish

SPECIFICATIONS

MODEL Harley-Davidson FLH Duo-Glide
CAPACITY 74cu. in. (1213cc)
POWER OUTPUT 55 bhp @ 7200 rpm
WEIGHT 670 lb (304 kg)
TOP SPEED 100 mph (161 km/h)
COUNTRY OF ORIGIN US

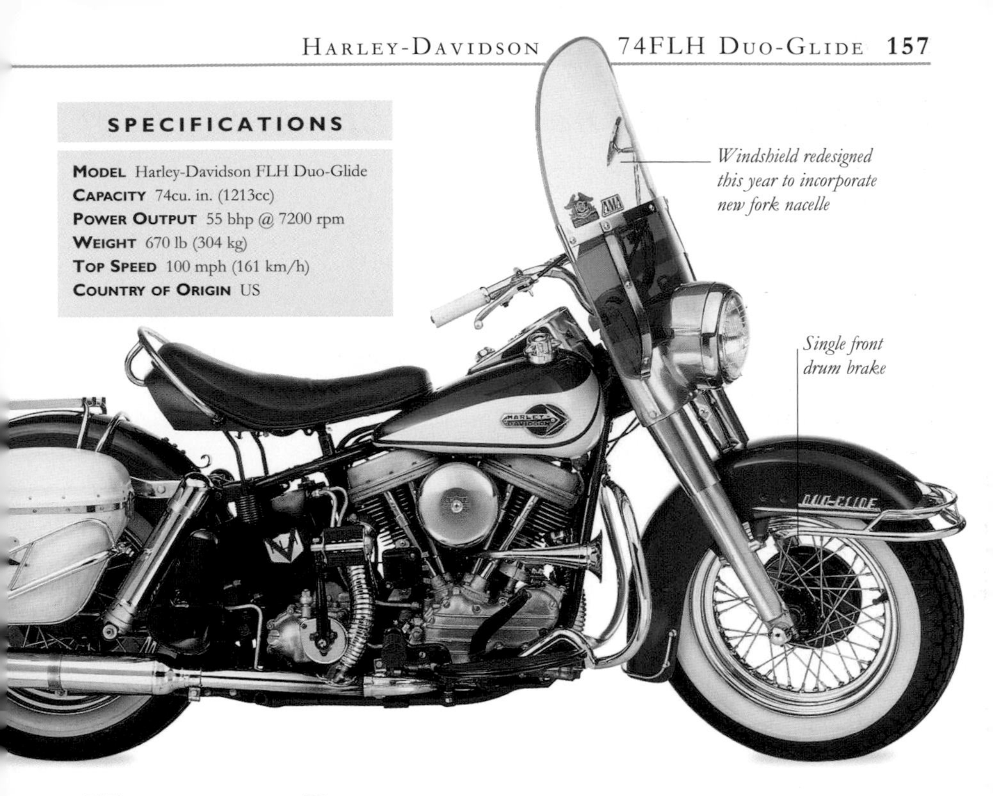

HARLEY-DAVIDSON *74FLH Duo-Glide*

BY THE LATE 1950S, Harley's big-twins had captured a section of the market for big, comfortable, large-capacity tourers. Weight wasn't an issue, but comfort and dependability were. In 1958 Harley finally added rear swingarm suspension to its "Panhead" big-twin and celebrated the addition with the Duo-Glide model name. By the time that this model was built in 1960, almost no two Harleys were the same, as owners added extra components in order to individualize the looks and improve the comfort and capabilities of their bike. Harley offered a range of accessory groups and color schemes, which meant that buyers could specify the extras they wanted on their machine when they ordered it from the dealer.

Harley-Davidson *KRTT*

Along with the introduction of the new K-series road bike in 1952 *(see p.155)*, Harley-Davidson also released a racing version, designated the KR. Its engine had all the tweaks you would expect in a competition power unit, while looking externally similar to the K. There were big valves, racing cams, and new bearings, as well as reshaped ports and a revised cylinder head. Riders like Brad Andres, Carroll Resweber, and Roger Reiman achieved many wins on the side-valve KR racers in the 1950s and '60s. Because of the variety of track surfaces found in American racing, both sprung and rigid versions of the KR's frame were produced.

Large fuel tank for long-distance races

Alloy wheel rim is lighter and stronger than the usual steel examples

SPECIFICATIONS

MODEL Harley-Davidson KRTT
CAPACITY 45cu. in. (748cc)
POWER OUTPUT 50 bhp
WEIGHT 320 lb (145 kg)
TOP SPEED 125 mph (233 km/h)
COUNTRY OF ORIGIN US

RACE LIFE
While the standard K-series road bike was discontinued for 1957, the racing KRs had continued success with this engine layout until the late 1960s.

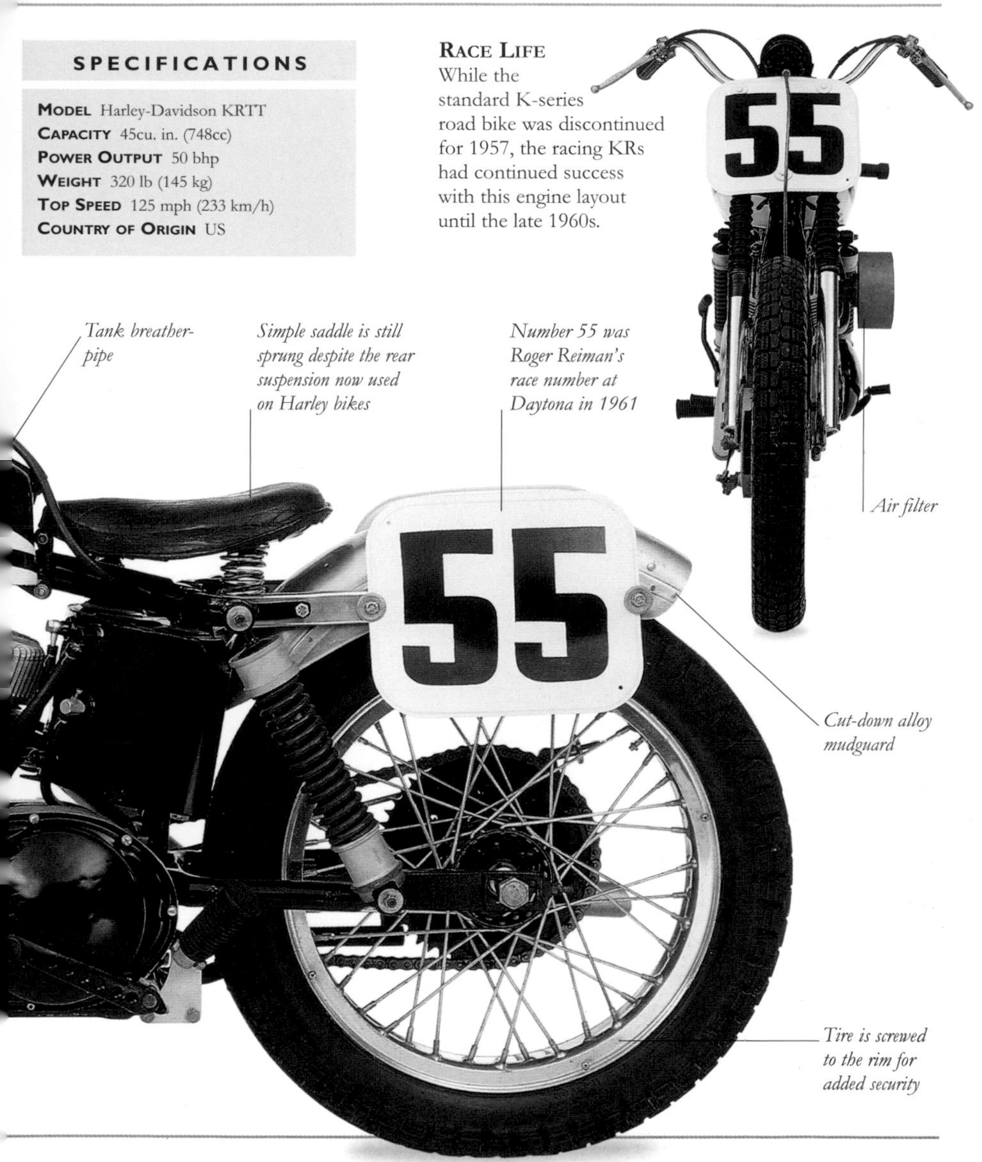

Tank breather-pipe

Simple saddle is still sprung despite the rear suspension now used on Harley bikes

Number 55 was Roger Reiman's race number at Daytona in 1961

Air filter

Cut-down alloy mudguard

Tire is screwed to the rim for added security

HARLEY-DAVIDSON *Sprint H*

HARLEY BOUGHT A SHARE in the Italian Aermacchi company in 1960, and this Harley-labeled Aermacchi 250 joined the line the following year under the Sprint moniker. It was unlike any other Harley-Davidson, and wary dealers treated the model with caution. The single-cylinder 246cc engine had wet sump lubrication, a cylinder positioned almost horizontally, and pushrod-operated valves. Unusually, the crankshaft rotated in the opposite direction to the wheels. Although the Sprint was a nice enough machine, it was pushed to keep up with comparable Hondas of the period. Production of the Italian four-strokes continued until 1974, by which time a 350cc model had joined the 250.

SPECIFICATIONS

MODEL Harley-Davidson Sprint H
CAPACITY 246cc
POWER OUTPUT 28 bhp
WEIGHT 280 lb (127 kg)
TOP SPEED 90 mph (145 km/h) (est.)
COUNTRY OF ORIGIN Italy

Frame brace reinforces the critical area between the swingarm and the suspension top-mounting

Hexagonal tank logo was introduced in 1966

Horizontal cylinder finning provided improved cooling

Italian Ceriani telescopic forks and front brakes were the best available components of the period

SPECIFICATIONS

MODEL Harley-Davidson CRTT
CAPACITY 248cc
POWER OUTPUT 35 bhp @ 10,000 rpm
WEIGHT 245 lb (111 kg)
TOP SPEED 115 mph (185 km/h)
COUNTRY OF ORIGIN Italy

Twin leading-shoe front drum brake

Elongated headlight shell contains the speedometer

HARLEY-DAVIDSON *CRTT*

THE BASIC DESIGN of this Italian-built, overhead-valve single was created by Alfredo Bianchi and was originally based on a 175cc unit that powered Aermacchi's distinctive Chimera road bike. Also known as the Ala D'Oro ("Golden Wing"), a number of racing versions of the bike were produced from 1961, in 250, 350, and 402cc formats. Although the layout was the same, the race bikes differed from the road bikes in many respects. Engine cases were sand-cast and incorporated the provision for a dry clutch and crankshaft-driven magneto ignition.

HARLEY-DAVIDSON *FX Super Glide*

THOUGH HARLEY-DAVIDSON frowned on the customizers who modified its machines in the 1960s—and didn't approve of the influence of the film *Easy Rider*—the company introduced its own tribute to customizing in 1971. The FX Super Glide joined a kick-start 74cu. in. (1213cc) engine with the forks and front wheel from a Sportster to give the bike a chopper-inspired look. The idea was to combine the grun of the big F-series engine with the lean looks of the X-series Sportsters. But although the Super Glide concept proved to be a winner in the long run, the unique bodywork on the 1971 model was too much for customers of the time, and '72 models came with a more conventional seat and mudguard.

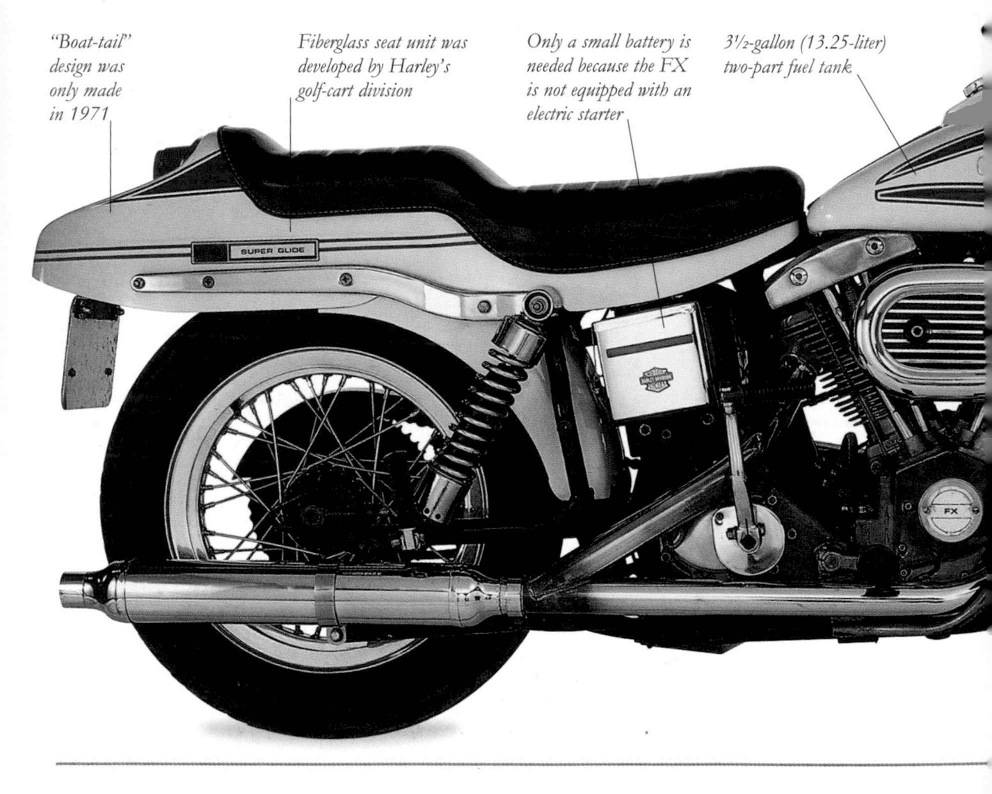

SPECIFICATIONS

MODEL Harley-Davidson FX Super Glide
CAPACITY 74cu. in. (1213cc)
POWER OUTPUT 65 bhp @ 5400 rpm
WEIGHT 549 lb (254 kg) (with ½ tank fuel)
TOP SPEED 108 mph (174 km/h)
COUNTRY OF ORIGIN US

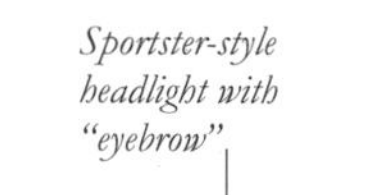

REAR PROFILE
The slim rear profile of the Super Glide is only compromised by the bulbous primary drive casing. Even so, the FX looked like no other bike Harley had produced.

HARLEY-DAVIDSON *XRTT*

ROAD-RACING WAS A RARITY in the US in the 1960s. Most racing action was on dirt tracks for which the XR and KR models were conceived, but Harley decided to build a road-race version of the XR. Unlike the flat-track bike, it was equipped with a front brake and a large-diameter four-leading shoe drum, which was combined with a rear disc like on the dirt bike. Most machines had the disc-drum combination the other way around. Although the XRTT was a useful machine, its success was in the most part due to its most famous rider, Cal Rayborn, who often beat machines of greater power on this XRTT. Ultimately, these road-race XR models did not have as much success as the dirt-track versions.

SPECIFICATIONS

MODEL Harley-Davidson XRTT
CAPACITY 45cu. in. (748cc)
POWER OUTPUT 90 bhp @ 8000 rpm
WEIGHT 324 lb (147 kg)
TOP SPEED 130 mph (209 km/h) (est.)
COUNTRY OF ORIGIN US

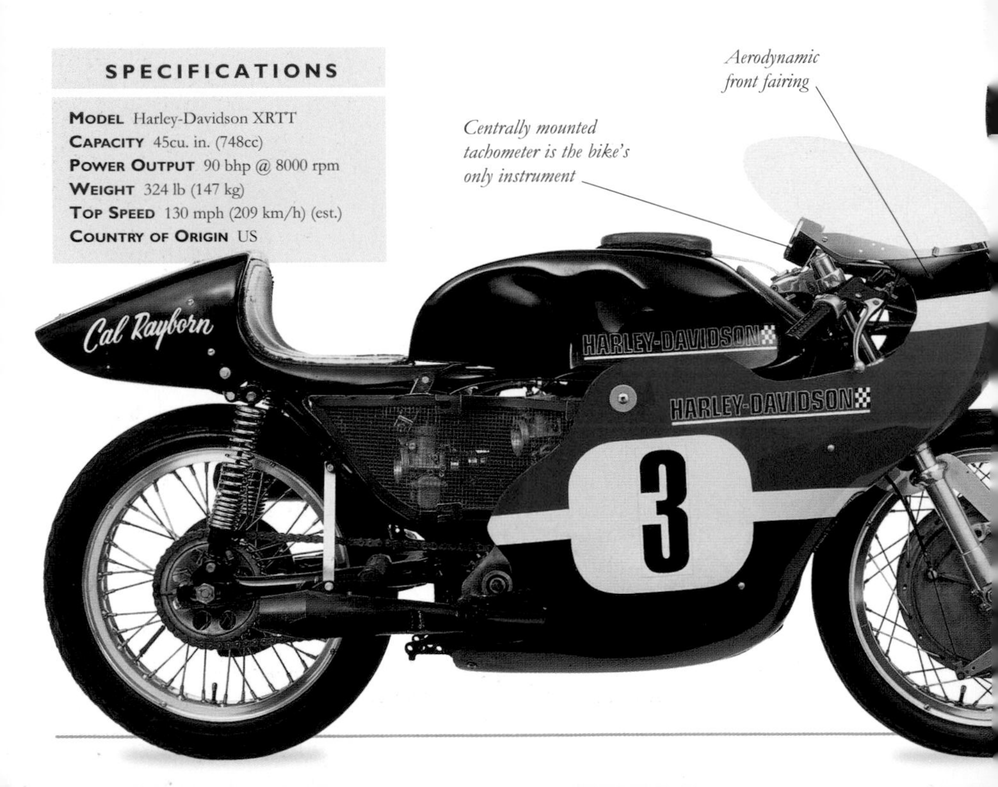

SPECIFICATIONS

MODEL Harley-Davidson 250SS
CAPACITY 243cc
POWER OUTPUT Not known
WEIGHT 245 lb (111 kg)
TOP SPEED 85 mph (137 km/h)
COUNTRY OF ORIGIN Italy

HARLEY-DAVIDSON *250SS*

IN THE MID-1970S Harley released a line of modern-looking, single-cylinder, two-strokes built in Italy at the Aermacchi factory. These replaced its aging lineup, which included the four-stroke Sprint *(see p.160)*. Offered in both street (SS) and trail (SX) styles, the bikes came in 125 and 175cc variants from 1974, and a 250cc model from 1975. It may have looked like a neat bike, but once again it couldn't match the Japanese competition of the time. The SX had some success and was produced until 1978, but only 1,417 SSs were sold in 1976, and this model was dropped soon after its release—along with the whole Aermacchi subsidiary, which Harley decided to relinquish in 1978.

Harley-Davidson *RR250*

Most traditional Harley riders may not realize that the company won a string of World Championships in the mid-1970s, and if they do they probably don't really care. The bikes that gave Harley the titles were as far removed from the traditional V-twin as it's possible to get. These high-revving two-stroke twins were developed in Italy to take on the Japanese manufacturers in international road-racing championships. Ridden by Italian ace Walter Villa, the twins won three straight 250cc World Championships in 1975, '76, and '77, and a 350cc version also took the World title in 1977.

SPECIFICATIONS

MODEL Harley-Davidson RR250
CAPACITY 246cc
POWER OUTPUT 53 bhp
WEIGHT 240 lb (109 kg)
TOP SPEED 140 mph (225 km/h)
COUNTRY OF ORIGIN Italy

BARE ESSENTIALS
Racing machines have minimal instrumentation, allowing the rider to concentrate on the job at hand, and the RR250 is no exception. The large, white-faced tachometer gives all the information needed to time gear-changes to perfection.

Tachometer mounting isolates the instrument from vibration

Road-racing grooves in tire

Lightweight mudguard

Borrani alloy wheel rims are flanged for extra strength

Expansion chamber exhaust pipe allows the two-stroke engine to realize its full power

Scarab front brake caliper

HARLEY-DAVIDSON *XLCR*

IN 1977 HARLEY INTRODUCED a new variation of the Sportster. The XLCR was designed by Willie G. Davidson, and the CR stood for café racer. Harley's model was a blend of 1960s café racer—a name given to stripped, souped-up hot rods used for blasting from bar to bar—with some of the styling cues of the XR flat-track racers. The frame and the exhaust pipes were new and would be used on the rest of the Sportster range the following year, but the engine was a stock XL1000. The really important stuff was the bodywork and the black finish. The XLCR looked great, but never sold in the numbers that were hoped for and was dropped after only two years.

SPECIFICATIONS

MODEL Harley-Davidson XLCR
CAPACITY 61cu. in. (1000cc)
POWER OUTPUT 55 bhp
WEIGHT 470 lb (213 kg)
TOP SPEED 105 mph (169 km/h)
COUNTRY OF ORIGIN US

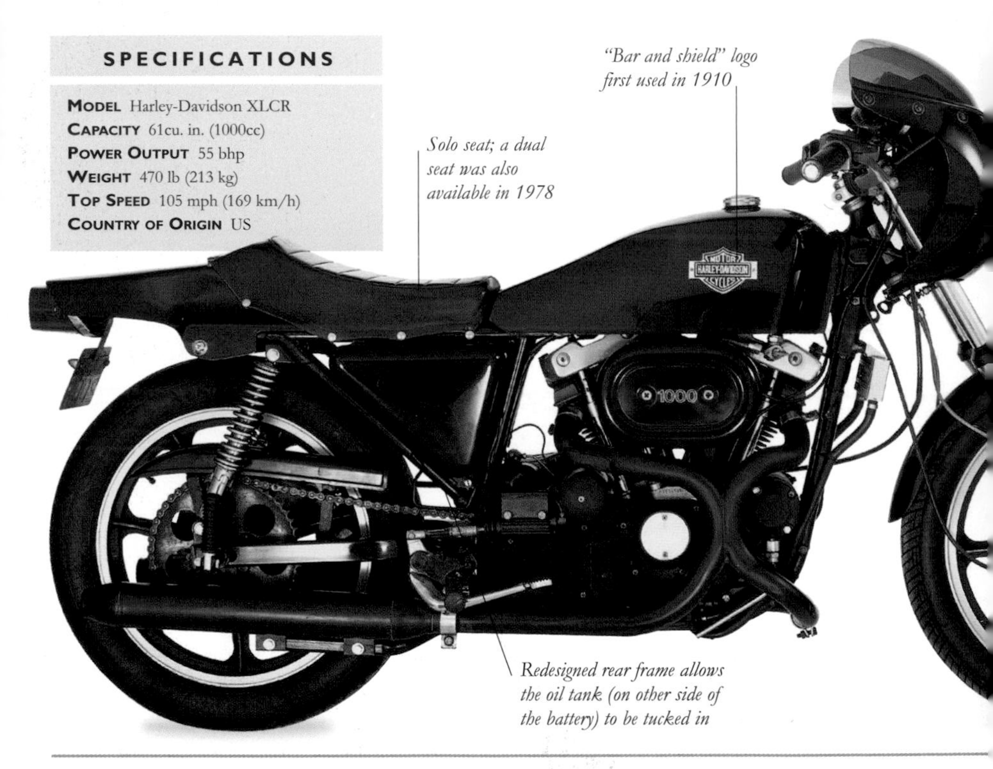

Solo seat; a dual seat was also available in 1978

"Bar and shield" logo first used in 1910

Redesigned rear frame allows the oil tank (on other side of the battery) to be tucked in

SPECIFICATIONS

MODEL Harley-Davidson RR500
CAPACITY 488cc
POWER OUTPUT 100 bhp (est.)
WEIGHT 265 lb (120 kg)
TOP SPEED 155 mph (250 km/h)
COUNTRY OF ORIGIN Italy/US

Large-capacity fuel tank is a prerequisite for thirsty two-stroke racers

Tubular frame was constructed by chassis specialists Bimota

Twin front discs are each gripped by two brake calipers

AMF Harley-Davidson

HARLEY-DAVIDSON *RR500*

HARLEY-DAVIDSON'S ITALIAN-BUILT 250 and 350cc twin-cylinder two-stroke racers bagged four World Championships in 1975, '76, and '77 *(see pp.166–67)*. Using the same formula to attempt similar success in the 500 class at the same time seemed like a good idea. The RR500 was a water-cooled, two-stroke twin, which was unusual in that each cylinder was fed by two carburetors. The model shown on this page is a late version that features a monoshock frame and cast-alloy wheels. The RR500 made a few appearances in Europe and America during the mid-Seventies, but never with any real success.

HARLEY-DAVIDSON *XR750*

THE XR750 IS THE most successful competition bike ever produced, though for many it's more famous as the bike that Evel Knievel used for his stunts. The early bikes were introduced in 1970 with iron barrels and heads that failed miserably, so a revised alloy engined version of the bike was introduced two years later. For the first time on production Harley V-twin, the rear cylinder had a forward-facing exhaust and rear-facing inlet port. The bike won the A.M.A. Grand National Championship in its first year, and is still winning races over 30 years later.

SPECIFICATIONS

Model Harley-Davidson XR750
Capacity 45cu. in. (748cc)
Power Output 90 bhp @ 8000 rpm
Weight 295 lb (134 kg)
Top Speed 115 mph (185 km/h) (est.)
Country of Origin US

Harley-Davidson *XR1000*

It seemed obvious. Harley's XR750 *(see pp.170–01)* was cleaning up in dirt-track racing, so why not offer a limited run of road-bike examples? In 1983 Harley produced the XR1000, attaching the alloy heads and twin Dell'Orto carburetors from the XR750 onto the bottom half of an XL1000 engine. The engine itself was put into a standard XLX chassis. The result was an engine that put out 10 percent more power than the stock Sportster in standard trim, though many owners souped up the bike even more. It created the fastest production bike Harley ever made, but buyers were not impressed—it looked almost identical to the cheapest bike in the range while costing a great deal more. The XR1000 didn't sell, but the bike is now a collector's item.

Nonstandard Corbin seat; the XR was supplied with a solo saddle only

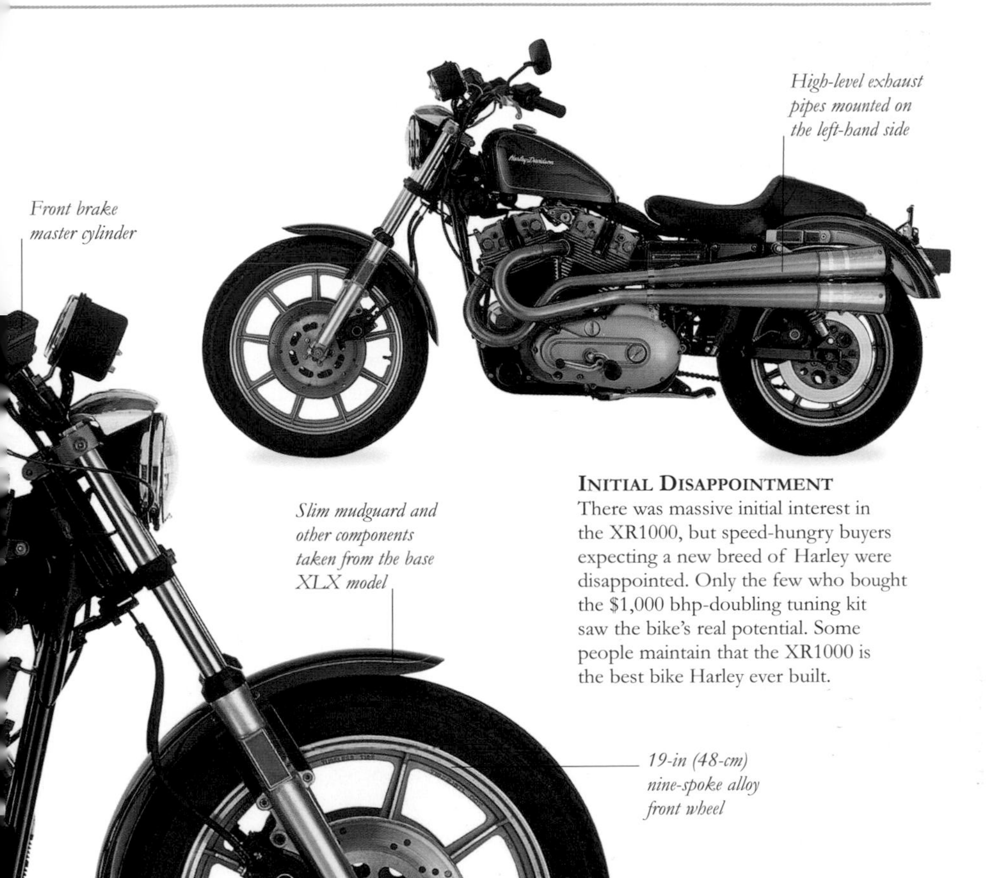

High-level exhaust pipes mounted on the left-hand side

Front brake master cylinder

Slim mudguard and other components taken from the base XLX model

19-in (48-cm) nine-spoke alloy front wheel

INITIAL DISAPPOINTMENT

There was massive initial interest in the XR1000, but speed-hungry buyers expecting a new breed of Harley were disappointed. Only the few who bought the $1,000 bhp-doubling tuning kit saw the bike's real potential. Some people maintain that the XR1000 is the best bike Harley ever built.

SPECIFICATIONS

MODEL Harley-Davidson XR1000
CAPACITY 61cu. in. (1000cc)
POWER OUTPUT 70 bhp @ 6000 rpm
WEIGHT 470 lb (213 kg) (est.)
TOP SPEED 120 mph (193 km/h) (est.)
COUNTRY OF ORIGIN US

HARLEY-DAVIDSON *FLHX Electra Glide*

AFTER 18 YEARS IN PRODUCTION the Shovelhead engine was replaced in 1984 by the new, but externally similar, Evolution engines. The FLHX was the swan song of the Shovelhead Electra Glides, a special limited-edition model (apparently only 1,250 were made) available in black or white with wire-spoked wheels and full touring equipment. Cynics would say that this was a good excuse to use up the last of the old-style engines, while others might argue that an engine with the reputation and life span of the Shovelhead deserved a celebratory parting shot. Either way it was the end of an era.

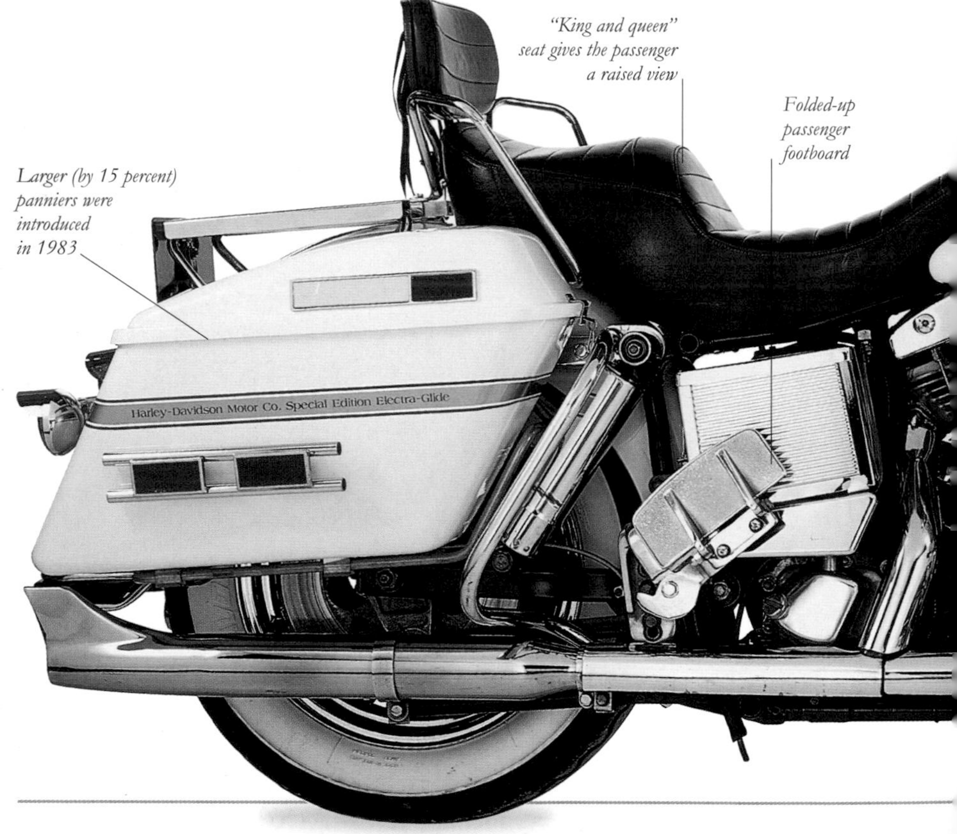

"King and queen" seat gives the passenger a raised view

Folded-up passenger footboard

Larger (by 15 percent) panniers were introduced in 1983

SPECIFICATIONS

Model Harley-Davidson FLHX Electra Glide
Capacity 80cu. in. (1312cc)
Power Output 65 bhp (est.)
Weight 752 lb (341 kg) (with ½ tank fuel)
Top Speed 90 mph (145 km/h)
Country of Origin US

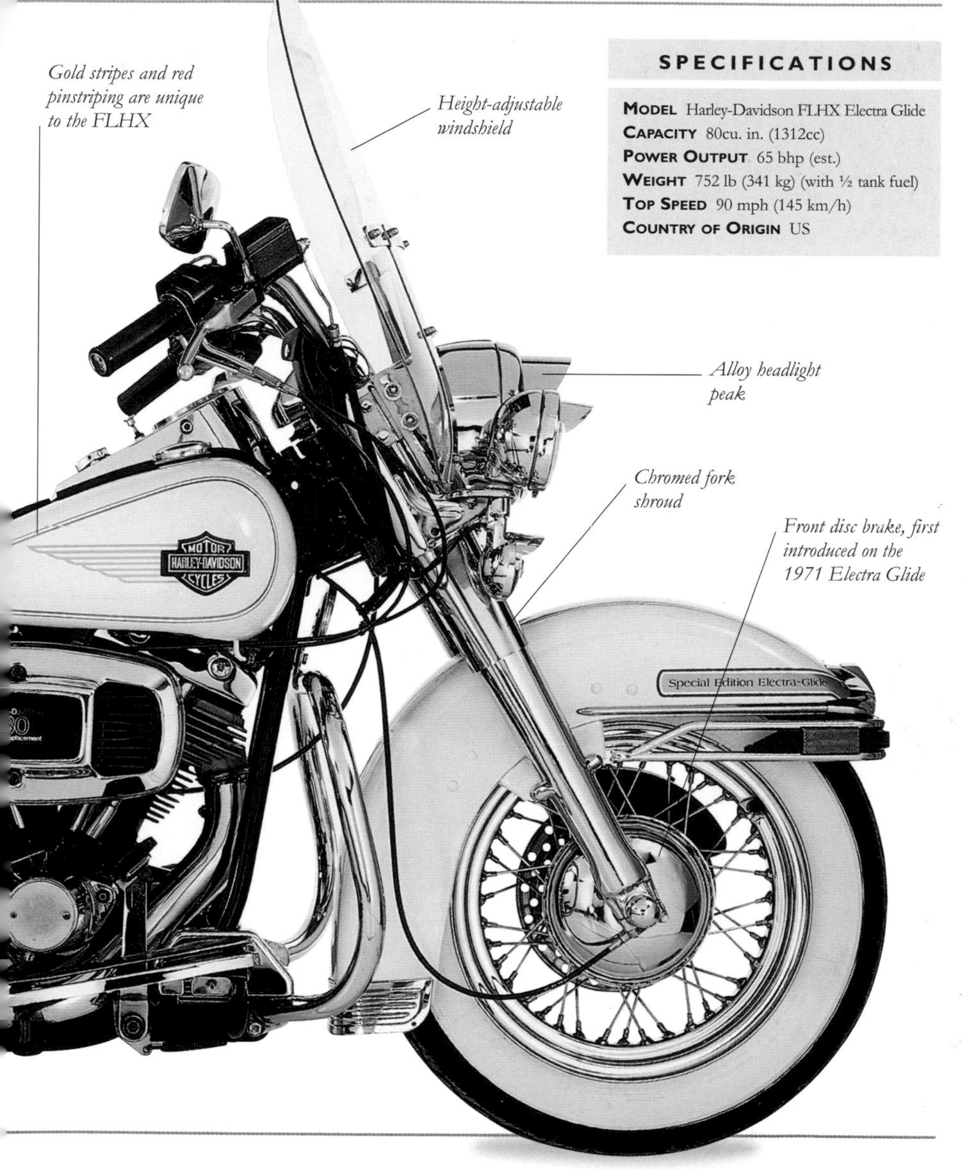

Harley-Davidson *VR1000*

There are probably several reasons why, in 1994, Harley decided to develop a totall new racing bike. Corporate pride, the necessity to familiarize itself with new technology, and a desire to appeal to a new type of customer are among them. Whatever, the VR10 made its debut under the spotlight at America's most prestigious road race—the 1994 Daytona 200-mile (322-km) Superbike. However, it wasn't a fairy-tale debut as the bike was off the pace and then blew up. Five years on, the VR1000 had still to achieve significant success despite swallowing large amounts of money and development time.

SPECIFICATIONS

Model Harley-Davidson VR1000
Capacity 61cu. in. (996cc)
Power Output 140 bhp @ 10,400 rpm
Weight 355 lb (161 kg)
Top Speed 190 mph (306 km/h)
Country of Origin US

uick-detach fairing steners allow a placement to be sed during a race

Inverted upside-down telescopic fork provides maximum rigidity for minimum weight

Race Rules

Limited numbers of the VR1000 were offered for sale to the public to comply with Superbike racing rules that stated that a number of production versions of the competing bikes had to be produced. This was the fourth bike off the production line.

Large-diameter twin front brake discs

HARLEY-DAVIDSON *FLHRI Road King*

BY THE MID-1990S IT WAS quite obvious that there wasn't going to be any significant "new idea" that had a Harley-Davidson logo on the tank. Harley knew what its customers wanted, and it knew what it was good at building. Hence the regular reappearance of old ideas such as the Road King, a middleweight tourer first introduced for the 1995 model year. This was a return to traditional values for the Electra Glide in much the same way that the FLHS had been a decade earlier. The Road King combined improvements to the Harley package—like electronic sequential port fuel injection on the FLHRI—with the looks of the traditional Electra Glide such as whitewall tires, spoked wheels, and plenty of chrome.

RIDING STANCE
The handlebars were positioned high and wide, which was good for short- to middle-distance touring but not so practical for longer journeys. It was a fun bike to ride on twisting roads.

Specifications

Model Harley-Davidson FLHRI Road King
Capacity 80cu. in. (1312cc)
Power Output 69 bhp
Weight 692 lb (314 kg)
Top Speed 96 mph (155 km/h)
Country of Origin US

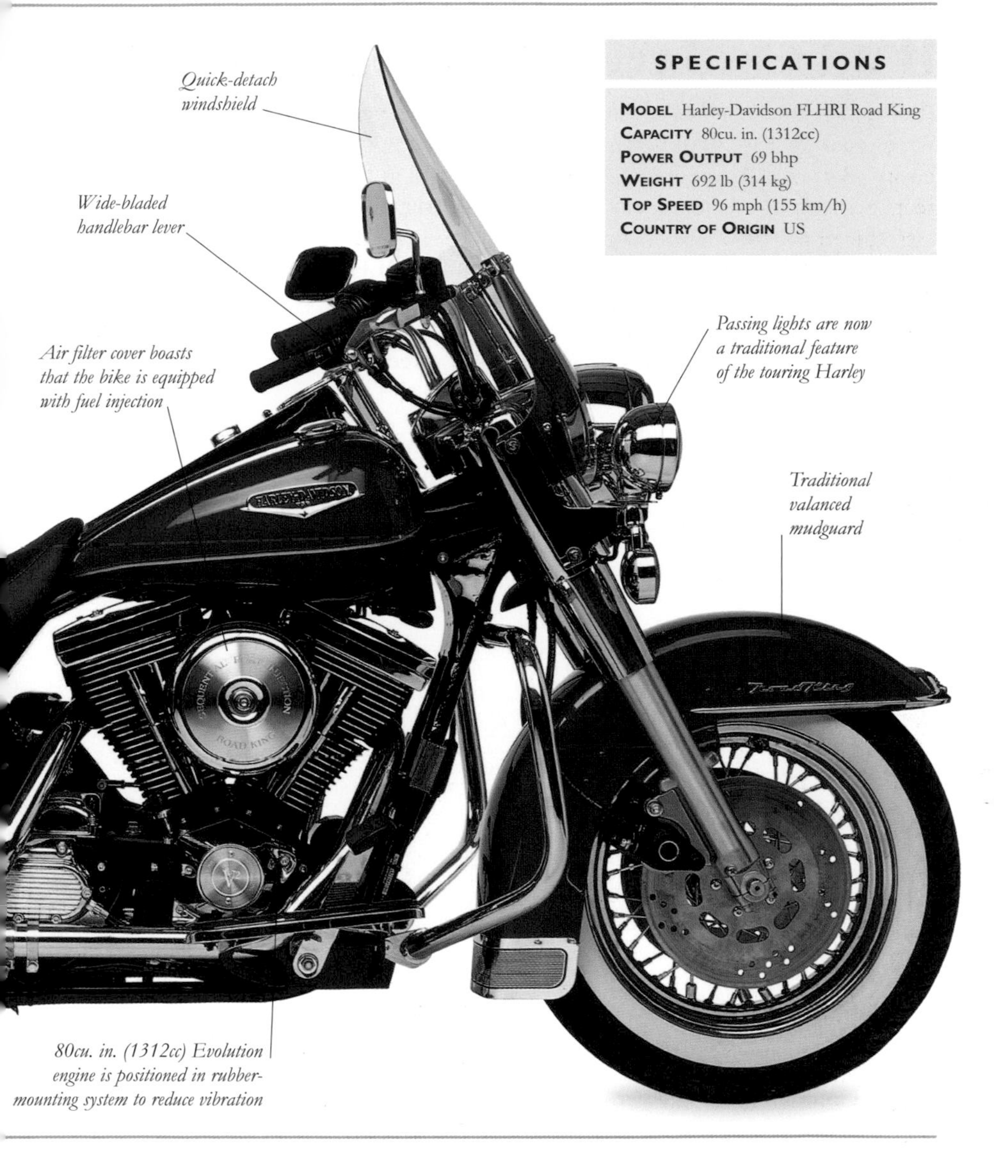

HARLEY-DAVIDSON *XL1200S*

THOUGH THE SPORTSTER engine has been through various guises and capacities over the years, it has always remained true to its original layout. A 1200cc version of the alloy Evolution engine was introduced in 1988, and the range was expanded to include a variety of model styles in both 1200 and 883cc engine capacities. The Sportster Sport, introduced in 1996, harked back to the XLCH Sportster of the late 1950s and early 1960s, a bare-boned bike intended for fast fun. It had uprated suspension and improved power output over the basic model to justify its S designation. Other models in the range were developed in a variety of custom and cruiser styles to attract a wider range of customers to the Harley brand.

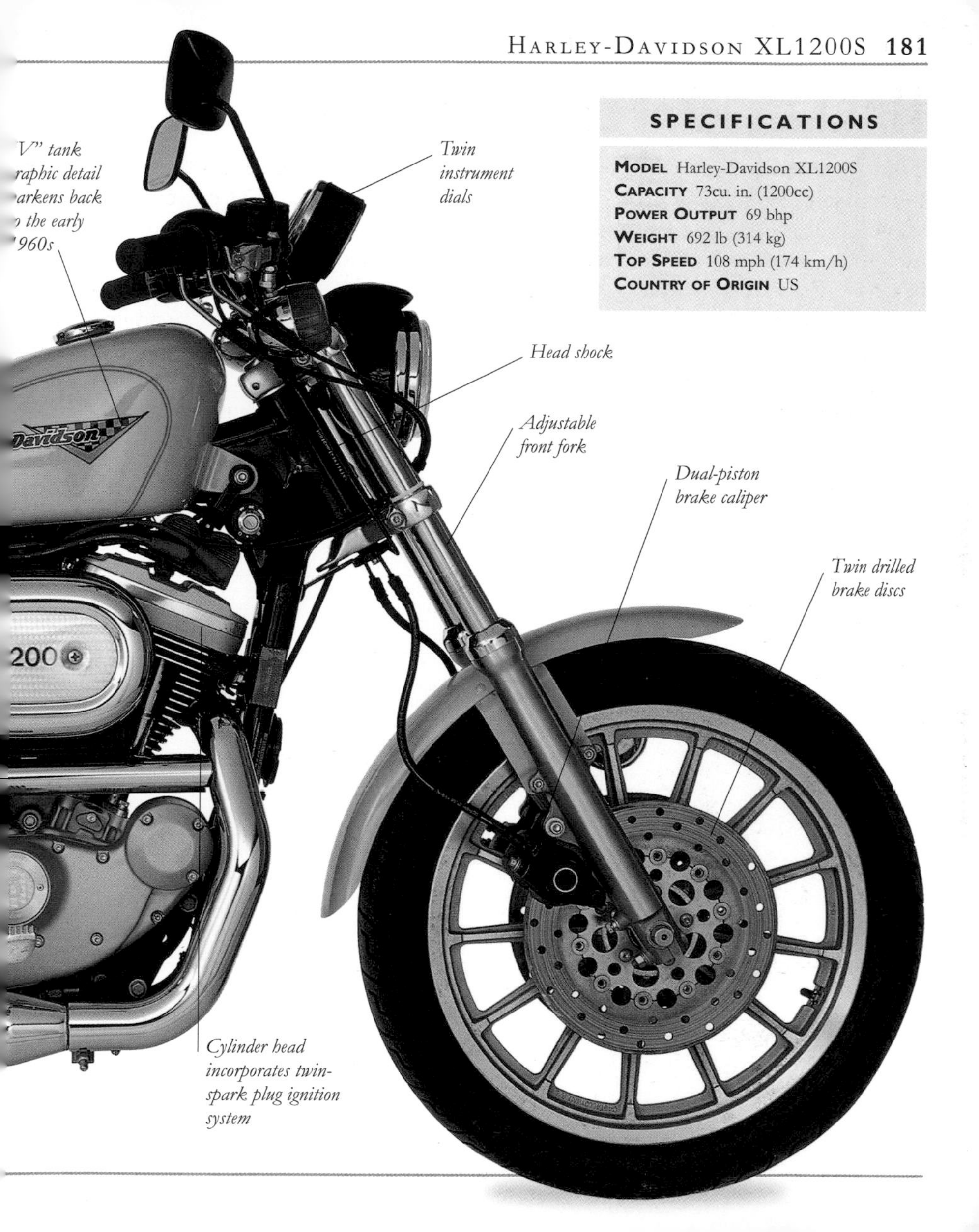

SPECIFICATIONS

MODEL Harley-Davidson XL1200S
CAPACITY 73cu. in. (1200cc)
POWER OUTPUT 69 bhp
WEIGHT 692 lb (314 kg)
TOP SPEED 108 mph (174 km/h)
COUNTRY OF ORIGIN US

HARLEY-DAVIDSON *CVO Softail*

CUSTOMIZING HAS ALWAYS BEEN part of the Harley-Davidson scene. Owners individualize their bikes by modifying them, tuning them, changing components, and painting them to their own taste. Harley had always supplied accessories, but in 1999 they offered customers the opportunity to buy bikes that had been personalized to their own specifications by the company's Custom Vehicle Operations division. A limited number of bikes were reworked with performance parts and other accessories as well as custom paint finishes. CVO bikes typically receive larger capacity engines the season before they are offered on standard bikes. This is a 2010 CVO Softail model.

Chromed rails allow panniers to be attached and bike to be converted into a tourer—hence the name

Leather solo saddle; the pillion makes do with a tiny pad

Speedometer and ignition switch are mounted in a tank top console

Chromed fork shrouds are inspired by 1950s Harley style

SPECIFICATIONS

Model Harley-Davidson CVO Softail
Capacity 1804cc
Power Output 85 bhp
Weight 724 lb (328 kg)
Top Speed 115 mph (185 km/h)
Country of Origin US

1804cc Twin Cam engine was Harley's biggest when this bike was built

Deeply valanced mudguards are styled after those fitted to Harley's traditional touring models

HENDERSON

ALTHOUGH THE FIRST US production four-cylinder motorcycle was the 1909 Pierce, the figure most identified with the layout in the US was Bill Henderson. In partnership with his brother Tom, he began motorcycle production at Detroit in 1912, although prototype machines were running the previous year. The motorcycle shown here is a 1912 model from the first production year. Early Henderson machines had a number of bizarre features, including a 65-in (165-cm) wheelbase and an optional passenger seat that mounted on the fuel tank in front of the rider. The valances on the mudguard were there to help keep ladies' skirts out of the wheel. The in-line four-cylinder engine had a four-bearing crankshaft and an i.o.e. valve layout.

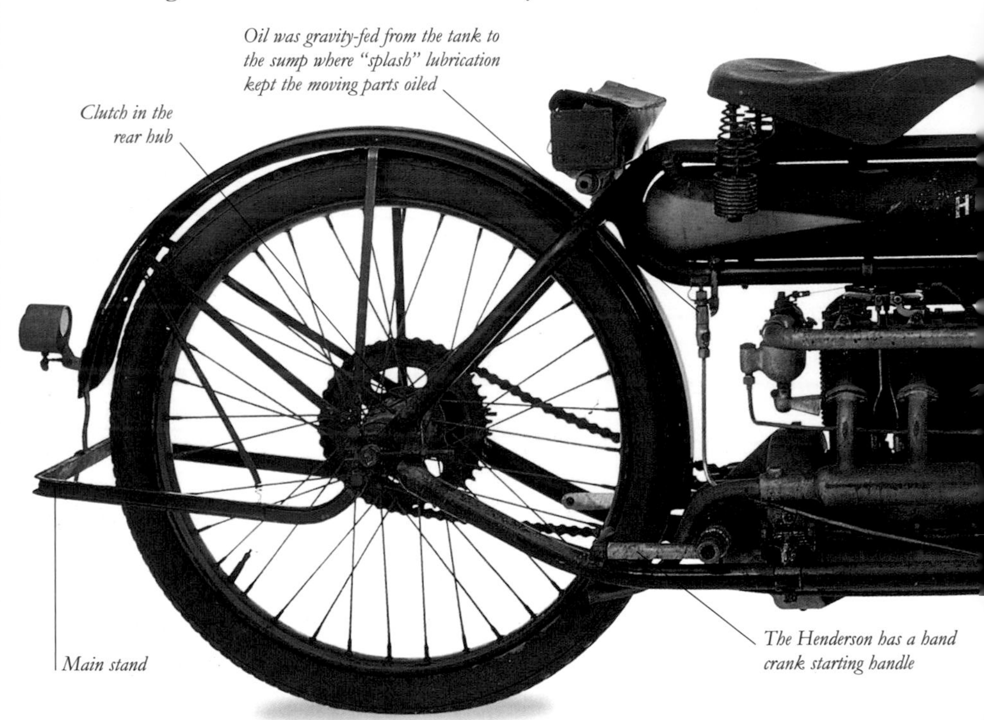

SPECIFICATIONS

MODEL Henderson
CAPACITY 56cu. in. (920cc)
POWER OUTPUT 8 bhp (est.)
WEIGHT 295 lb (134 kg)
TOP SPEED 60 mph (96 km/h)
COUNTRY OF ORIGIN US

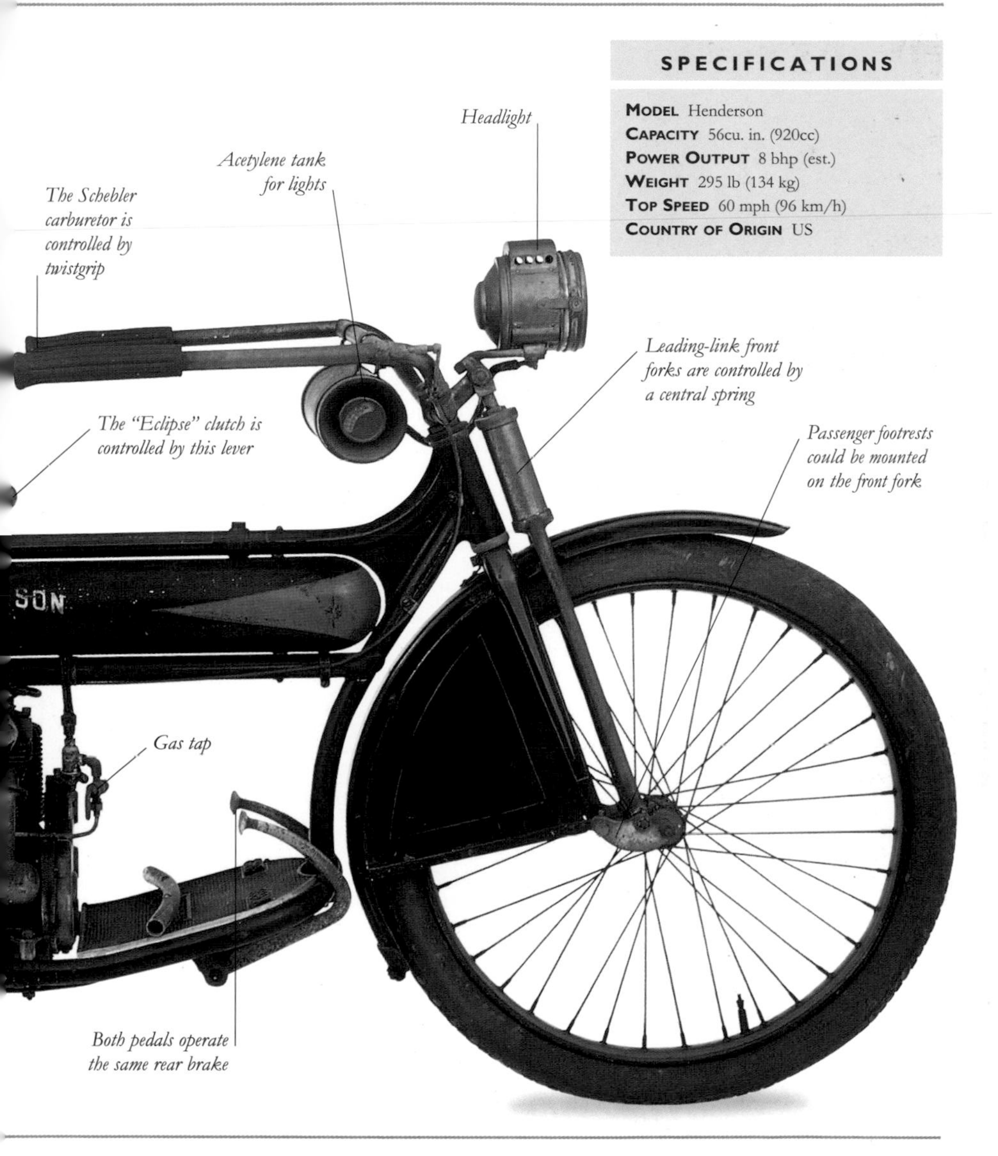

HENDERSON *KJ*

FOR 1929 HENDERSON INTRODUCED a new machine called the KJ. Designed by Arthur Constantine, it retained the 79cu. in. (1300cc) capacity of the earlier K series models. The crankshaft now had five main bearings and alloy pistons were used. Overheating of the rear cylinders was reduced through improved air cooling. In addition, a Schebler carburetor was now incorporated. New frames allowed lower seats and "streamlined" fuel tanks that included an instrument panel. However, the KJ was short-lived. Ignatz Schwinn, the manufacturer of bicycles and the Excelsior motorbike, who had bought the Henderson company in 1917, pulled out of the motorcycle business. The last Hendersons were made in 1931.

SPECIFICATIONS	
MODEL	Henderson KJ
CAPACITY	79cu. in. (1300cc)
POWER OUTPUT	40 bhp
WEIGHT	500 lb (227 kg) (est.)
TOP SPEED	100 mph (161 km/h)
COUNTRY OF ORIGIN	US

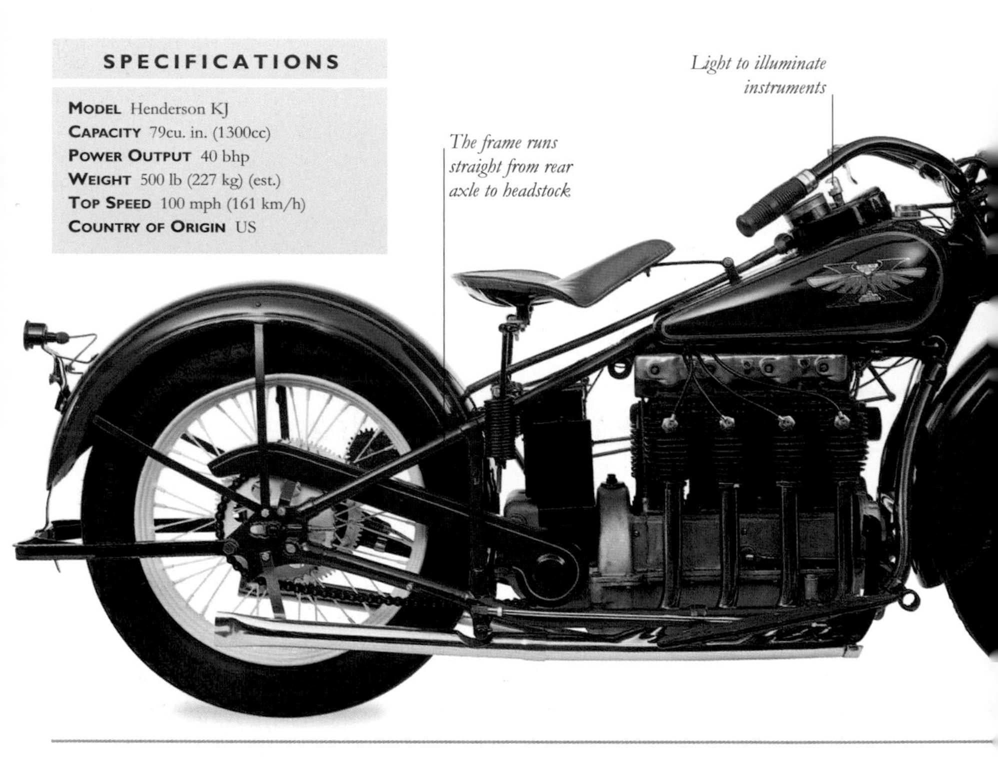

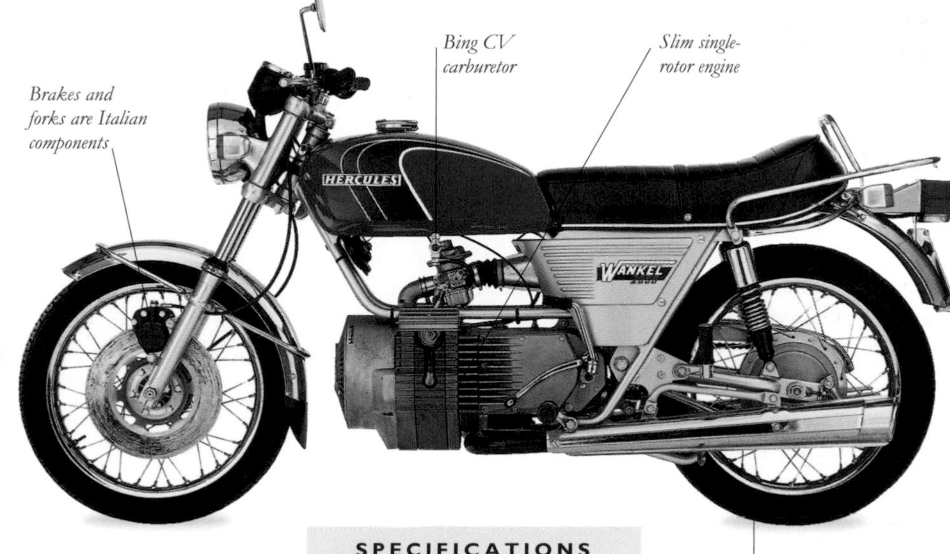

Only 2,000 W2000s were produced

SPECIFICATIONS

MODEL Hercules W2000
CAPACITY 294cc
POWER OUTPUT 27 bhp @ 6500 rpm
WEIGHT 386 lb (175 kg)
TOP SPEED 91 mph (147 km/h)
COUNTRY OF ORIGIN Germany

HERCULES *W2000*

SEVERAL motorcycle companies developed Wankel-engined prototypes in the early 1970s. The Hercules was the first to reach the market, with production versions appearing in late 1974; the W2000 was sold in some markets as a DKW. Engine rotation was in-line with the frame, and the drive had to be turned through 90° before it reached the six-speed gearbox. The W2000 was not a success: it was expensive, ugly, and unproven.

HODAKA *Super Rat*

HODAKA'S LIGHTWEIGHT off-road motorcycles first appeared in the US in 1964. The original machine was a 90cc two-stroke single; capacity later grew to 100cc and then 125cc. Hodaka motorcycles were simple, rugged, and reliable. During the 1960s, as the American dirt-bike market was developing, Hodaka machines proved very popular and gained a cult following. The machine shown on these pages is a 1971 Super Rat Motocross, but enduro and trail models were similar and had names like Combat Wombat and Dirt Squirt.

19-in (48-cm) front wheel

BEATEN BY THE COMPETITION
Production ended in 1978 when Hodaka was unable to compete with the technology of major Japanese manufacturers.

Front drum brake

SPECIFICATIONS

MODEL Hodaka Super Rat
CAPACITY 98cc
POWER OUTPUT 16 bhp @ 7250 rpm
WEIGHT Not known
TOP SPEED Not known
COUNTRY OF ORIGIN Japan

HONDA *RC160*

HONDA'S FIRST FOUR-CYLINDER motorcycle was the 1959 RC160, shown here. It was built in time for the All-Japan Championships of 1959. It did not compete in the TT races that year, but the design was revised and taken to Europe the following season. In 1961, Mike Hailwood won the 250cc World Championship for Honda on an RC162. The spine frame used the engine as a stressed frame member. Early models like this one had a vertically mounted engine with shaft-driven cams; later the engine had central gear drive to the twin camshafts and was forward inclined. This helped cooling and lowered the center of gravity for improved handling.

SPECIFICATIONS

MODEL Honda RC160
CAPACITY 249cc
POWER OUTPUT 35 bhp @ 14,000 rpm
WEIGHT 273 lb (124 kg)
TOP SPEED 125 mph (201 km/h)
COUNTRY OF ORIGIN Japan

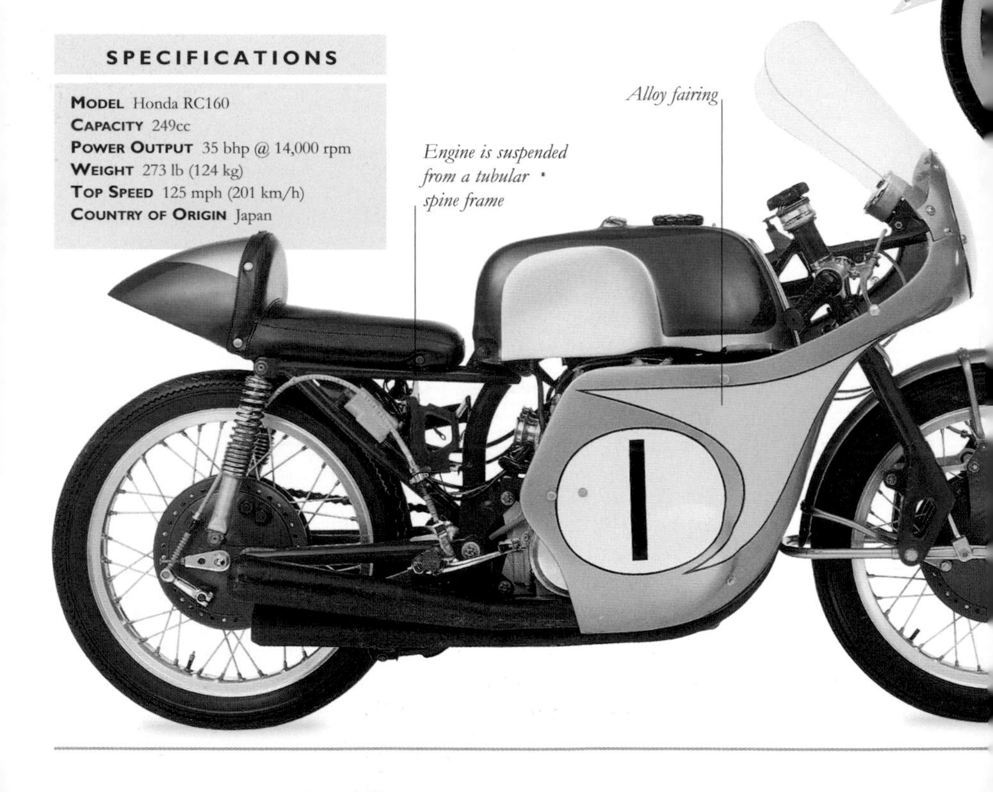

Enclosed chain

Deep-valanced front mudguard

O.h.c. parallel-twin engine

The RC160 could hit speeds of up to 125 mph (201 km/h)

SPECIFICATIONS

MODEL Honda Dream 300
CAPACITY 305cc
POWER OUTPUT 21 bhp @ 7000 rpm
WEIGHT 348 lb (158 kg)
TOP SPEED 80 mph (129 km/h)
COUNTRY OF ORIGIN Japan

HONDA *Dream 300*

HONDA ENTERED THE 1960S with a line of four-stroke twins. The Dream 300 was the touring model of the lineup, and included the luxury of electric start and fully enclosed final-drive chain. The chassis was pressed steel with leading-link forks and twin oil-damped rear shocks. The wheels were small—41 cm (16 in) front and back—and had 3¼-in (8.25-cm) tires. Ultimately, although the Dream offered potential buyers what they wanted—a reliable nonleaking engine with good electrics—it did not handle well.

Honda *CB92 Benly*

The CB92 Benly was perhaps the most radical of the early Hondas that were exported to the West. No other 125cc machines came close to the performance, specification, or quality of construction offered by the Benly. The design was typical period Honda with pressed-steel frame and forks. The twin-cylinder engine featured a single o.h.c., breathing through a single carburetor and driving via a four-speed gearbox. It was equipped with an electric starter, large-diameter drum brakes, and 18-in (46-cm) rims. The machine shown here is a 1960 model. Naturally this specification came at a price, and the Benly wasn't cheap. Within a few years Honda had dropped the ungainly pressed-steel frames and leading-link forks in favor of a crisper look featuring telescopic forks and tubular frames.

Specifications

Model Honda CB92 Benly
Capacity 124cc
Power Output 15 bhp @ 10,500 rpm
Weight 220 lb (100 kg)
Top Speed 70 mph (113 km/h)
Country of Origin Japan

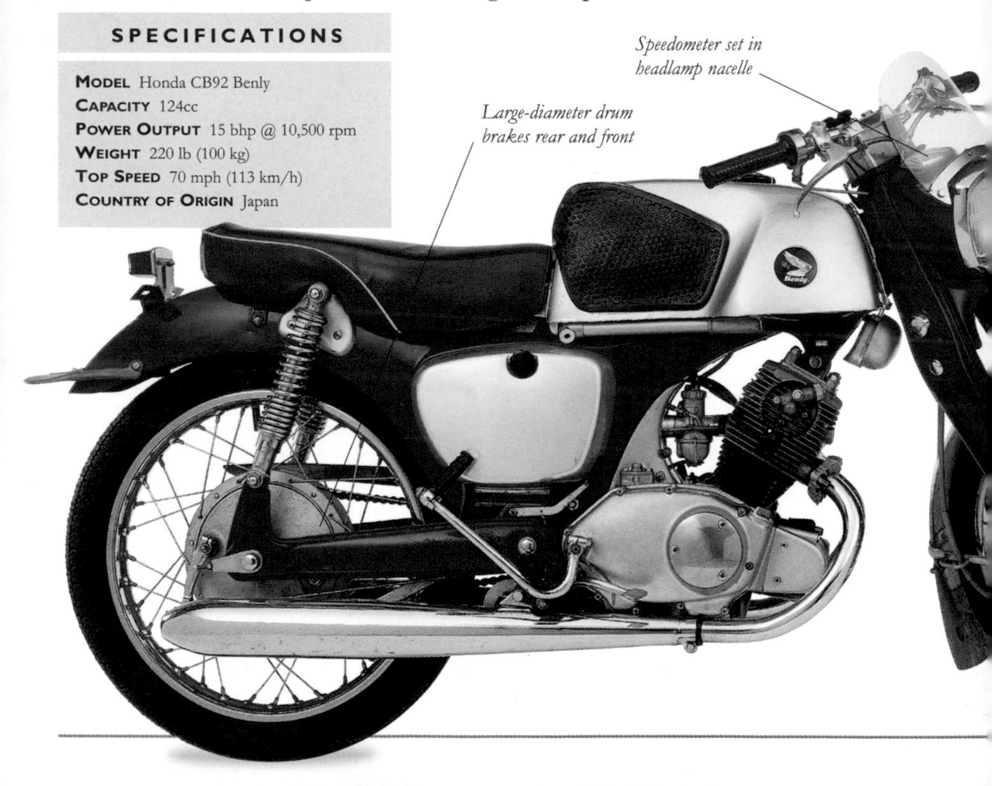

SPECIFICATIONS

MODEL Honda RC164
CAPACITY 249cc
POWER OUTPUT 45 bhp @ 14,000 rpm
WEIGHT 232 lb (105 kg)
TOP SPEED 140 mph (225 km/h)
COUNTRY OF ORIGIN Japan

HONDA *RC164*

THE 1964 RC164 WAS the result of five years of development of the earlier RC160 *(see p.190)*. It was 10 bhp more powerful and almost 20 mph (32 km/h) faster than the earlier machine. The dated leading-link forks were replaced with more modern telescopics. The 1964 250 RC164 was a lighter, lower version of the well-tried four, but still not quick enough to match the fast-improving Yamaha two-stroke twins. A new six-cylinder 250 model was introduced at the end of 1964.

HONDA *CL72 Scrambler*

HONDA'S SPORTY CL72 street scrambler was introduced in 1962. Pictured here is a 1964 model. It had the CB72's 249cc 180°, four-stroke motor, a derivative of the 1960's C70 with recently introduced wet-sump engine. In tune with its street scrambler image, it had twin, high-level exhaust pipes that exited to the left of the engine. It also had a tubular-steel cradle frame with telescopic forks and a hydraulic steering shock absorber. It was the forerunner of the modern trail bike and in many ways exemplifies why Honda was so successful in the 1960s. The CL was reliable and easy to ride, but it was also good looking and fun. At the time few other manufacturers could boast all four attributes.

SPECIFICATIONS

MODEL Honda CL72 Scrambler
CAPACITY 249cc
POWER OUTPUT 24 bhp
WEIGHT 337 lb (153 kg)
TOP SPEED 80 mph (128 km/h)
COUNTRY OF ORIGIN Japan

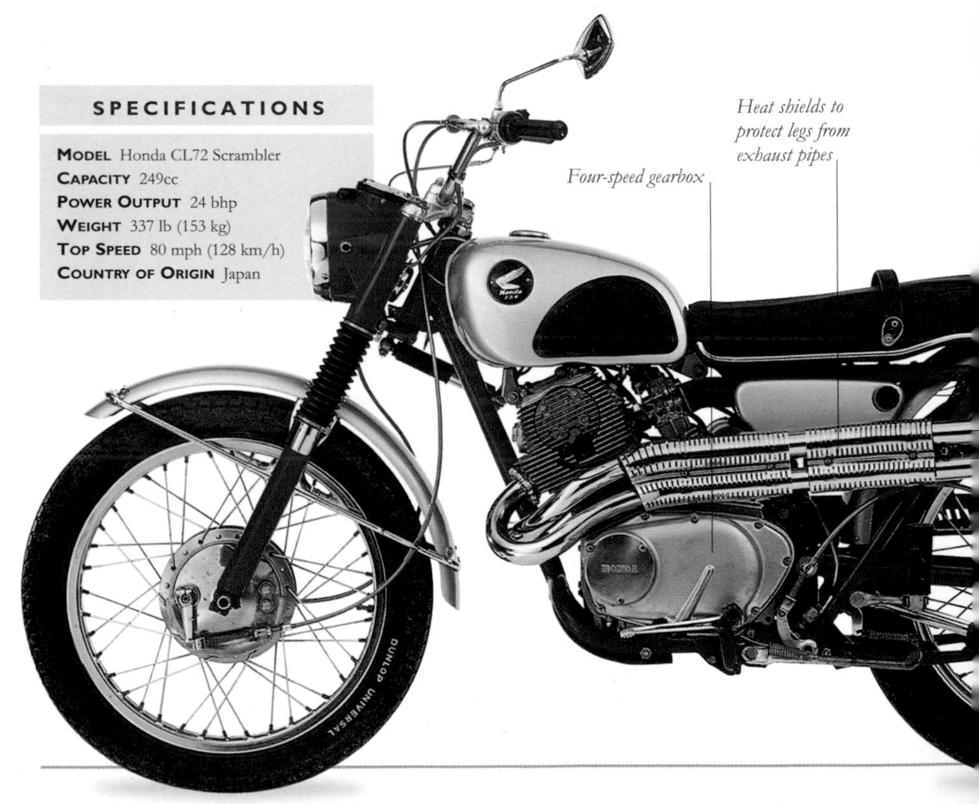

Twin carburetors

Speedometer and tachometer are combined in one unit

Electric start motor

SPECIFICATIONS

MODEL Honda CB77 Super Hawk
CAPACITY 305cc
POWER OUTPUT 27.5 bhp @ 9000 rpm
WEIGHT 353 lb (160 kg)
TOP SPEED 103 mph (166 km/h)
COUNTRY OF ORIGIN Japan

HONDA *CB77 Super Hawk*

HONDA'S CB77 Super Hawk (Super Sport in Britain) was an important model in the 1960s. It proved the Japanese could build stylish, powerful motorcycles—not just small-capacity commuters—and at an affordable price. The 305cc was the biggest bike Honda produced at the time. The tubular frame used the motor as a stressed member. The engine itself was essentially a bored-out 250 motor. Introduced in 1964, the model shown here is a 1966 version.

HONDA *CB750*

IF ONE MACHINE CHANGED the course of motorcycle development, it was the CB750. Four-cylinder engines are common today, as are disc brakes, electric starters, and 124 mph (200 km/h) performance. However, when the CB750 was launched in 1969 such a high level of specification was not usually considered by designers of production motorcycles. The CB750 set new standards for performance, practicality, and reliability in the big bike class, heralding the era of the production superbike. This was Honda's first attempt at a big bike, and it was immediately successful. The CB750 was not revolutionary, but it was well equipped with state-of-the-art technology.

Plenty of chrome as well as a candy-red color with gold stripes

A 210-watt alternator is mounted at the left end of the crankshaft, with the ignition points at this end

SPECIFICATIONS

MODEL Honda CB750
CAPACITY 736cc
POWER OUTPUT 67 bhp @ 8000 rpm
WEIGHT 485 lb (220 kg)
TOP SPEED 124 mph (200 km/h)
COUNTRY OF ORIGIN Japan

Handlebar-mounted hydraulic fluid reservoir for front disc brake

Engine tachometer and speedometer are positioned for maximum visibility

Telescopic front forks

Indicator

A five-speed gearbox is used; Honda later produced a semiautomatic version

FOUR PIPES

The impressive arrangement of four exhaust pipes—one per cylinder—assisted the engine's performance and also made the bike look powerful. Honda replaced the eight-valve s.o.h.c. four with a 16-valve d.o.h.c. model in 1978.

Unusually for Honda, it made the 736cc engine undersquare at 2.4 × 2.5 in (61 × 63 mm)

The CB750 was the first production bike to be equipped with a hydraulic disc brake

HONDA *Goldwing GL1000*

HONDA'S ORIGINAL 1975 GOLDWING was worlds away from today's heavily accessorized successor *(see pp.208–09)*. The brief was for it to be "the king of motorcycles," to beat the Z1 *(see pp.258–59)*, and to regain the glory Honda had lost to Kawasaki. When launched, it was Japan's first water-cooled four-stroke: a massive grand tourer that was an immediate hit in the US. The British press was less enthusiastic. However, the Goldwing's smooth cruising ability made it a favorite with long-distance riders. Many Goldwings were modified with panniers and fairing, which prompted Honda to later supply them as standard. The bike had a dummy fuel tank; the real one was under the seat.

SPECIFICATIONS

MODEL Honda Goldwing GL1000
CAPACITY 999cc
POWER OUTPUT 80 bhp @ 7500 rpm
WEIGHT 571 lb (259 kg)
TOP SPEED 120 mph (193 km/h)
COUNTRY OF ORIGIN Japan

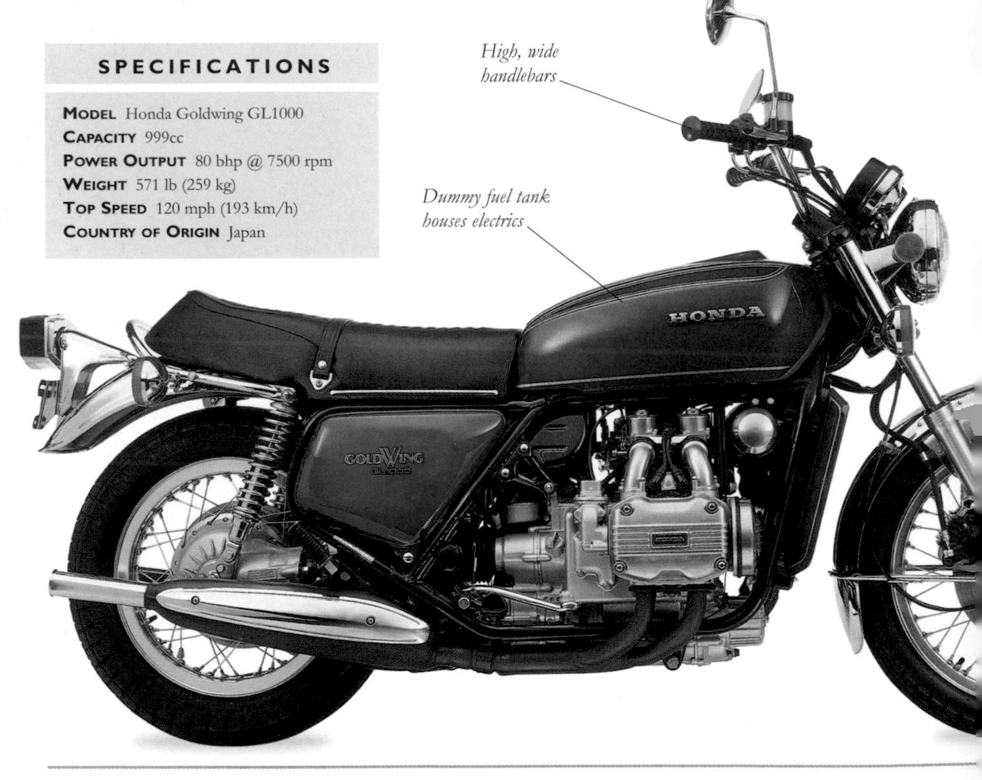

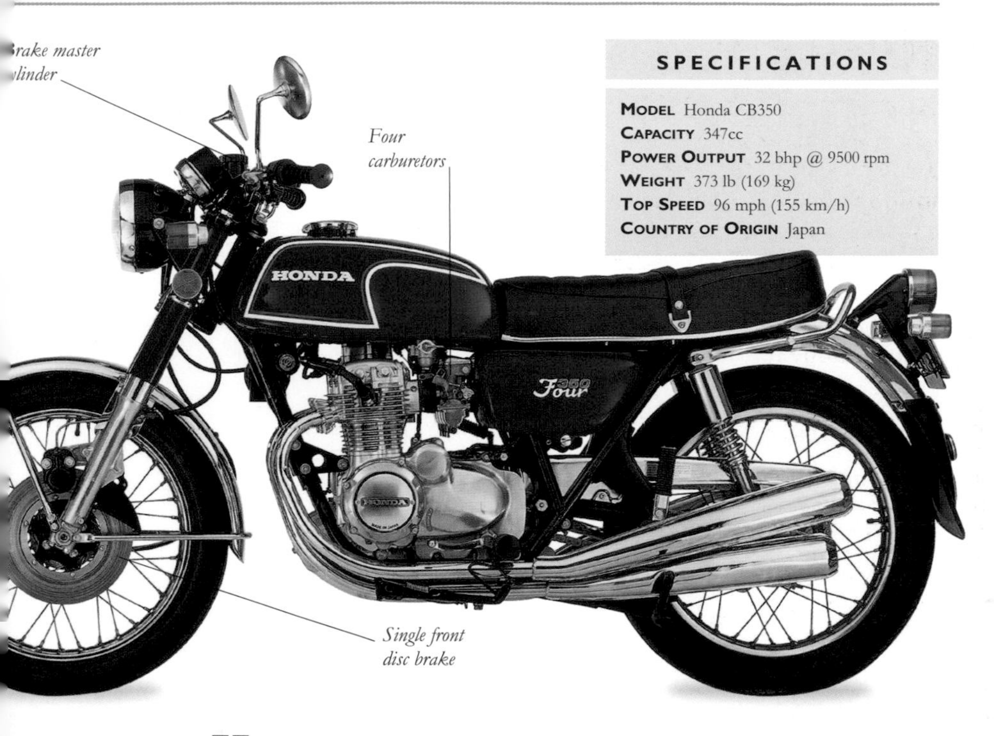

SPECIFICATIONS

MODEL Honda CB350
CAPACITY 347cc
POWER OUTPUT 32 bhp @ 9500 rpm
WEIGHT 373 lb (169 kg)
TOP SPEED 96 mph (155 km/h)
COUNTRY OF ORIGIN Japan

HONDA *CB350*

THE CB350 WAS the smallest of the 1970s' four-cylinder, s.o.h.c. road machines. Never officially imported into Britain, and failing to gain support in the US, Honda looked to other countries in Europe, where some countries' import tax laws favor sub-350cc bikes. Police versions were offered in France and the Netherlands. Even so, the CB350 had a short life. Introduced in 1972, it was replaced in the markets by the CB400F, which appeared in 1974. Capacity of the 400 four was increased to 408cc. It was restyled and given a four-into-one exhaust pipe. The 400 was a big success and was produced until 1979.

HONDA *Elsinore*

THE ELSINORE WAS BUILT in a bid for greater success in US motocross when Honda went against its "four-stroke only" policy. Honda employee Soichiro Miyakoshi taught himself everything about two-stroke technology, and the air-cooled CR250 Elsinore was the result. Its equipment included upside-down, remote-reservoir rear shock absorbers, electronic ignition, leading axle, air-assisted front forks, and box-section swingarm. This 1980 model was the first Honda motorcycle built in the US.

Small-capacity fuel tank

Radial fins help cool the engine

Aluminum wheel with magnesium front and rear brake hubs

Rubber fork gaitors protect the sliders from damage

Expansion chamber exhaust is routed over the engine for increased ground clearance

SPECIFICATIONS

MODEL Honda Elsinore
CAPACITY 247cc
POWER OUTPUT 40 bhp
WEIGHT 224 lb (102 kg)
TOP SPEED Not known
COUNTRY OF ORIGIN Japan

REAR VIEW
Grip is vital for success in motocross. The rear tire tread pattern is changed for differing conditions and is run at low pressure to increase traction. The rear suspension keeps the wheel on the ground as much as possible.

Honda *CBX1000*

In 1968 Honda changed the face of motorcycling with the CB750 *(see pp.196–97)*. A decade later, it astounded the world once again with an air-cooled, across-the-frame six. Early impressions of the awesome six-cylinder stunned journalists. A few were said to have arrived back at the CBX launch pits pale, after experiencing high-speed weaves. Handling improved later, when Honda upgraded the swingarm. The CBX1000's engine was suspended from a tubular spine frame and canted forward 30° to aid cooling. It had four valves per cylinder and six carburetors. This is a 1980 model; monoshock rear suspension, a full fairing, and more sober styling came later.

SPECIFICATIONS

MODEL Honda CBX1000
CAPACITY 1047cc
POWER OUTPUT 105 bhp @ 9000 rpm
WEIGHT 556 lb (252 kg)
TOP SPEED 135 mph (217 km/h)
COUNTRY OF ORIGIN Japan

PUBLIC DOUBTS
Although journalists may have been astounded by the CBX, it was too much for public taste; despite its superb engine and looks, it never won favor.

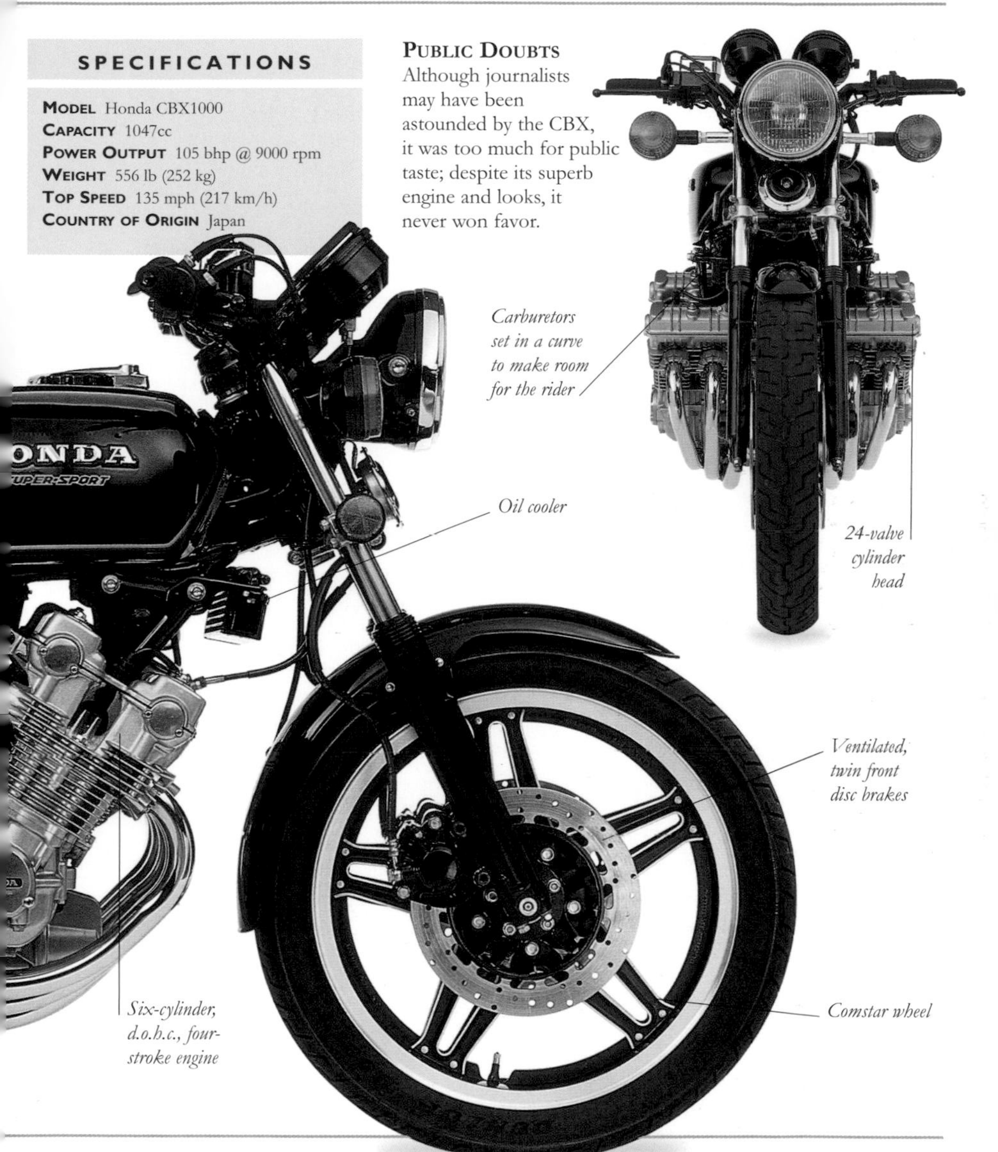

Carburetors set in a curve to make room for the rider

Oil cooler

24-valve cylinder head

Ventilated, twin front disc brakes

Comstar wheel

Six-cylinder, d.o.h.c., four-stroke engine

HONDA *RS500R*

HONDA HAD FOUND IN THE 1970s that to stay competitive on the track it had to turn to two-stroke technology. The firm's first two-stroke racer was the NS500—not only a two-stroke but also a 90° V3. The central cylinders faced forward while the two outer cylinders were more vertical. Though Honda made its Grand Prix comeback in 1979 with a four-stroke bike, it went on to win the 1983 500cc World Championship with this superb two-stroke bike. "Fast Freddie" Spencer went on to win many 500cc Grand Prix races for Honda in 1985. The RS500 had an alloy frame and lightweight racing Comstar wheels. This RS500 was one of a limited number of NS500 replicas sold to nonfactory riders in 1983.

SPECIFICATIONS

MODEL Honda RS500R
CAPACITY 499c
POWER OUTPUT 130 bhp @ 11,500 rpm
WEIGHT 270 lb (122 kg) (est.)
TOP SPEED 175 mph (282 km/h)
COUNTRY OF ORIGIN Japan

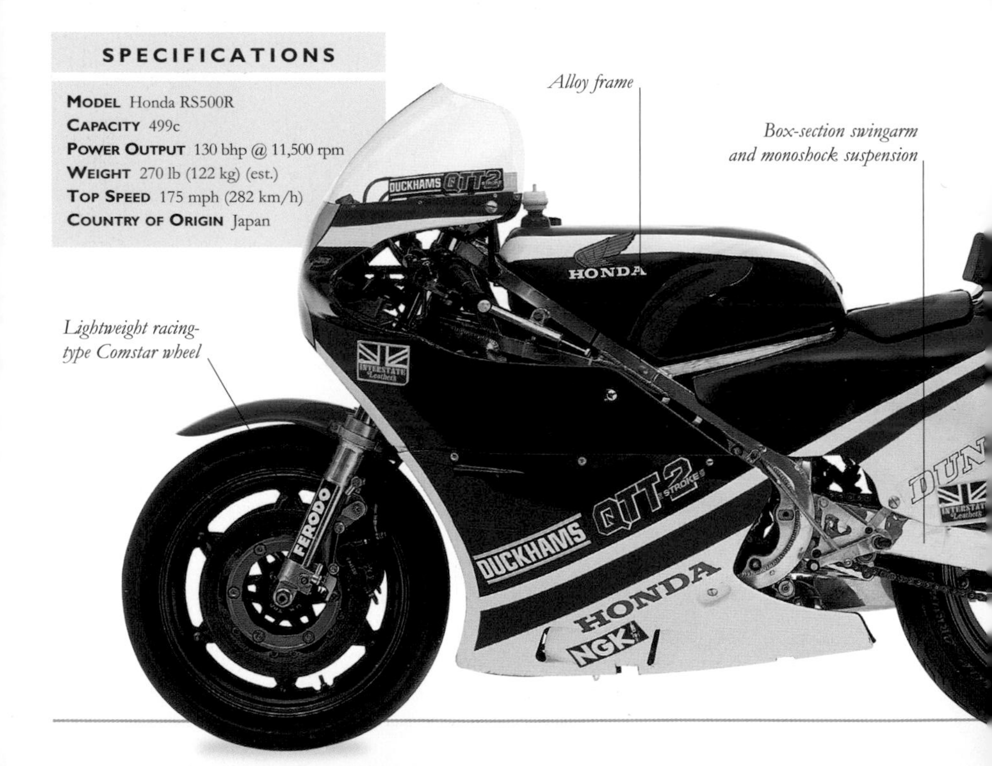

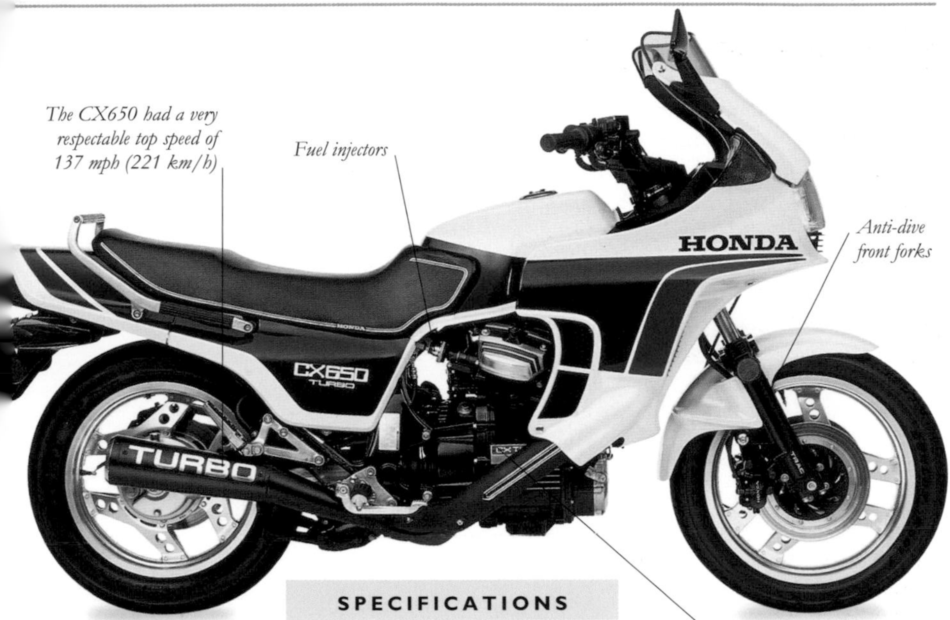
The CX650 had a very respectable top speed of 137 mph (221 km/h)
Fuel injectors
Anti-dive front forks
O.h.v., transverse twin motor with turbocharger
HONDA
CX650
TURBO
TURBO

SPECIFICATIONS

MODEL Honda CX650 Turbo
CAPACITY 673cc
POWER OUTPUT 100 bhp @ 8000 rpm
WEIGHT 518 lb (235 kg)
TOP SPEED 137 mph (221 km/h)
COUNTRY OF ORIGIN Japan

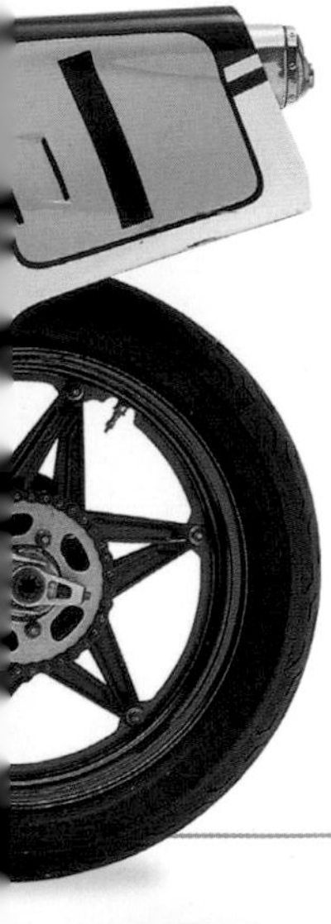

HONDA *CX650 Turbo*

HONDA STARTED A SHORT craze for turbocharging when it built the CX500 Turbo in 1981. It was based on the humble o.h.v. shaft-drive V-twin CX500. Capacity was increased from 496cc to 673cc in 1983 to create the equally short-lived CX650 Turbo. Both models had electronically controlled fuel injection, a single-shock rear suspension system, and an aerodynamic fairing. This is a 1983 model.

HONDA *VF750F*

THE PRODUCTION VF750F WAS INTRODUCED in 1983. The new model had a single suspension unit and a box-section swingarm, but it suffered from reliability problems as Honda continued to struggle to make the V-four engine layout work. This highly modified superbike racer was ridden to victory in the 1985 Daytona 200 by Freddie Spencer. It was a good year for Spencer, who also won the 250 and 500cc World Championships on Grand Prix Honda two-strokes.

Cutdown seat reduces comfort but allows the rider to sit lower on the bike

Box-section tubing

SPECIFICATIONS

MODEL Honda VF750F
CAPACITY 748cc
POWER OUTPUT 125 bhp (est.)
WEIGHT 398 lb (180 kg)
TOP SPEED 160 mph (258 km/h)
COUNTRY OF ORIGIN Japan

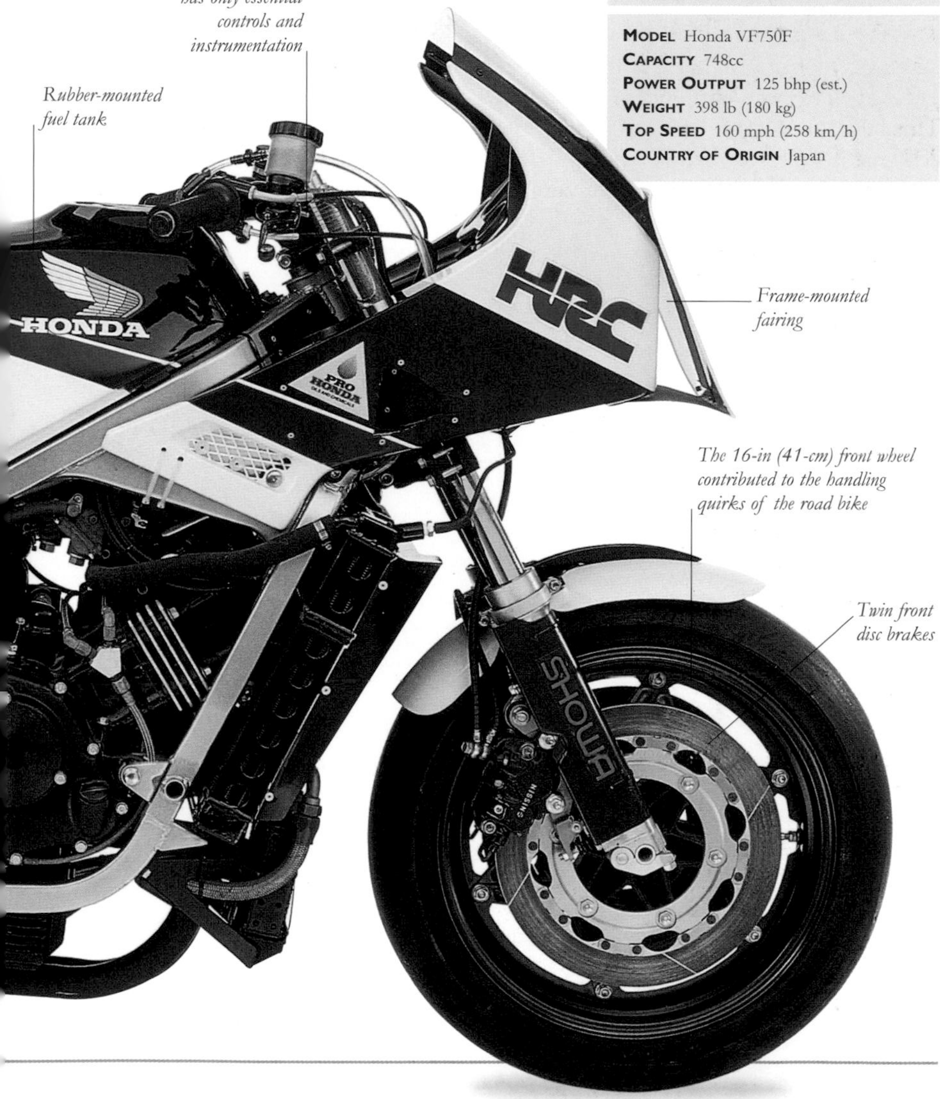
The racing bike has only essential controls and instrumentation
Rubber-mounted fuel tank
Frame-mounted fairing
The 16-in (41-cm) front wheel contributed to the handling quirks of the road bike
Twin front disc brakes
HONDA
HRC
PRO HONDA
SHOWA

HONDA *Goldwing GL1500*

THE SYMBOL OF THE GOLDEN wing was already a familiar Honda trademark when the model was christened in 1975. At the time of its launch, the 1000cc four-cylinder machine was simply the biggest, most complex motorcycle ever produced in Japan. By the 1980s, Honda was an established auto manufacturer, with car and motorcycle assembly plants around the world. In 1981, Goldwing production moved from Japan to Ohio; the biggest market for the bike was in the US, but it was exported all over the world, including Japan. A new 1520cc, six-cylinder Goldwing—the largest, most complex bike ever built in the US—was unveiled in 1988.

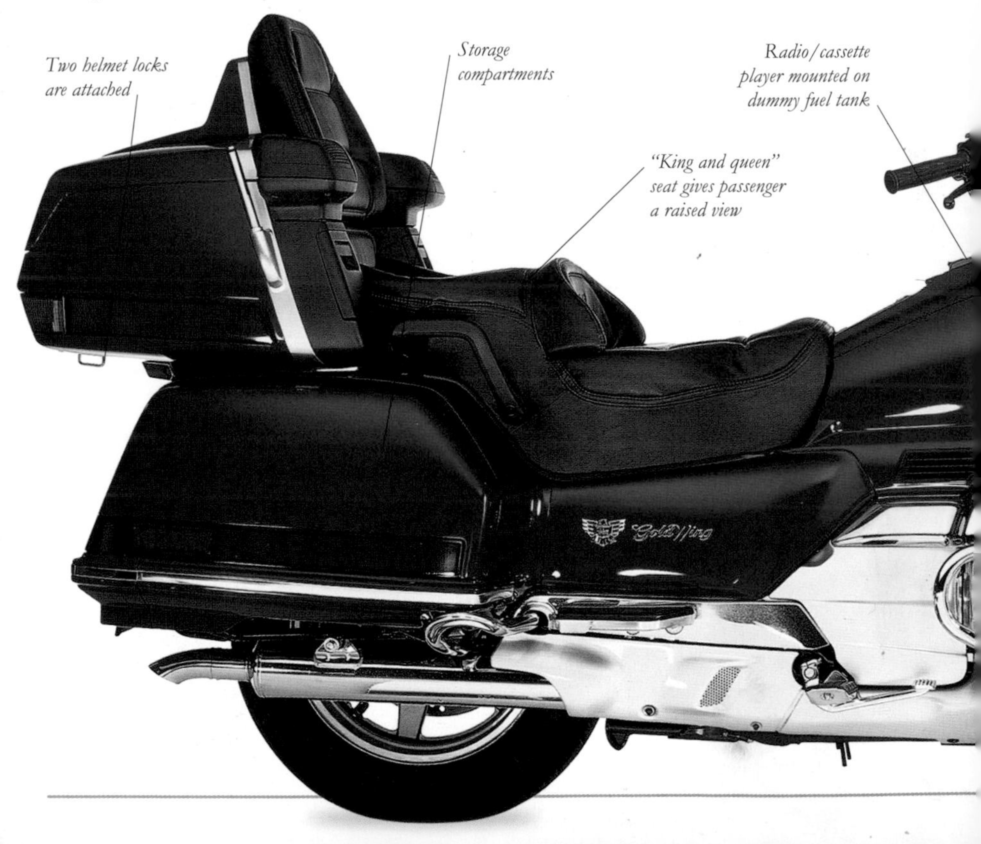

DIVIDED OPINION

This machine inspires either love or hate: for some people, the Goldwing is the ultimate two-wheeled luxury motorcycle, for others it is an overweight, ugly, and expensive monolith.

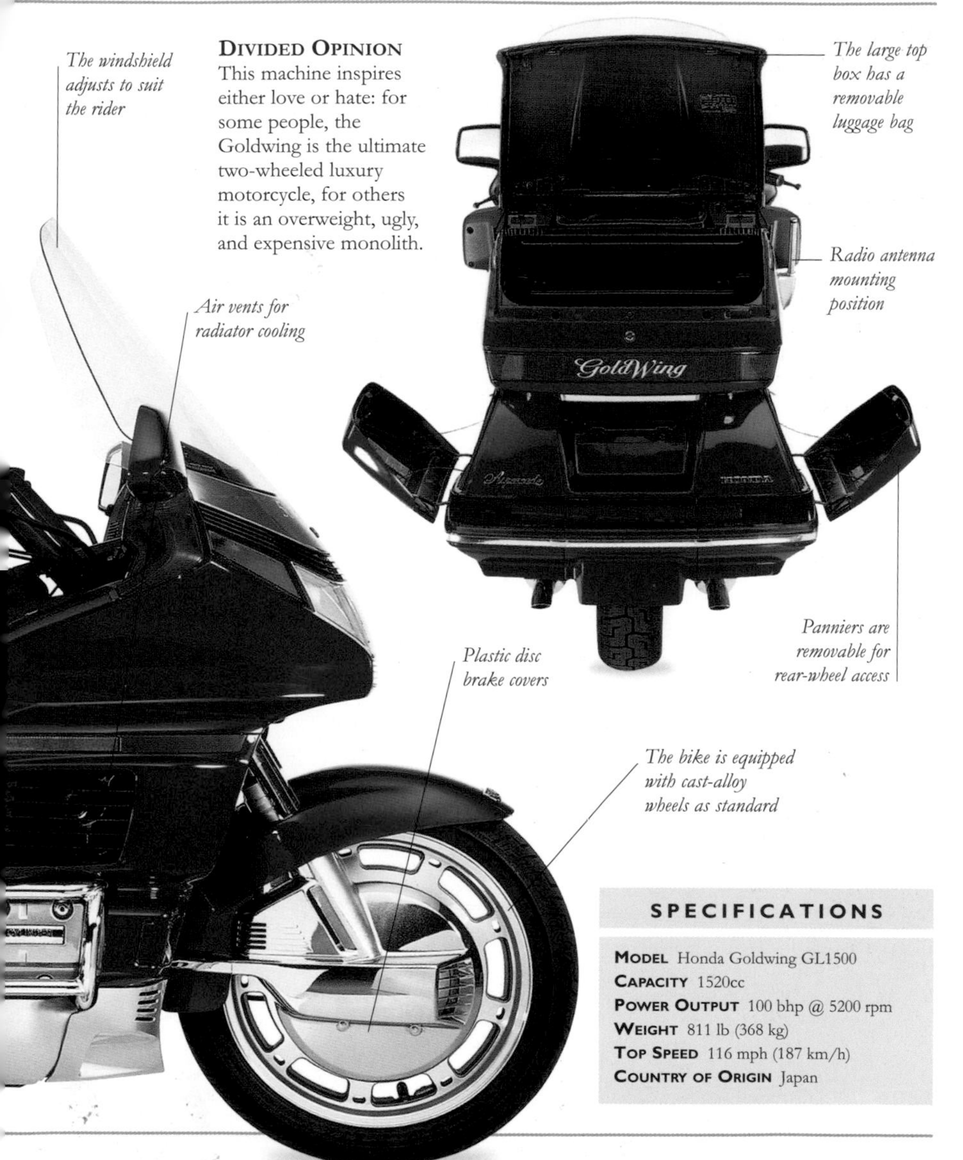

SPECIFICATIONS

MODEL Honda Goldwing GL1500
CAPACITY 1520cc
POWER OUTPUT 100 bhp @ 5200 rpm
WEIGHT 811 lb (368 kg)
TOP SPEED 116 mph (187 km/h)
COUNTRY OF ORIGIN Japan

HONDA *XRV750*

HONDA BUILT EIGHT MACHINES to compete in the 1991 Paris-Dakar Rally. Only two of them completed the course. This one failed to finish after its rider, John Watson-Miller, was injured three days' ride from the finish. The "Marathon" class in which they are entered allows the use of lightly modified production machines with changes permitted only to the tank, rear shock absorber, and exhaust system. Long-distance competitions have a dedicated following in Spain, France, and Italy, where there is a big market for replicas of this type of machine.

SPECIFICATIONS

MODEL Honda XRV750
CAPACITY 742cc
POWER OUTPUT 59 bhp @ 5500 rpm
WEIGHT 463 lb (210 kg)
TOP SPEED 115 mph (185 km/h)
COUNTRY OF ORIGIN Japan

Navigation notes are contained in the "Road Book"

The Honda is dominated by its huge fuel tank

Long-distance suspension

DESERT EXTRAS

Among the extras found on this bike in contrast to a normal trail bike include an additional rear fuel tank and a computerized compass mounted in a slot behind the seat.

The tires are filled with puncture-resistant mousse

The bashplate protects the bottom of the engine

HONDA *CBR900RR Fireblade*

THERE WAS NO NEW technology on the Fireblade when it was launched in 1992, but it was a genuine leap forward in motorcycle design. So much so, that it wasn't until the arrival of the Yamaha R1 *(see pp.462–63)* six years later that the opposition caught up. The secret of the 'Blade's success was that it combined a 900cc power with 600cc weight and 400cc dimensions – and it could easily outhandle its overweight rivals. Construction quality and design detailing were superb; every component on the 'Blade had been considered for performance, weight, and size. It was a racer for the road.

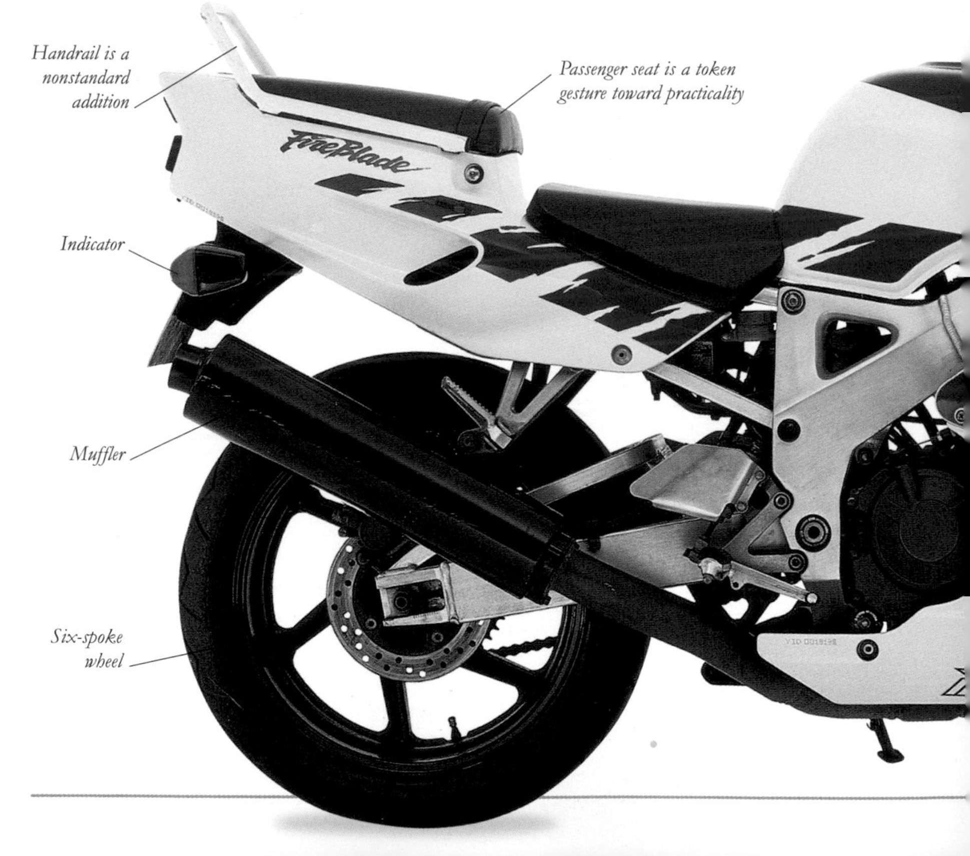

SPECIFICATIONS

MODEL Honda CBR900RR Fireblade
CAPACITY 893cc
POWER OUTPUT 122 bhp
WEIGHT 408 lb (185 kg)
TOP SPEED 160 mph (258 km/h)
COUNTRY OF ORIGIN Japan

Perforated fairing

24° steering head angle makes for superfast steering

Despite giving away 20 bhp to 1000 and 1100cc opposition, the 'Blade had a similar power-to-weight ratio

Close-fitting fairing appears shrink-wrapped to the machine

IMPOSING FRONT

The 'Blade was compact and purposeful. The perforated fairing and the twin round headlights implied a racing pedigree that it didn't have. At least not to start with. The 16-in (41-cm) front wheel made handling frisky.

Nissin four-piston brake calipers

HONDA *CBR600F*

HONDA TOOK THE middleweight field by storm in 1987 with the CBR600F. With its revolutionary bodywork and over 140 mph (225 km/h) motor it soon ousted Kawasaki's GPZ600 from the top of the middleweight sales charts. Since then, the CBR600 has been an almost constant best-seller in its class. The design was completely revised in 1991 to keep it ahead of the competition, with power increased to 99 bhp and the chassis redesigned with a smaller frame and other minor refinements. This is a 1993 model.

Twin-spar steel frame

Behind the distinctive fully enclosed bodywork hides a water-cooled, 16-valve, d.o.h.c., four-cylinder engine

17-in (43-cm) wheels with twin-disc front brake

Road tire

SPECIFICATIONS

Model Honda CBR600F
Capacity 599cc
Power Output 99 bhp @ 12,000 rpm
Weight 408 lb (185 kg)
Top Speed Not known
Country of Origin Japan

Power Boost

There have been a few changes to the CBR600's specification since the original 83 bhp, H model. Engine changes in J and K models saw power output upped to 93 bhp.

HONDA *RC45*

THE RC45 IS HONDA'S ultimate V4 production bike. Built to comply with World Superbike racing regulations, the 1994 machine shown here replaced the earlier RC30. Much of the technology for the new machine was developed on the factory's RVF racers. The RC45 had an all-new engine with the gear drive for the camshafts taken from the end of the crank. Electronic fuel injection replaced the RC30's carburetors. Honda's optional race kit increased power to 150 bhp but doubled the bike's price.

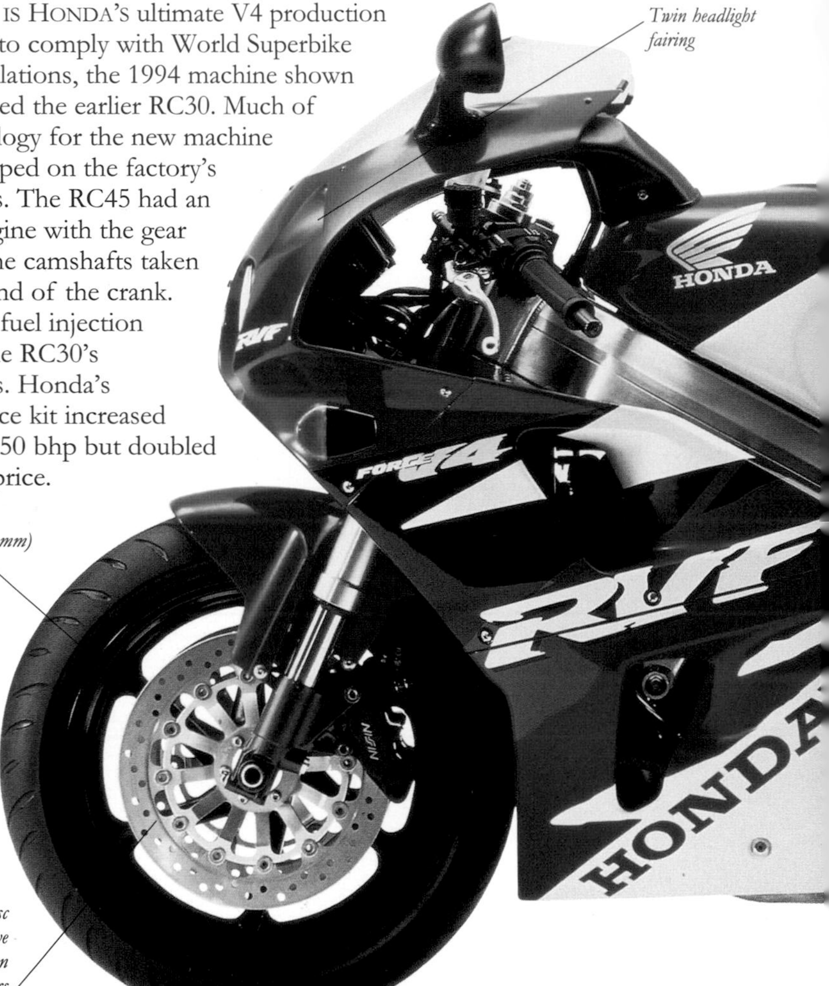
Twin headlight fairing
16-in (406-mm) front wheel
Twin front disc brakes have four-piston calipers
HONDA
RVF
HONDA

SPECIFICATIONS

Model Honda RC45
Capacity 749cc
Power Output 118 bhp @ 12,000 rpm
Weight 417 lb (189 kg)
Top Speed Not known
Country of Origin Japan

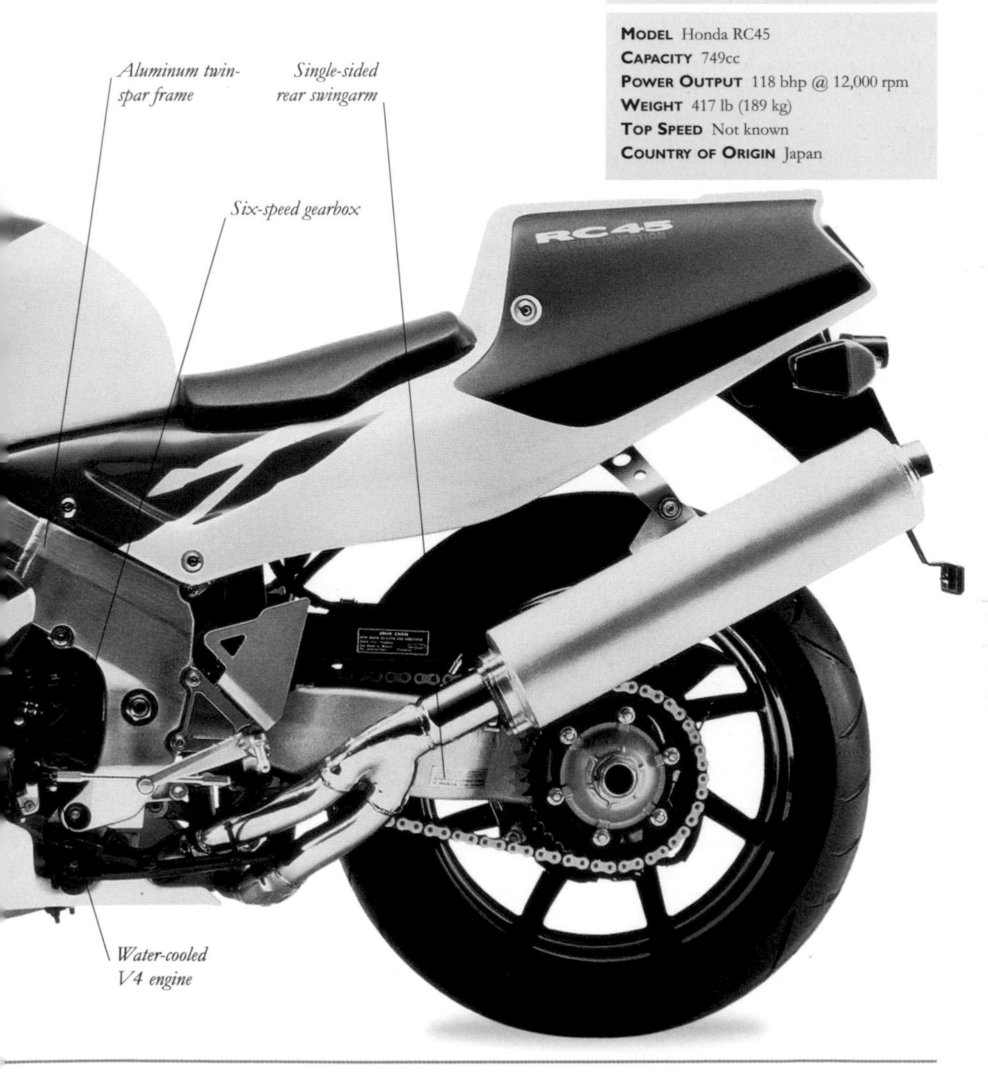

Honda *CBR1100 Blackbird*

Motorcycle makers have always wanted to make the fastest bike on the market. Kawasaki claimed the title in 1990 with its ZZ-R1100 *(see pp.266–67)*, and it wasn't until six years later that Honda was able to respond. Placing a 164 bhp engine in an aerodynamic bike weighing 531 lb (241 kg) resulted in a top speed of 177 mph (285 km/h). However, the Blackbird is also docile, easy to ride at low speeds, comfortable, and has surprisingly agile handling. It combines the punch of a boxer with the manners of an English butler and has become a popular bike for two-person sports touring use. Inevitably the Blackbird had a limited life at the top of the performance chart; in 1999 it was overshadowed by the Suzuki Hayabusa *(see pp.398–99)*.

SPECIFICATIONS

MODEL Honda CBR1100 Blackbird
CAPACITY 1137cc
POWER OUTPUT 164 bhp @ 10,000 rpm
WEIGHT 531 lb (241 kg)
TOP SPEED 177 mph (285 km/h)
COUNTRY OF ORIGIN Japan

SLEEK AND FAST
Aerodynamics become very important when a bike is as fast as the Blackbird. The headlights are stacked to make them narrower, and allow the front to take a more pointed form. The indicators are incorporated into the mirrors.

HONDA *VTR Firestorm*

FROM THE LATE 1960S, Honda was committed to developing impressive four-cylinder superbikes. However, by the 1990s, Ducati had achieved success, winning sales, acclaim, and races with its good-looking, idiosyncratic V-twin machines. Inspired by the Italian opposition, Honda soon built its own V-twin. The Firestorm was a capable and, for Honda, quirky bike that found a ready market among buyers who wanted the handling and power of a lightweight V-twin without the suspect reliability of Italian machines.

SPECIFICATIONS

MODEL Honda VTR Firestorm
CAPACITY 996cc
POWER OUTPUT 108 bhp @ 9000 rpm
WEIGHT 423 lb (192 kg)
TOP SPEED 150 mph (241 km/h)
COUNTRY OF ORIGIN Japan

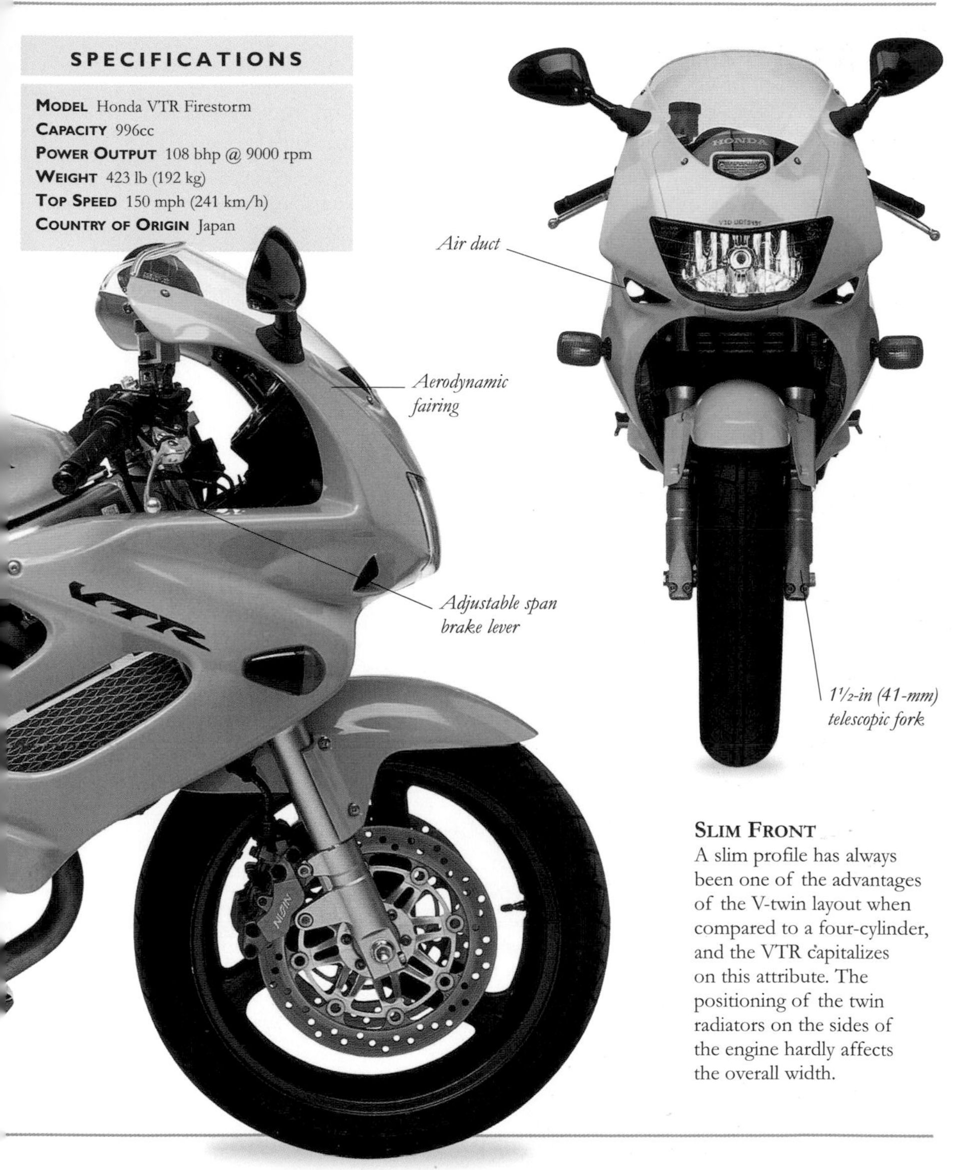

SLIM FRONT
A slim profile has always been one of the advantages of the V-twin layout when compared to a four-cylinder, and the VTR capitalizes on this attribute. The positioning of the twin radiators on the sides of the engine hardly affects the overall width.

HONDA *RC30*

THE RC30 APPEARED IN 1987. It was a road bike designed for the racetrack. The World Superbike championship for production-based bikes was kicking off the following year, and Honda wanted to win. It did, clocking up Isle of Man TT, World Endurance, and hundreds of national race wins to boot. The V-four engined bike was compact, clever, and beautifully engineered. The single-sided swingarm, intricate four-into-one exhaust system, and gear-driven camshafts were things of beauty. And if it hadn't been for the prohibitive price tag it would have been an almost perfect road bike as well as racer. Its race pedigree, road manners, and evocative droning exhaust note make the RC30 a classic.

The RC30 was known as VFR750R in Europe

SPECIFICATIONS

MODEL Honda RC30/VFR750R
CAPACITY 748cc
POWER OUTPUT 112 bhp
WEIGHT 408 lb (185 kg)
TOP SPEED 155 mph (250 km/h)
COUNTRY OF ORIGIN Japan

RACE MODIFICATIONS
Many RC30s had their lights removed and fairing changed for racing, where they inevitably suffered damage and abuse. Road bikes with original bodywork are rare and sought after.

HONDA *RC51*

IN FIVE YEARS OF RACING the V-four cylinder RC45 in the World Superbike Championship, Honda only won the title once—compared to four wins for Ducati. In a blatant case of "if you can't beat them join them," Honda abandoned the V-four layout and switched to a Ducati-style 90° V-twin for the 2000 season. The RC51 (SP-1 in Europe) was the road bike that made the racer possible and the effect was immediate. Colin Edwardes won the WSB title in the RC51 's maiden year. The RC51 featured an alloy twin-spar chassis and distinctive triangular air-intake. The power, performance, and handling of the RC51 reflects the fact that the bike is designed for the track.

Alloy swingarm is braced for increased rigidity

SPECIFICATIONS

MODEL Honda RC51
CAPACITY 998cc
POWER OUTPUT 118 bhp
WEIGHT 441 lb (200 kg)
TOP SPEED 168 mph (270 km/h)
COUNTRY OF ORIGIN Japan

FUNCTIONAL INTAKE
The central air intake pressurizes the airbox that feeds the fuel injection system. The faster the bike is going, the higher the pressure, and the greater the power.

Honda *CB1000R*

The CB1000R combines the attitude and technology of a sports bike, but in a package that is more suited to road riding and an urban environment. The four-cylinder engine comes from the Honda Fireblade *(see pp.212–13)* sports bike, but for this model its peak power is reduced to 123 bhp. That's not a problem on an unfaired machine where performance is limited by the lack of wind protection, and the benefit is that the re-tuned engine is stronger and more usable at lower rpm. The CB1000R is very punchy and easy to ride, at least until the wind pressure tells you that it is time to slow down. It's designed to be a good-looking bike and an engaging ride, though not ideal for covering long distances. The bike shown here is a 2012 model.

Frame wraps over the top of the engine

Passenger accommodation is minimal

Exhaust system is kept short and stubby to centralize the mass of the bike to improve handling

SPECIFICATIONS

MODEL Honda CB1000R
CAPACITY 998cc
POWER OUTPUT 123 bhp @ 10,000 rpm
WEIGHT 478 lb (217 kg)
TOP SPEED 140 mph (225 km/h)
COUNTRY OF ORIGIN Japan

Four-cylinder engine was developed for the Honda Fireblade sports bike

Dash panel has three LED displays for speed, engine revs, and other information

Front forks have adjustable preload and damping

HOREX *Imperator*

THE FIRST HOREX MOTORCYCLES were 248cc o.h.v. singles from 1923. Production of the new Imperator model began in 1954, although twin-cylinder prototypes and racing machines had been produced earlier. The new machine's o.h.c. engine was mounted in a twin-loop cradle frame with swingarm rear suspension. This 1955 model has Schnell front forks; these were once optional but became standard in 1955. But the German market was in steep decline and Horex production ended in 1957.

SPECIFICATIONS

MODEL Horex Imperator
CAPACITY 398cc
POWER OUTPUT 24 bhp @ 5650 rpm
WEIGHT 386 lb (175 kg)
TOP SPEED 84 mph (135 km/h)
COUNTRY OF ORIGIN Germany

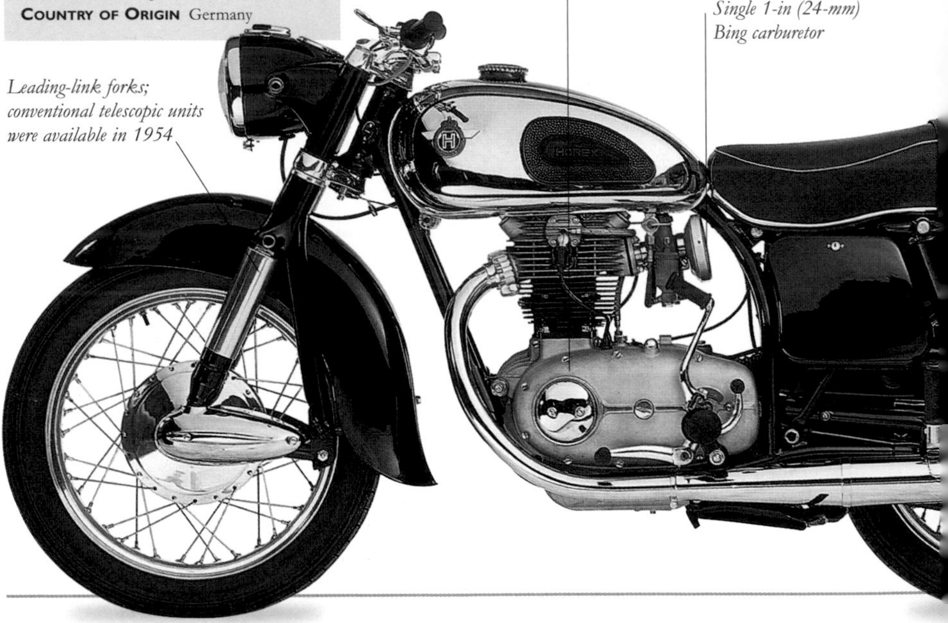

Unit-construction four-speed gearbox

Single 1-in (24-mm) Bing carburetor

Leading-link forks; conventional telescopic units were available in 1954

O.h.v. JAP engine has twin exhaust ports

Fully enclosed final-drive chain

Front drum brake

SPECIFICATIONS

MODEL HRD
CAPACITY 344cc
POWER OUTPUT 20 bhp (est.)
WEIGHT 300 lb (136 kg) (est.)
TOP SPEED 85 mph (137 km/h) (est.)
COUNTRY OF ORIGIN UK

HRD

HOWARD DAVIES' FIRST claim to fame was as the only person to win the 500cc Senior TT on a 350cc machine, a feat he achieved on an AJS in 1921. His second claim to fame was as one of the few people to have won a TT race on a machine of his own design and manufacture when, on an HRD, he won the 1925 Senior race, having already repeated his 1921 performance of second in the Junior race. The company went into liquidation in 1927, but not before Freddie Dixon gave it another Junior TT victory.

HUSQVARNA *V-Twin Racer*

INTRODUCED IN 1932 AND steadily developed until the company withdrew from racing after the 1935 season (the year of the model shown), the Husqvarna V-twin never quite made the grade, although it showed the vulnerability of the highly developed single-cylinder opposition to the challenge of a simple twin. High spots during its time were three consecutive victories in the Swedish Grand Prix and Stanley Woods' record lap in the 1934 Senior TT before he ran out of gas. Extensive use of light alloys kept the weight low, but despite being very fast, the handling of the machine was always something of a handicap. Stanley Woods switched to a Moto Guzzi to win the 1935 Senior TT. He also won the Junior in 1938 and 1939 aboard Velocettes.

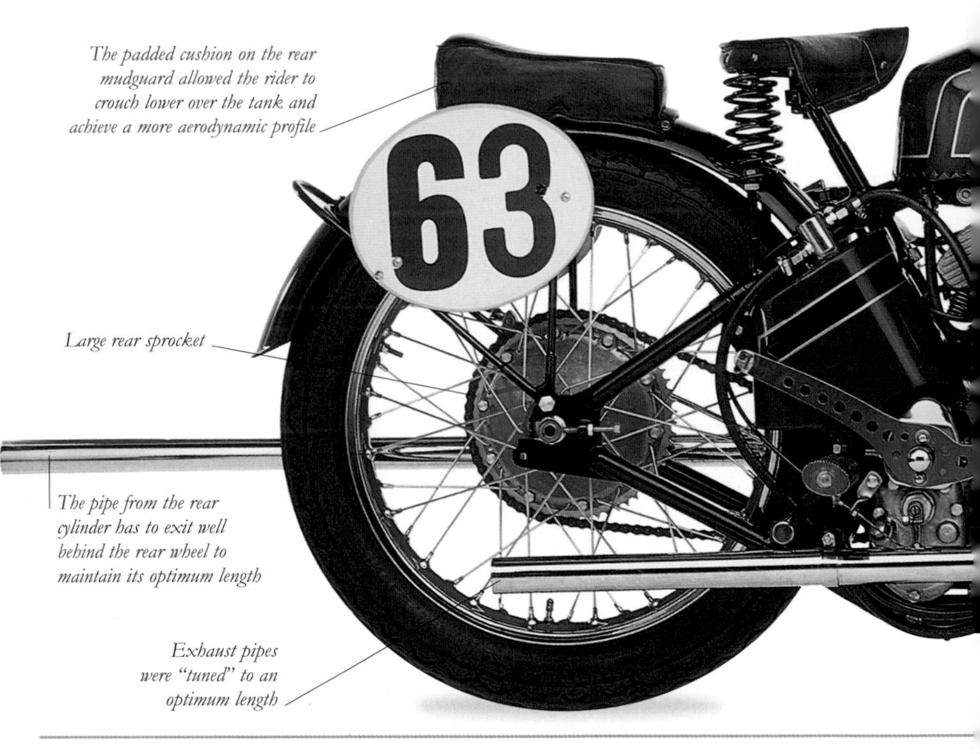

The padded cushion on the rear mudguard allowed the rider to crouch lower over the tank and achieve a more aerodynamic profile

Large rear sprocket

The pipe from the rear cylinder has to exit well behind the rear wheel to maintain its optimum length

Exhaust pipes were "tuned" to an optimum length

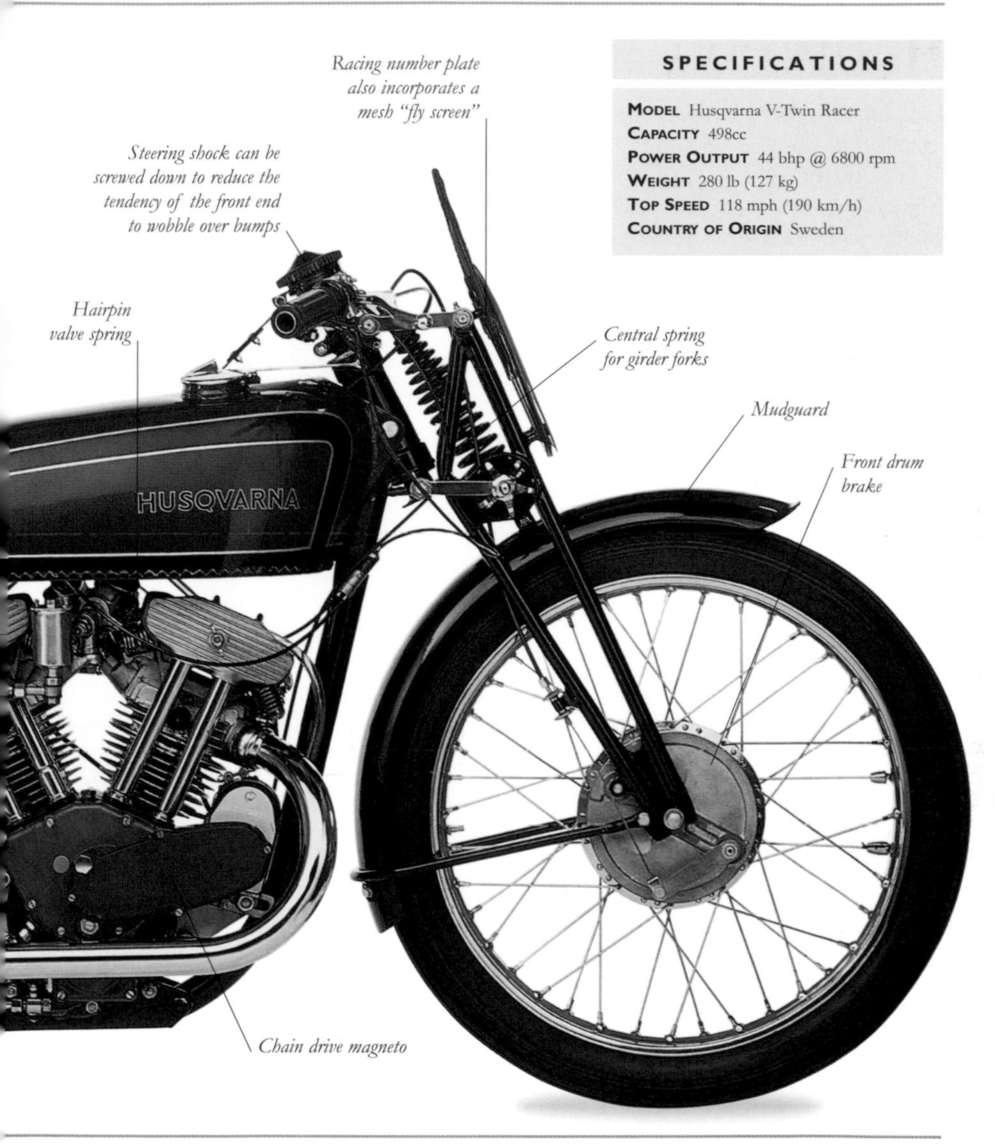

SPECIFICATIONS

Model Husqvarna V-Twin Racer
Capacity 498cc
Power Output 44 bhp @ 6800 rpm
Weight 280 lb (127 kg)
Top Speed 118 mph (190 km/h)
Country of Origin Sweden

HUSQVARNA *250*

DEVELOPED FROM A MID-FIFTIES 175cc three-speed road model, by 1963 the Husqvarna was the best 250cc motocross machine available, and for a while Torsten Hallman and his "Huskie" were virtually unbeatable in 250cc World Championship events. The engine was a full 250, joined with a four-speed gearbox, and combined power with reliability. It was used as a stressed member in the simple tubular frame. A single, large-diameter tube connected the headstock to the swingarm pivot. Husqvarna continued to develop its air-cooled two-stroke motocross and enduro machines, which were the first choice for many private competitors in the late Sixties and into the Seventies. Among them was actor Steve McQueen, who was an accomplished off-road competitor.

SPECIFICATIONS

MODEL Husqvarna 250
CAPACITY 245cc
POWER OUTPUT 22 bhp @ 6000 rpm
WEIGHT 208 lb (94 kg)
TOP SPEED Depends on gearing
COUNTRY OF ORIGIN Sweden

Norton forks were standard but some owners preferred to use Italian Cerianis

Heat shield on high-level exhaust pipe

Light alloy, conical brake hubs

Gold anodized wheel rims

Equipped with larger fuel tank, lights, and a stand for enduro events

Spark arresting muffler

Six-speed engine/ gearbox unit

SPECIFICATIONS

MODEL Husqvarna 390WR
CAPACITY 384cc
POWER OUTPUT Not known
WEIGHT 250 lb (113 kg) (est.)
TOP SPEED Depends on gearing
COUNTRY OF ORIGIN Sweden

HUSQVARNA *390WR*

CONTINUED DEVELOPMENT of Husqvarna's two-stroke off-road machines resulted in bikes such as this 1979 390WR model. The 390 was produced in three versions for motocross, enduro, and desert racing. All were based around the same air-cooled, single-cylinder, two-stroke engine and steel tubular frame. The WR version was equipped with lights to meet the regulations for enduro competitions. Bikes like this were very popular in the US as off-road machines.

HUSQVARNA *TC610*

RENEWED INTEREST IN the four-stroke single for off-road competition use prompted Husqvarna to go out and develop new machines. This 1992 TC610 uses a typically high-tech engine that features d.o.h.c., four valves, and water cooling. Power is transmitted via a six-speed gearbox. A 349cc version of the machine was also produced. The Husqvarana was the most popular four-stroke motocrosser until the arrival of the Yamaha YZ400F in the late 1990s.

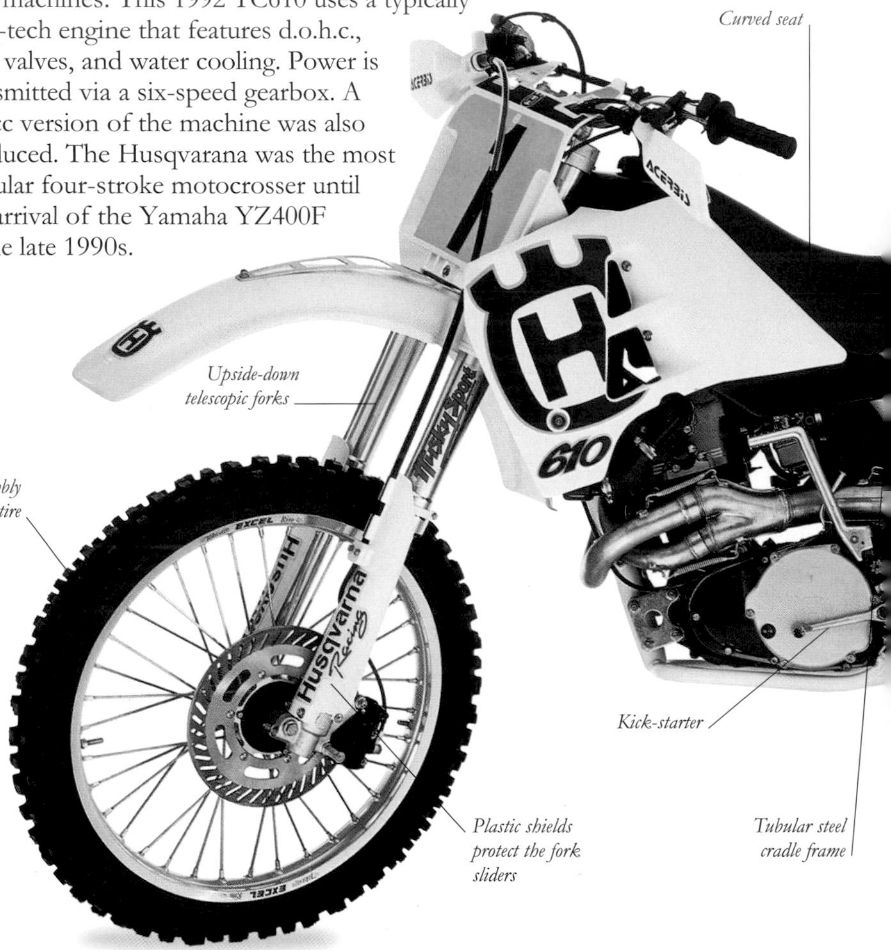
Curved seat
Upside-down telescopic forks
Knobbly front tire
Kick-starter
Plastic shields protect the fork sliders
Tubular steel cradle frame
610
Husqvarna Racing
EXCEL

SPECIFICATIONS

MODEL Husqvarna TC610
CAPACITY 577cc
POWER OUTPUT 50 bhp (est.)
WEIGHT 258 lb (117 kg)
TOP SPEED Depends on gearing
COUNTRY OF ORIGIN Sweden

ITALIAN SALE
The Swedish Husqvarna company sold its motorcycle division to the Italian Cagiva group in 1986, and production was moved to Italy.

IMME *R100*

BRAINCHILD OF NORBERT RIEDEL, the Imme (so named because it was made in Immenstadt) was unconventional in almost every respect. The combined engine and gearbox was in the form of a "power egg" and mounted on a forward extension of the rear suspension arm, which extended back beyond the rear wheel. It also formed the exhaust system. This arm was pivoted behind the engine to the lower end of the curved backbone-frame tube. This bike dates from 1948.

SPECIFICATIONS

MODEL Imme R100
CAPACITY 99cc
POWER OUTPUT 4 bhp @ 5800 rpm
WEIGHT Not known
TOP SPEED Not known
COUNTRY OF ORIGIN Germany

Front suspension uses parallel links and a spring but the wheel is mounted on a single arm

Complete engine and rear-wheel assembly pivots to provide rear suspension

Single front fork

SPECIFICATIONS

MODEL Indian Single
CAPACITY 17.59cu. in. (288cc)
POWER OUTPUT 2 bhp
WEIGHT 115 lb (52 kg)
TOP SPEED 30 mph (48 km/h)
COUNTRY OF ORIGIN US

Battery container

Coil for "constant loss" ignition system

INDIAN *Single*

INDIAN WAS FOUNDED by George Hendee and Oscar Hedstrom in Springfield, Massachusetts, in 1901 to make what they called "motocycles." The early models established the firm's reputation for quality. The 1904 2 bhp single shown here is typical of Indians made between 1901 and 1908. Designed by Hedstrom, the 288cc a.i.v. engine formed the saddle downtube of the bicycle-style diamond frame. The exhaust valve was mechanically operated while the automatic inlet valve was actuated by suction. Ignition was a total-loss system with three rechargeable two-volt dry-cell batteries.

INDIAN *1914 V-Twin*

INDIAN BUILT ITS FIRST V-twins in 1907. A "V" was constructed at 42°, with cylinders and barrels similar to those of Indian singles. The 1914 61cu. in. (1000cc) V-Twin used the cradle spring frame introduced the previous year with the standard leaf-sprung front forks. There was an electric-start version known as the Hendee Special, but only a handful were built because battery technology was not sufficiently developed to make the system reliable. The machine here is equipped with the two-speed gearbox driven through a dry multi-plate clutch. The i.o.e. V-twins were superseded by new side-valve engines in 1916.

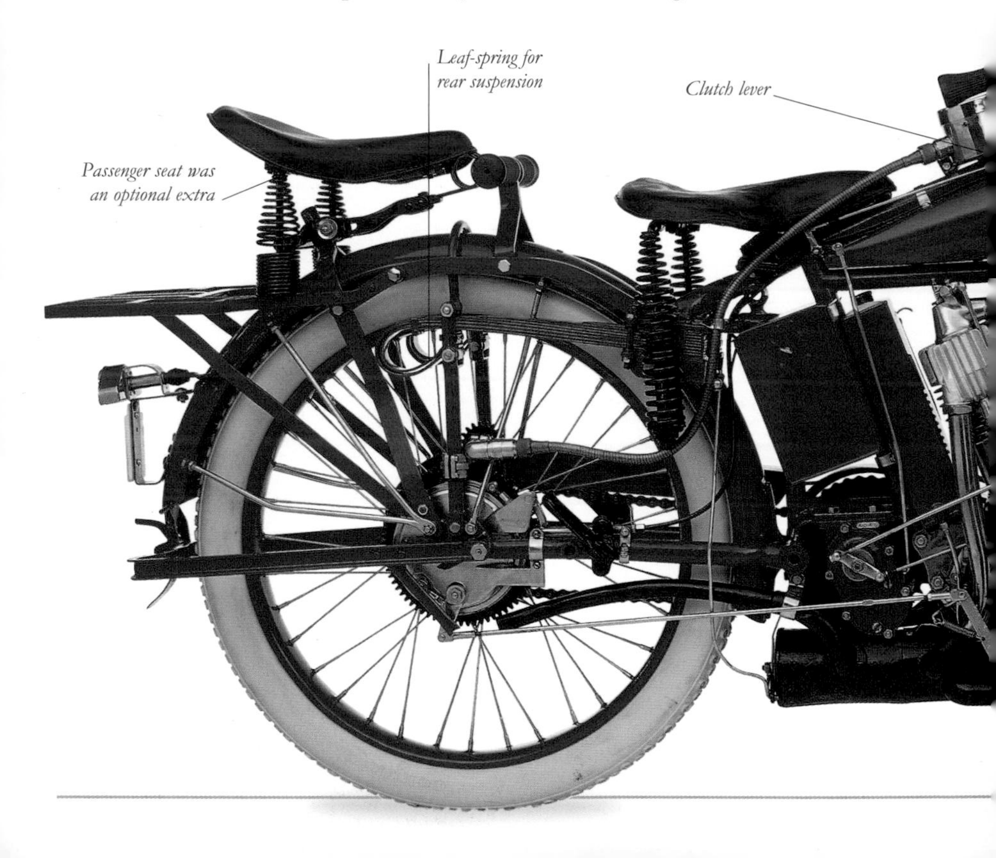

SPECIFICATIONS

MODEL Indian 1914 V-Twin
CAPACITY 61cu. in. (1000cc)
POWER OUTPUT 7 bhp
WEIGHT 400 lb (181 kg)
TOP SPEED 55 mph (89 km/h)
COUNTRY OF ORIGIN US

DECENT RIDE

The leaf-sprung rear suspension combined with the sprung saddle ensured a more comfortable ride than on most other contemporary bikes. The machine's electric horn, speedometer, and passenger seat were optional extras.

Complex rod linkages were soon to be replaced by cables

White rubber tires were common for the period

INDIAN *Model H*

THE DOMINANT FORM OF motorcycle sport in the US before 1920 was board-track racing. The minimalist machines that raced on the banked circuits had no gears, no brakes, no suspension, and no concessions to rider comfort since anything superfluous to the business of speed was not attached. Eight-valve Indian racers had first appeared in 1910, but most dated from between 1916 and the early 1920s. Performance of these machines was astonishing—typical average lap speeds were 90 mph (145 km/h). This is a 1920 bike.

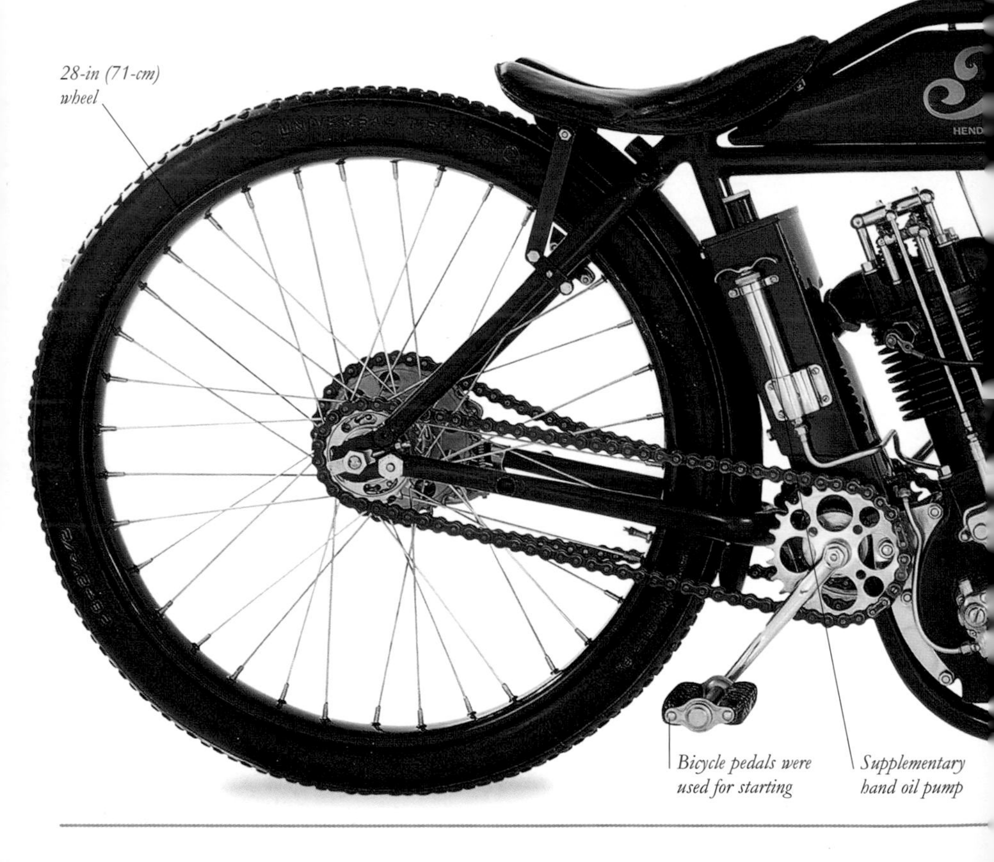

28-in (71-cm) wheel

Bicycle pedals were used for starting

Supplementary hand oil pump

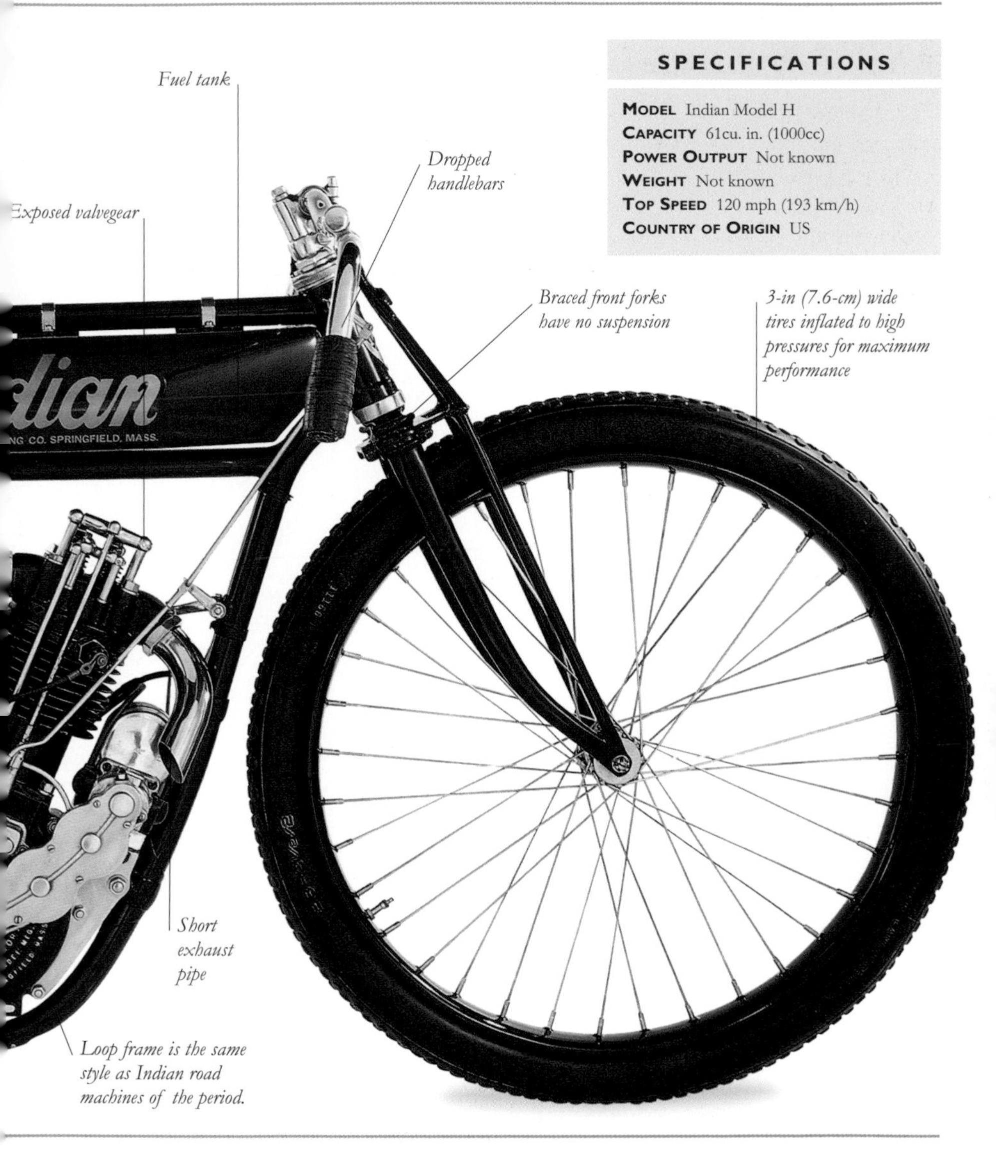

SPECIFICATIONS

MODEL Indian Model H
CAPACITY 61cu. in. (1000cc)
POWER OUTPUT Not known
WEIGHT Not known
TOP SPEED 120 mph (193 km/h)
COUNTRY OF ORIGIN US

INDIAN *Scout*

THE FIRST OF INDIAN'S famous Scout models appeared in 1920. More popular than its larger sibling, the Chief *(see pp.244–45)*, the Scout was the work of Indian's Irish-born designer Charles B. Franklin. The engine layout followed Indian's traditional 42°, V-twin configuration and, like the larger Powerplus, featured side valves. The three-speed gearbox was bolted to the back of the engine and driven by an apparently indestructible, if rather noisy, helical-gear primary drive that ran in a cast-aluminum oil bath case. The machine shown here is a 1930 model.

SPECIFICATIONS

MODEL Indian Scout
CAPACITY 45cu. in. (750cc)
POWER OUTPUT 18 bhp
WEIGHT 370 lb (168 kg)
TOP SPEED 100 mph (161 km/h)
COUNTRY OF ORIGIN US

101 frames were slightly longer than earlier models, making handling more stable

Detachable cylinder heads; early models had one-piece cylinders

18-in (46-cm) wheel

ALL-AROUND BIKE

Although Scouts were offered with the full array of extras for touring, their low center of gravity and excellent handling meant they were popular with racers, hill climbers, and trick riders.

INDIAN *Chief*

THE CHIEF MADE ITS DEBUT in 1922. Advanced design allowed the Chief, together with its famous stablemate the Scout *(pp.242–43)*, to dominate the marketplace for over 20 years. Color options, tank, mudguards, and tires for the Chief varied throughout its long production run. The later Chiefs (such as this fine example) were particularly handsome machines. Valanced mudguards (or "fenders") and girder forks gave the 1947 model a look of elegance and streamlined luxury, although its performance was rather sluggish—because of its weight—and its speed and acceleration could not match the equivalent Harley-Davidson. The Chiefs were the last true Indians.

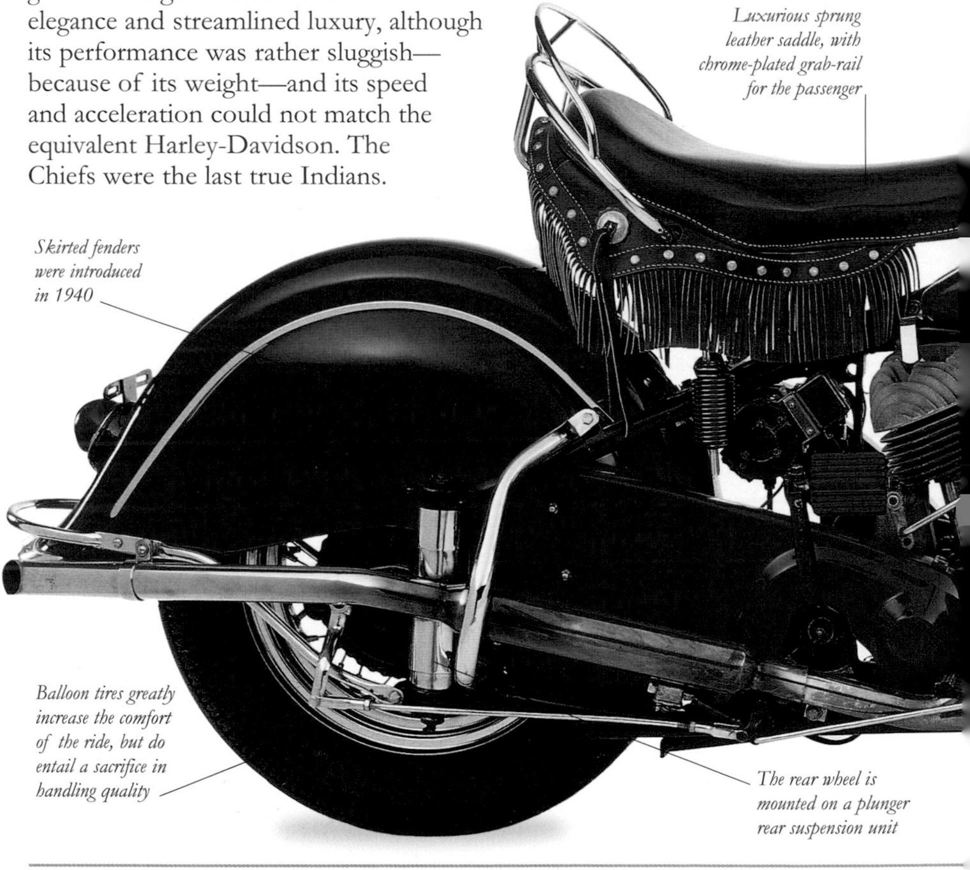

Luxurious sprung leather saddle, with chrome-plated grab-rail for the passenger

Skirted fenders were introduced in 1940

Balloon tires greatly increase the comfort of the ride, but do entail a sacrifice in handling quality

The rear wheel is mounted on a plunger rear suspension unit

The right-hand twistgrip controls the ignition timing, not the throttle (Indian practice since 1901)

Caps for oil tank and reserve fuel tank

Chromed headlight and additional sidelights

SPECIFICATIONS

MODEL Indian Chief
CAPACITY 74cu. in. (1213cc)
POWER OUTPUT 40 bhp (est.)
WEIGHT 550 lb (249 kg)
TOP SPEED 85 mph (137 km/h)
COUNTRY OF ORIGIN US

Girder forks replaced the traditional leaf-sprung design

The small drum brake is barely sufficient to stop a 550 lb (249 kg) motorcycle

Indian

Gearchange linkage (the lever is on the other side)

INDIAN *440*

INDIAN'S FOUR-CYLINDER machines continued to evolve during the 1930s, but were expensive to produce and never really sold in large numbers. Some observers credit Indian's obsession with the layout as contributing to its decline. After an experiment with an exhaust-over-inlet-valve cylinder head in 1936 and 1937, the company reverted to the traditional layout in 1938. For 1940, the fours got Indian's famous skirted mudguards and plunger rear suspension. Optional extras on the 440 included luggage rack, rearview mirror, and crash bars. Production continued until 1942.

SPECIFICATIONS

MODEL Indian 440
CAPACITY 77.21cu. in. (1265cc)
POWER OUTPUT 40 bhp
WEIGHT 568 lb (258 kg)
TOP SPEED 90 mph (145 km/h)
COUNTRY OF ORIGIN US

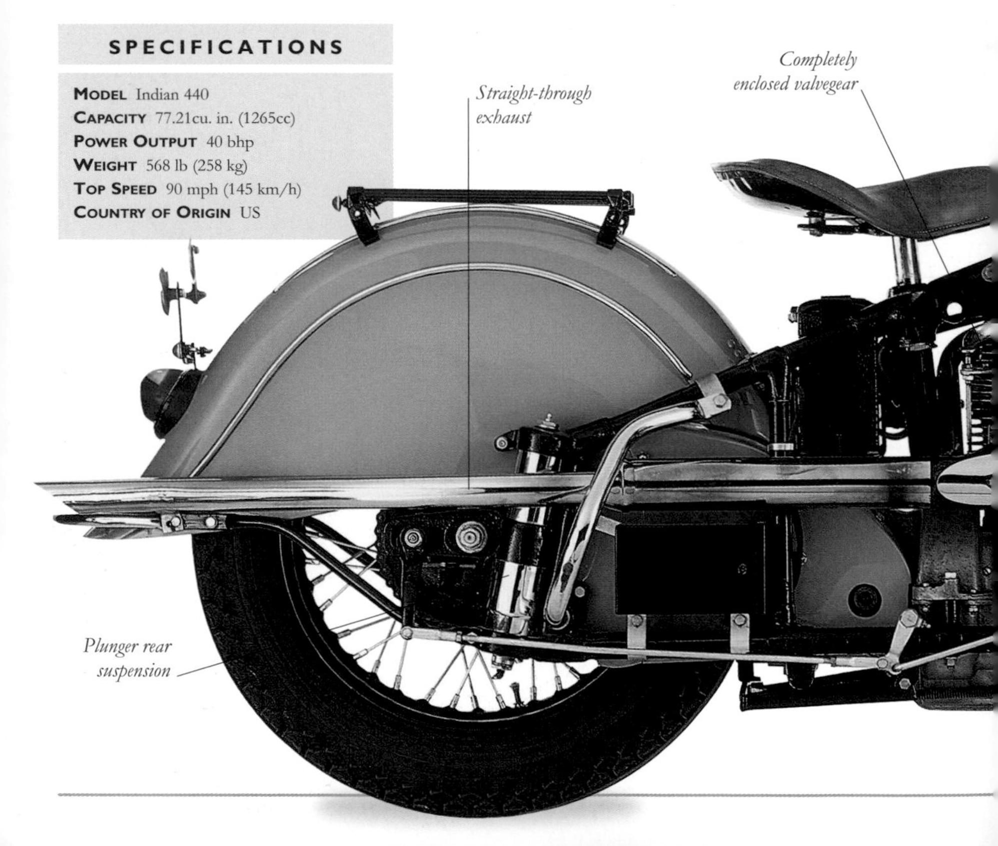

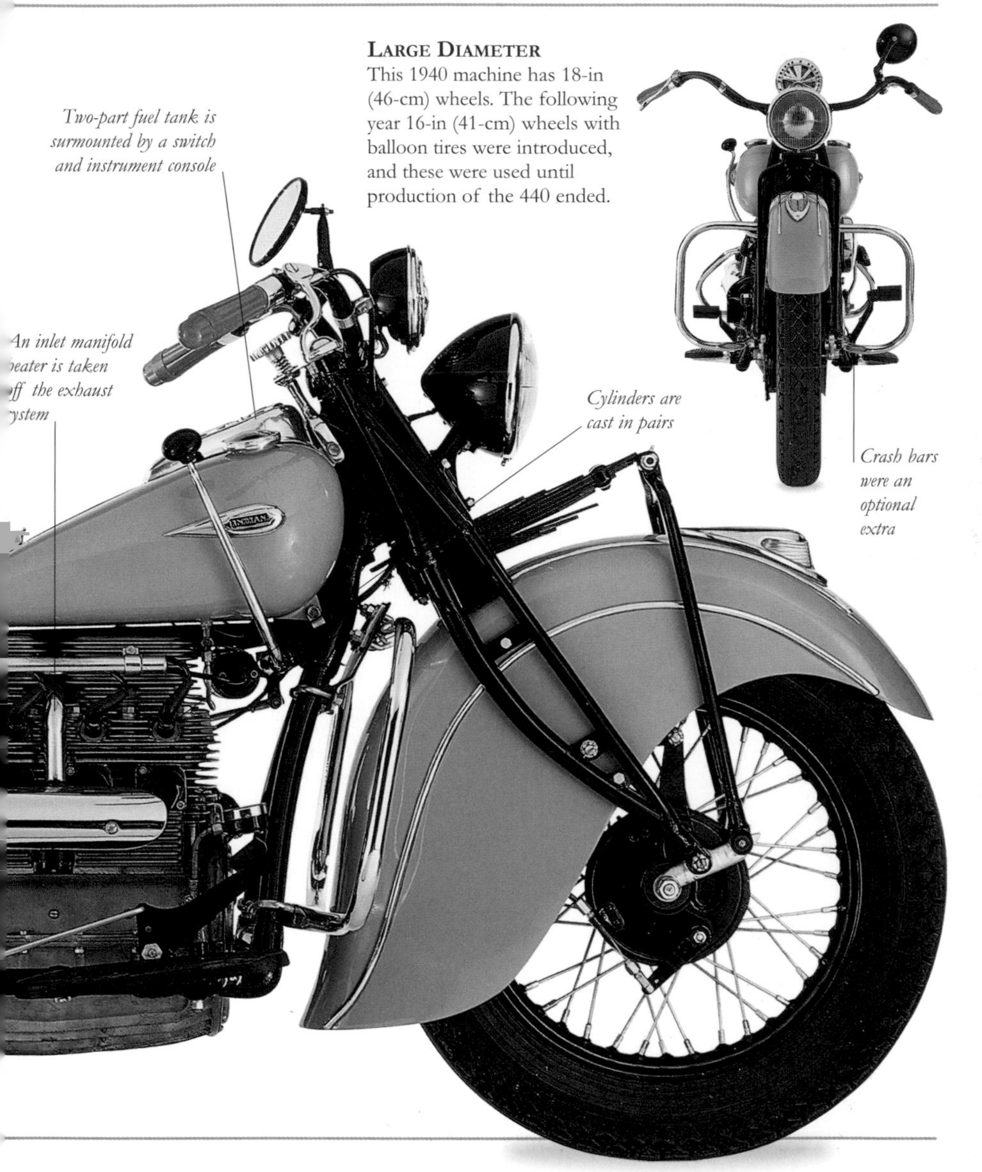

LARGE DIAMETER
This 1940 machine has 18-in (46-cm) wheels. The following year 16-in (41-cm) wheels with balloon tires were introduced, and these were used until production of the 440 ended.

INDIAN *Junior Scout*

A NEW SMALL-CAPACITY V-twin was introduced into the Indian range in 1932. The machine was the cheapest twin on the US market at the time of the Depression. Nicknamed the "Thirty-fifty" to reflect its cylinder capacity, it was sold as a Pony Scout but later became known as the Junior Scout. The new machine had a scaled-down V-twin engine in the frame and the running gear of the single-cylinder Prince. The Junior Scout followed the style of the bigger models and it got skirted mudguards along with the rest of the range in 1940 when the model shown was built. Production ended in 1941 and was not restarted after the war. The model here dates from 1940.

SPECIFICATIONS

MODEL Indian Junior Scout
CAPACITY 30.5cu. in. (500cc)
POWER OUTPUT 15 bhp (est.)
WEIGHT 340 lb (154 kg) (est.)
TOP SPEED 70 mph (113 km/h) (est.)
COUNTRY OF ORIGIN US

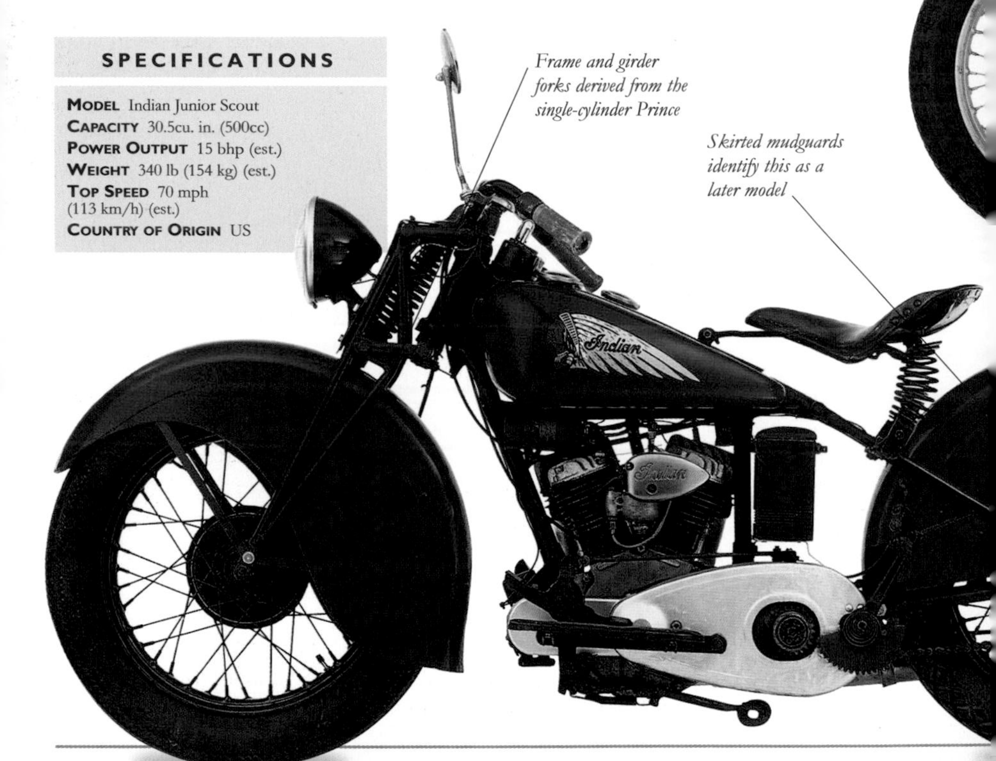

Cylinder finning was improved for this 1940 model

Triplex-chain primary drive

SPECIFICATIONS

MODEL Indian Sports Scout
CAPACITY 45.44cu. in. (745cc)
POWER OUTPUT 38 bhp
WEIGHT 380 lb (172 kg)
TOP SPEED 105 mph (169 km/h)
COUNTRY OF ORIGIN US

INDIAN *Sports Scout*

INTENDED FOR THE sports-bike market, many machines were converted into racers for use in the production-based Class C. Indian offered a line of performance parts and racing accessories to the budding private racer. They were equipped with Daytona engines, so named after Ed Kretz's spectacular Daytona 200 win in 1937. The 2.9 x 3.5-in (73 x 89-mm) side-valve engine had racing camshafts, high compression pistons, and polished ports generating 38 bhp. Coil and distributor ignition was standard. Indian and Harley-Davidson fought for Class C supremacy with their 45cu. in. (738cc) side-valve V-twins throughout the 1930s, with neither gaining a significant advantage.

INDIAN *648 Scout*

INDIAN HAD INTRODUCED the Sport Scout model in 1934 and it had been regularly updated and improved ever since. Although the engine was less powerful than the larger capacity Chief models *(see pp.252–53)*, this was the most lively machine that Indian built in prewar years, but the 1948 model 648 was to be the last of the line. The company changed hands in 1945, and the new owners were developing a new line of single and parallel-twin cylinder models. The 648 was a limited production stopgap to maintain Indian's presence in racing. Only 50 648 Scouts were made and the bike on these pages was ridden to victory in the prestigious Daytona 200 by Floyd Emde.

SPECIFICATIONS

MODEL Indian 648 Scout
CAPACITY 45cu. in. (738cc)
POWER OUTPUT 38 bhp
WEIGHT 182 lb (401 kg) (est.)
TOP SPEED 115 mph (185 km/h) (est.)
COUNTRY OF ORIGIN US

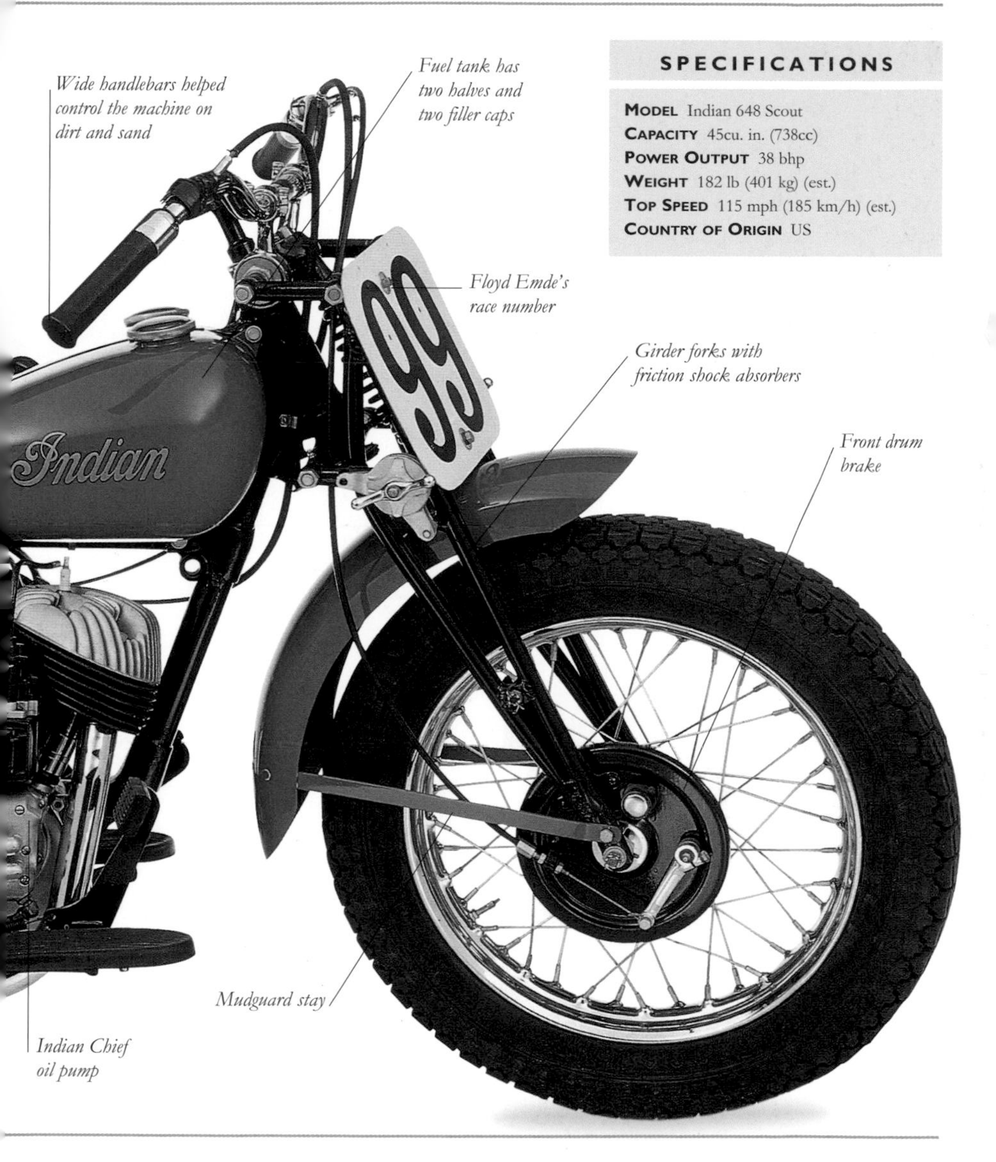

INDIAN *Chief Roadmaster*

INDIAN'S FLAGSHIP model underwent various changes in 1950 (the year of the model shown here), the most obvious of which was the use of hydraulically damped telescopic front forks instead of the girders that had replaced the old leaf-spring design after World War II. The 74cu. in. (1213cc) side-valve engine was enlarged to 80cu. in. (1311cc) by increasing the stroke to a massive 4⅘ in (122 mm). Generating 50 bhp at a lazy 4800 rpm, the engine drove a standard three-speed handshift gearbox by triplex chain through a wet, multiplate clutch. The handshift moved to the left and the throttle to the right, bringing Indian into line with common practice after 49 years.

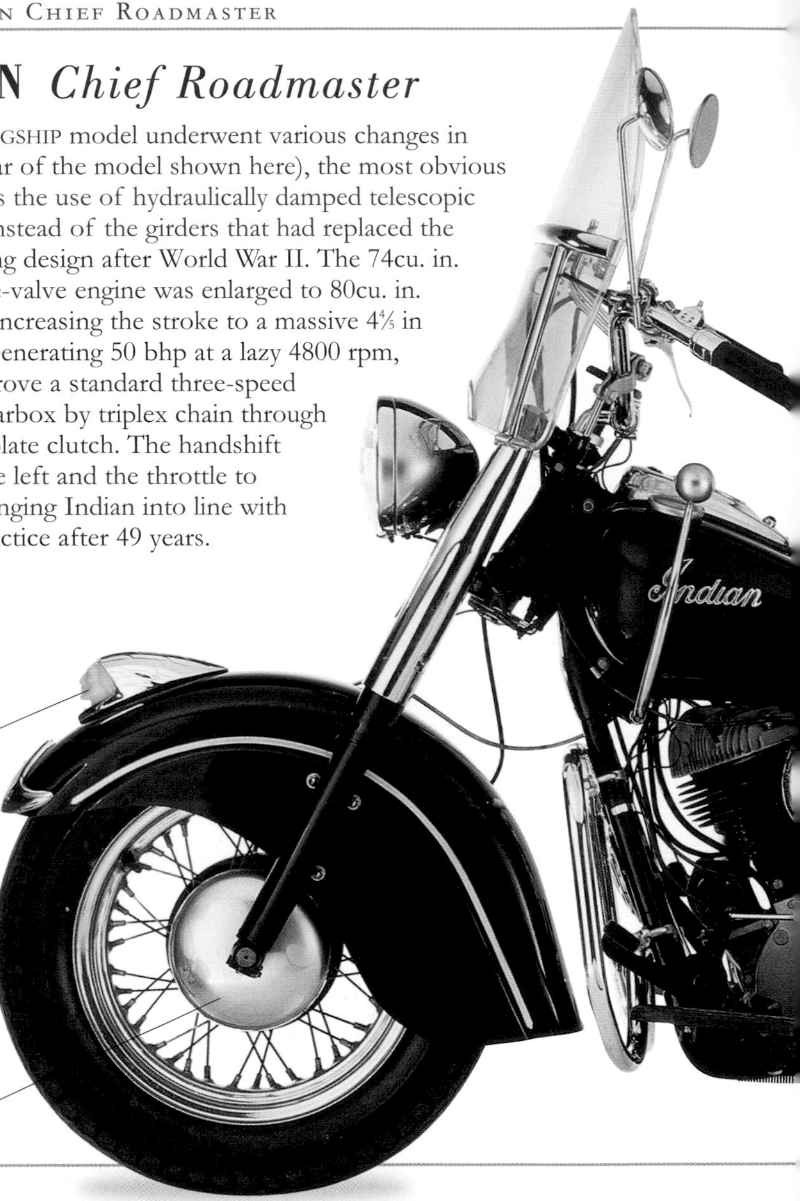

Postwar models had a distinctive "Indian Head" running light on the mudguard

Chrome dome wheel trim

SPECIFICATIONS

MODEL Indian Chief Roadmaster
CAPACITY 80cu. in. (1311cc)
POWER OUTPUT 50 bhp @ 4800 rpm
WEIGHT 570 lb (259 kg) (est.)
TOP SPEED 85 mph (137 km/h)
COUNTRY OF ORIGIN US

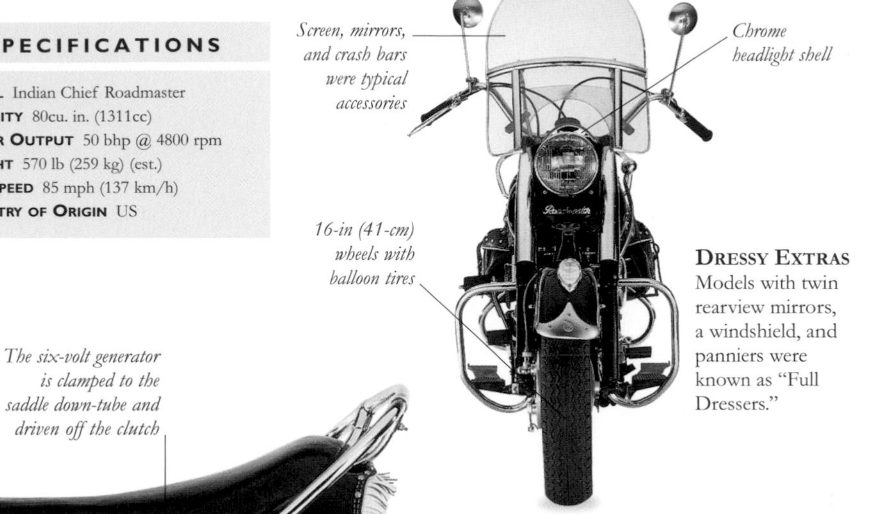

DRESSY EXTRAS
Models with twin rearview mirrors, a windshield, and panniers were known as "Full Dressers."

INDIAN *Velocette*

AROUND ONE HUNDRED OF THESE Anglo-Italian hybrids were built during 1969 and 1970. The bike used an overhead-valve single-cylinder Velocette engine and gearbox in a twin-loop frame, with Italian suspension and brakes. Around this time the Velocette company was on the verge of extinction, yet the machine did not enthuse admirers of either brand. The man responsible for the Indian Velocette was former West Coast Indian distributor and motorcycle magazine publisher Floyd Clymer, who dreamed of reviving the famous name. He died in 1970 having succeeded only in attaching the label to a selection of unlikely machines.

SPECIFICATIONS

MODEL Indian Velocette
CAPACITY 500cc
POWER OUTPUT 37 bhp @ 6200 rpm
WEIGHT 345 lb (156 kg)
TOP SPEED 101 mph (163 km/h)
COUNTRY OF ORIGIN US

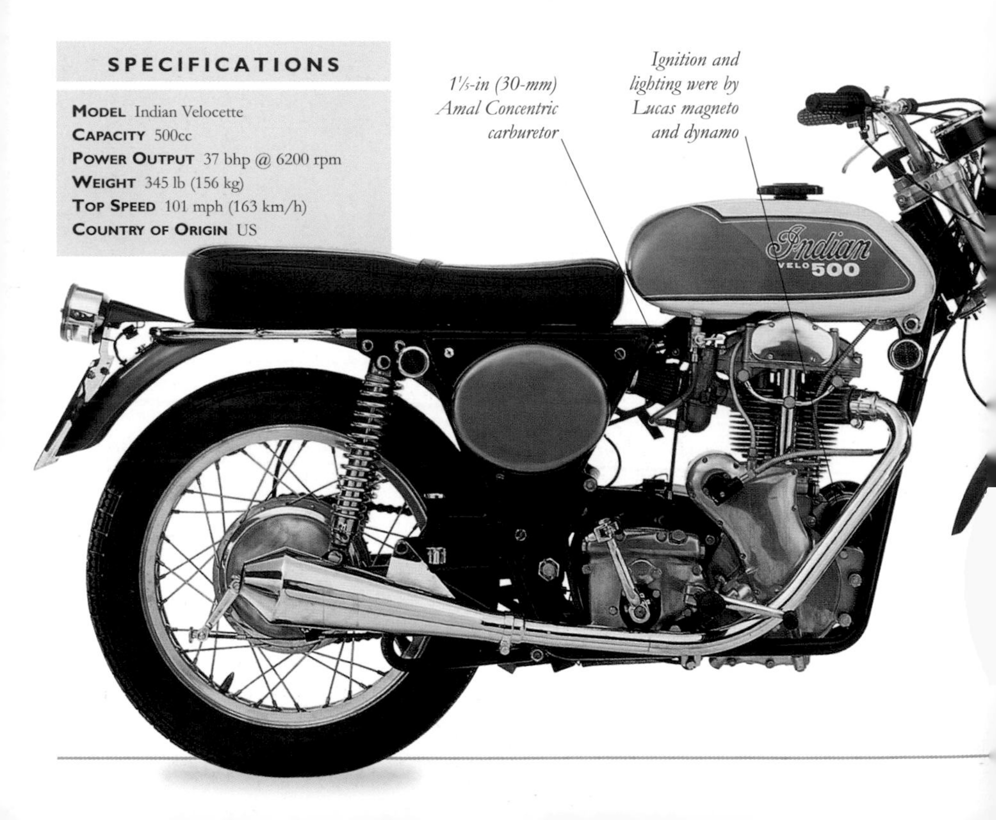

SPECIFICATIONS

MODEL Kawasaki Commander
CAPACITY 624cc
POWER OUTPUT 53 bhp @ 7000 rpm
WEIGHT 476 lb (216 kg)
TOP SPEED 108 mph (174 km/h)
COUNTRY OF ORIGIN Japan

KAWASAKI *Commander*

IN 1965 THE Commander was the biggest capacity bike built in Japan. Effectively a copy of BSA's 650cc A10, it had a four-stroke, air-cooled parallel-twin engine. Kawasaki had high hopes for the bike and expected good US sales. Despite healthy home sales, the Commander did poorly in the US because it was often compared to the popular BSAs, and the Kawasaki often came off worse. Kawasaki retreated from the US big-bike market to rethink its strategy. It returned four years later with the Mach III *(see pp.256–57)*.

KAWASAKI *Mach III*

THE KAWASAKI TRIPLE was launched in 1969 and soon gained a reputation for awesome performance and marginal handling. One of the last incarnations of the 500cc Mach III Kawasaki was this 1973 model. So successful was the 60 bhp two-stroke triple engine that minor changes only were made. The capacitor discharge ignition system was replaced by a more conventional battery and coil arrangement. The cycle parts were also modified as Kawasaki fought to contain the power of the triple. The bike was eventually forced out of production by stringent American emissions laws that Kawasaki could not meet, despite a relatively sophisticated lubrication system.

Fuel is fed to the engine by three carburetors

Specifications

Model Kawasaki Mach III
Capacity 498cc
Power Output 60 bhp @ 8000 rpm
Weight 395 lb (179 kg)
Top Speed 119 mph (191 km/h)
Country of Origin Japan

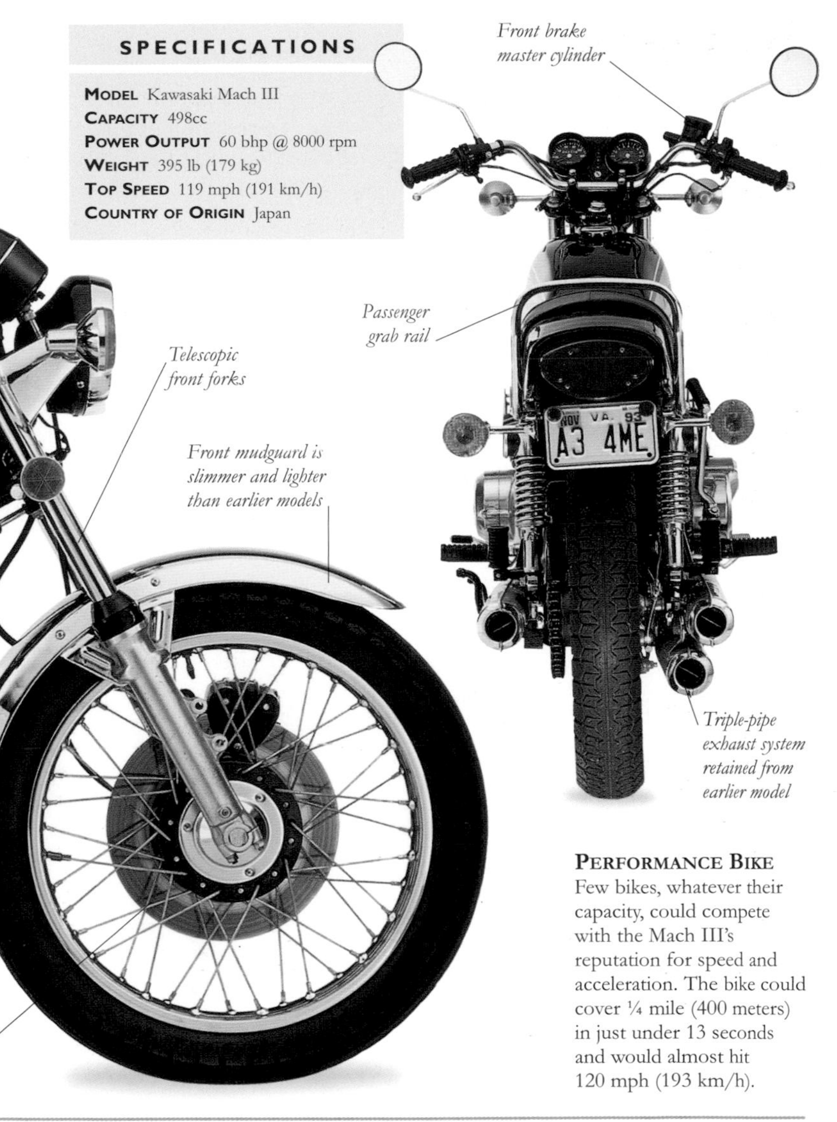

Performance Bike
Few bikes, whatever their capacity, could compete with the Mach III's reputation for speed and acceleration. The bike could cover ¼ mile (400 meters) in just under 13 seconds and would almost hit 120 mph (193 km/h).

KAWASAKI *Z1*

IF HONDA'S 750 FOUR created a sensation on its introduction in 1969, then Kawasaki's answer, 1972's 903cc Z1, was a reply worth waiting for. Like the Honda, the Z1 was also an air-cooled, four-stroke in-line four, but where the Honda had a single o.h.c., the Z1 had two. Soon after its introduction, the Z1 was entered in races with great success. In March 1973, the Z1 established a new 24-hour speed and endurance record at Daytona, Florida. The Z1 proved extremely popular with those riders looking for out-and-out performance. Often accused of poor handling and being under-braked, it was nevertheless exciting, impressive, and impossible to ignore. The 900 model was replaced by the 1015cc Z1000 in late 1976. The machine illustrated is a 1973 Z1.

The four-pipe exhaust proved too expensive to produce and was discontinued on later models

SPECIFICATIONS

MODEL Kawasaki Z1
CAPACITY 903cc
POWER OUTPUT 82 bhp @ 8500 rpm
WEIGHT 229.5 kg (506 lb)
TOP SPEED 211 km/h (131 mph)
COUNTRY OF ORIGIN Japan

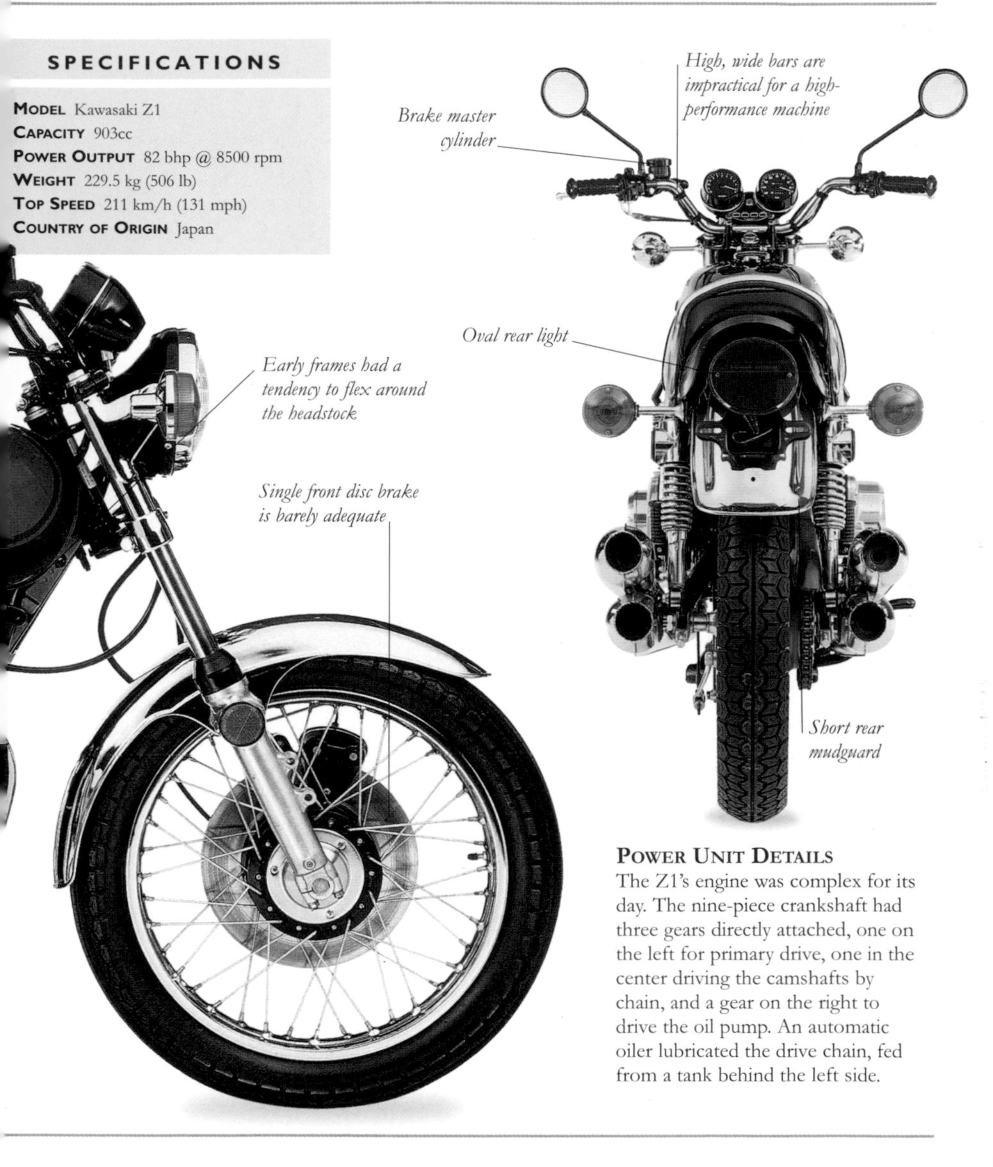

POWER UNIT DETAILS

The Z1's engine was complex for its day. The nine-piece crankshaft had three gears directly attached, one on the left for primary drive, one in the center driving the camshafts by chain, and a gear on the right to drive the oil pump. An automatic oiler lubricated the drive chain, fed from a tank behind the left side.

KAWASAKI *KR250*

ALTHOUGH KAWASAKI IS the smallest of the leading Japanese manufacturers, it has always had a reputation for building powerful, innovative machines. This KR250 is no exception. Designed for world-class racing, it is a rotary-valved, two-stroke twin, but instead of having the two cylinders arranged conventionally side by side, and operating a single crankshaft, they are placed one behind the other, and two crankshafts are used. The benefit is a narrower engine, as slim as a single-cylinder machine, with big improvements in aerodynamics.

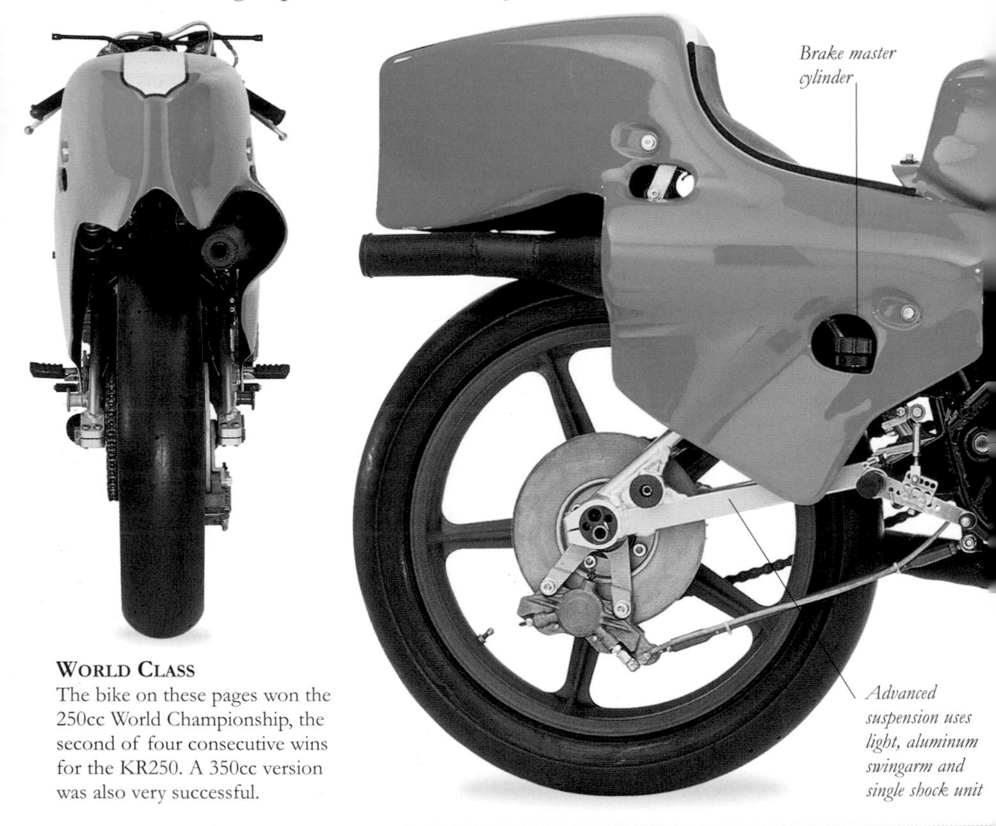

WORLD CLASS
The bike on these pages won the 250cc World Championship, the second of four consecutive wins for the KR250. A 350cc version was also very successful.

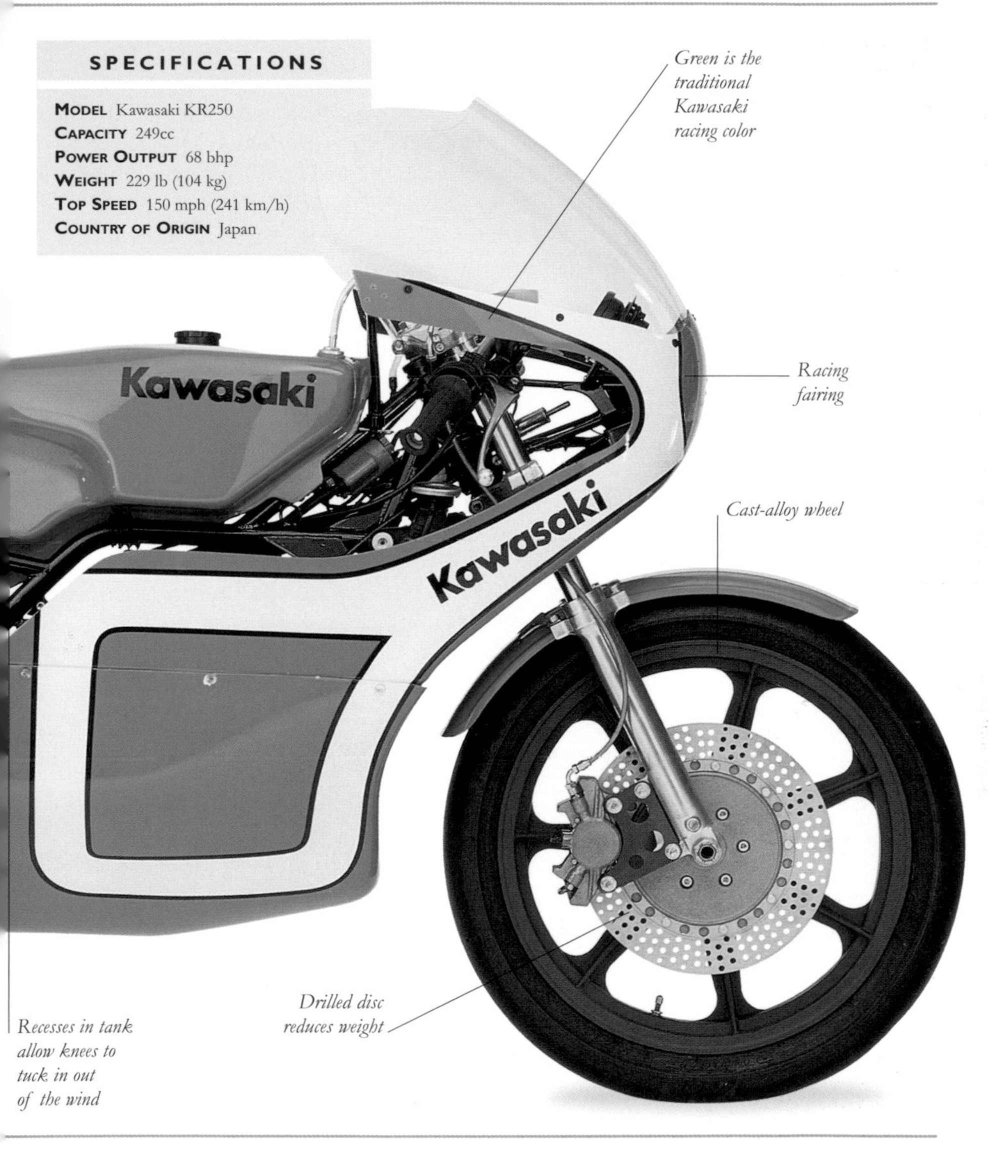

Specifications

Model Kawasaki KR250
Capacity 249cc
Power Output 68 bhp
Weight 229 lb (104 kg)
Top Speed 150 mph (241 km/h)
Country of Origin Japan

Green is the traditional Kawasaki racing color

Racing fairing

Cast-alloy wheel

Drilled disc reduces weight

Recesses in tank allow knees to tuck in out of the wind

KAWASAKI *ZX750*

THE SUPERBIKE CLASS FOR MODIFIED production machines originated in the US in the 1970s. This Kawasaki ZX750 won the American Championship in 1983 ridden by Wayne Rainey who later won three 500cc Grand Prix World Championships for Yamaha. The ZX750 was dubbed GPZ750 in Europe, and was a sports machine which used a conventional air-cooled, two-valve engine derived from the original Kawasaki Z1 *(see pp.258–59)*. From 1983 the ZX/GPZ was equipped with single-shock rear suspension.

Special fasteners allow the bodywork to be removed quickly

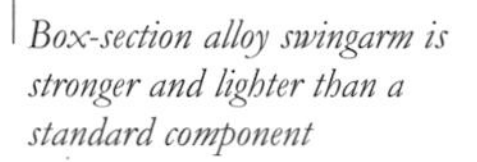

Box-section alloy swingarm is stronger and lighter than a standard component

SPECIFICATIONS

MODEL Kawasaki ZX750
CAPACITY 738cc
POWER OUTPUT 40 bhp
WEIGHT 493 lb (219 kg)
TOP SPEED 160 mph (257 km/h) (est.)
COUNTRY OF ORIGIN Japan

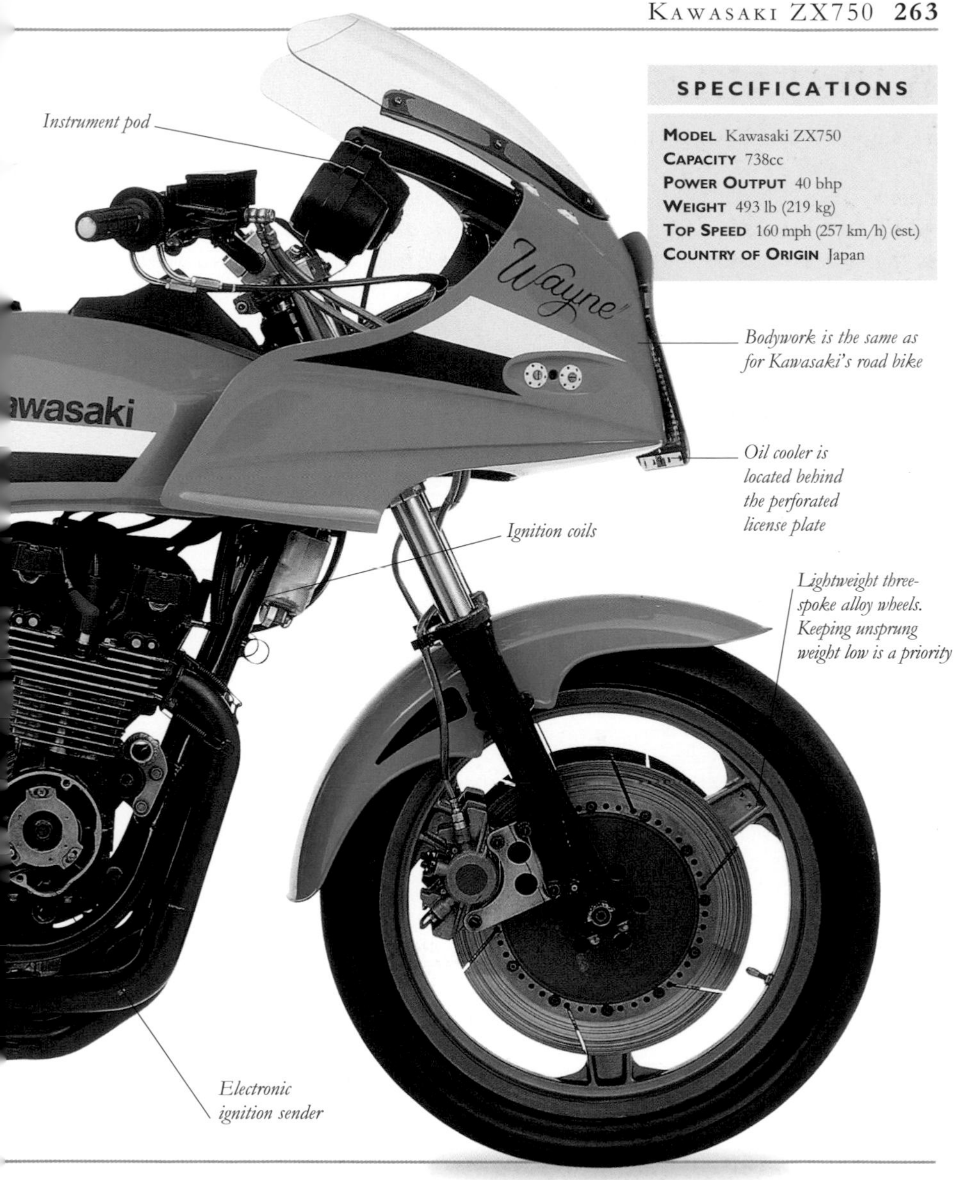
Instrument pod
Wayne
awasaki
Bodywork is the same as for Kawasaki's road bike
Oil cooler is located behind the perforated license plate
Ignition coils
Lightweight three-spoke alloy wheels. Keeping unsprung weight low is a priority
Electronic ignition sender

KAWASAKI *GPZ900R*

KAWASAKI'S REPUTATION FOR BUILDING groundbreaking sports bikes was further enhanced with the arrival of the GPZ900R in early 1984. Wet liners (the water flows directly against the outside of the cylinder liner rather than in an aluminum block) meant the cylinders could be positioned closely together, giving a very slim profile to the engine. The GPZ's performance and fine handling made it an immediate success and a future classic. It remained in production for 10 years.

Passenger grab rail

Single rear disc brake

Alternator is behind cylinder block and chain driven from the crankshaft

SPECIFICATIONS

MODEL Kawasaki GPZ900R
CAPACITY 908cc
POWER OUTPUT 119 bhp @ 10,900 rpm
WEIGHT 502 lb (228 kg)
TOP SPEED 154 mph (248 km/h)
COUNTRY OF ORIGIN Japan

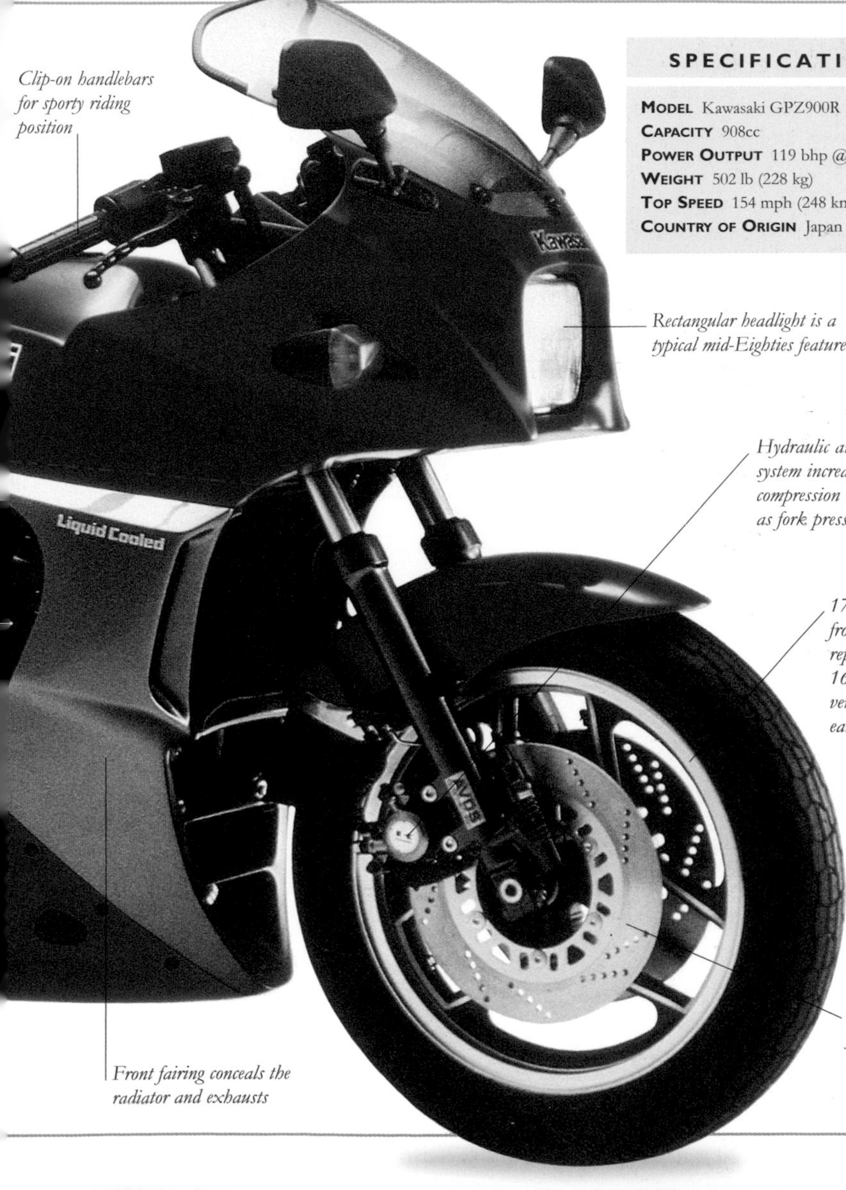

Clip-on handlebars for sporty riding position

Rectangular headlight is a typical mid-Eighties feature

Hydraulic anti-dive system increases compression damping as fork pressure builds

17-in (43-cm) front wheel replaced a 16-in (41-cm) version used on earlier models

Perforated front disc brake

Front fairing conceals the radiator and exhausts

KAWASAKI *ZZ-R1100*

BY THE TIME THE ZZ-R1100 premiered in 1990, Kawasaki had already built a long line of classic powerful in-line four-stroke fours, so the ZZ-R had a lot to live up to. Neither an out-and-out sportster nor a full tourer, the ZZ-R nevertheless found a niche among lovers of its hugely powerful yet silky-smooth engine. The bike could go from 20 mph (32 km/h) in top gear right up to 175 mph (282 km/h) in one strong surge. Riders found the bike easy to handle despite its weight, much of this being attributed to the huge, aluminum perimeter frame. Everything about the ZZ-R was big and impressive, from its wide 180-section rear tire to the huge 126-in (320-cm) twin front brake discs. In the early 1990s, as the true sports emphasis turned away from the large-capacity bikes toward the 600cc class, the ZZ-R lost its impact.

Passenger grab rail

Tank is sculpted to accept rider's legs, keeping wind drag to a minimum

Rear subframe is welded to main, twin-beam perimeter frame

SPECIFICATIONS

MODEL Kawasaki ZZ-R1100
CAPACITY 1052cc
POWER OUTPUT 125 bhp @ 9500 rpm
WEIGHT 513 lb (233 kg)
TOP SPEED 175 mph (282 km/h)
COUNTRY OF ORIGIN Japan

The red line on the tachometer is at an impressive 11,500 rpm

Indicators are integral to the design of the fairing

Forks are adjusted at top for preload and rebound shock absorption

Front mudguard is contoured for aerodynamic efficiency

Mufflers sheathed in sheet aluminum

UNBEATABLE SPEED

A power output of 125 bhp and a top speed of 175 mph (282 km/h), made the Kawasaki ZZ-R1100 the fastest production motorcycle of its era. But its extra weight and length meant that the handling was not as agile as some contemporary machines.

Opposed, four-piston, hydraulic caliper gives powerful, yet predictable braking

KAWASAKI *ZX-7R*

IN AN AGE WHEN MODEL CHANGES are introduced regularly, the Kawasaki ZX-7R had been, by 2001, almost unchanged for five years. It's even more astonishing because almost as soon as it was introduced, it was dismissed as old fashioned, underpowered, and overweight. Road riders knew differently. The conservative chassis geometry makes the 7R stable and well mannered. And with 163 mph (262 km/h) performance, it's plenty fast enough. The fact that it is so good looking is a bonus. The ZX-7R was developed from the earlier ZXR750, itself no spring chicken. Despite its apparent disadvantages, the 7R has also managed some impressive racing results.

Four-into-two-into-one exhaust system is designed to increase midrange power

SPECIFICATIONS

MODEL Kawasaki ZX-7R
CAPACITY 748cc
POWER OUTPUT 106 bhp
WEIGHT 505 lb (229 kg)
TOP SPEED 163 mph (262 km/h)
COUNTRY OF ORIGIN Japan

BLUNT FRONT

The two huge scoops dominate the blunt front of the ZX-7R. Ducts direct the air into the airbox from which the four 1½-in (38-mm) Keihin carburetors suck.

KAWASAKI *ZX10R*

BLISTERINGLY FAST AND GREAT-LOOKING, the Kawasaki ZX10R is a well-equipped sports bike. A redesign for 2011 gave the bike a new engine and chassis, but followed the usual formula for sports bikes, making it lighter, faster, and more powerful. However, the addition of rider aids, such as traction control and an engine management system that allows the rider to choose a softer power delivery to calm the 175 bhp engine for road riding or wet weather use, have made it more usable.

Fuel tank capacity is 4½-gallon (17-liter) giving a range of around 140 miles (225 km) on the road

Rigid fabricated swingarm is controlled by horizontal shock absorber

SPECIFICATIONS

MODEL Kawasaki ZX10R
CAPACITY 998cc
POWER OUTPUT 175 bhp @ 13,500 rpm
WEIGHT 436 lb (198 kg)
TOP SPEED 184 mph (296 km/h)
COUNTRY OF ORIGIN Japan

The airbox that feeds the engine is located in front of the fuel tank, under a dummy tank cover

Dash console includes a lap timer for track use

Showa forks have 1¾-in (43-mm) stanctions that resist flex under braking forces

12¼-in (310-mm) wavy-edged discs are gripped by four-piston brake calipers

KREIDLER *Renn Florett*

IN THIRTY YEARS OF MOTORCYCLE, MANUFACTURING, Kreidler only made 50cc machines. Therefore, the creation of a 50cc racing class in the late 1950s was a natural opportunity for Kreidler to develop racing machines. The firm's first racing bikes—built in time for the 1961 season—were heavily based on production Florett models. By the time of this 1963 model, the bikes had become more sophisticated. Disc valve induction replaced the piston port, and the four-speed gearbox was complemented by an externally mounted three-speed overdrive. When combined, the two sets offered riders 12 gears. A number of Kreidler's racing machines actually broke world speed records for their engine size.

SPECIFICATIONS

MODEL Kreidler Renn Florett
CAPACITY 49cc
POWER OUTPUT 14 bhp @ 15,000 rpm
WEIGHT 131 lb (59 kg) (est.)
TOP SPEED 106 mph (171 km/h)
COUNTRY OF ORIGIN Germany

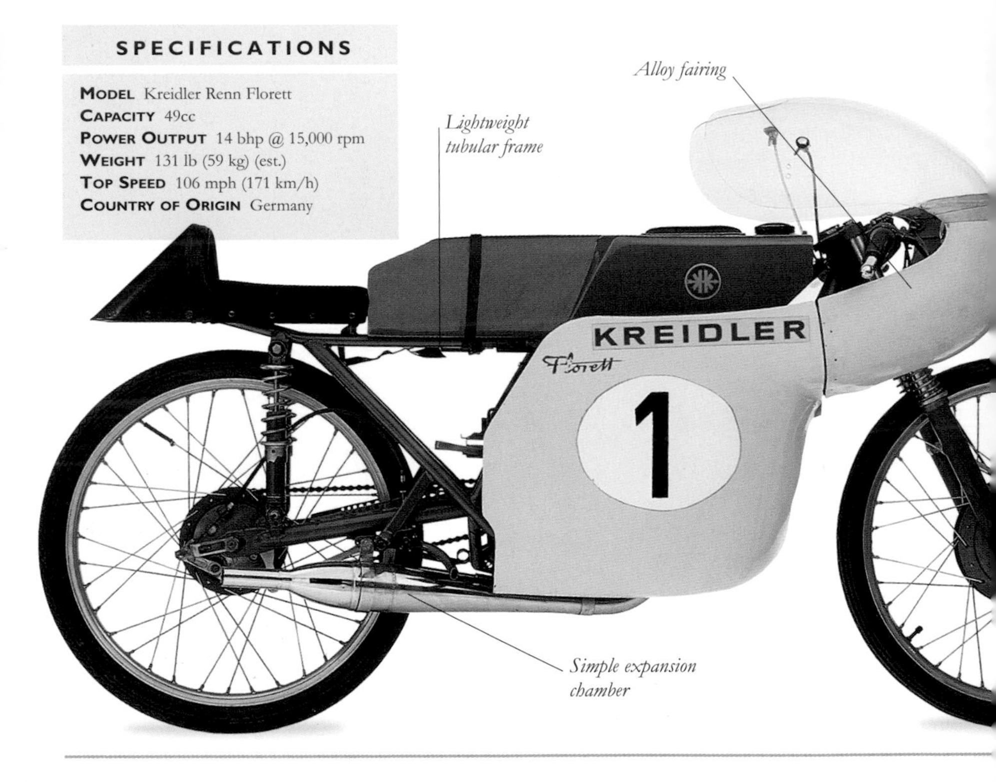

Radiators hidden behind plastic cowls

Plastic bodywork

SPECIFICATIONS
Model KTM 125 Motocross
Capacity 125cc
Power Output Not known
Weight 193 lb (88 kg)
Top Speed Not known
Country of Origin Austria

KTM *125 Motocross*

During the 1970s the Austrian company KTM became famous for its off-road competition machines and won its first World Championship in 1977 when Soviet rider Gennady Moisseyev took the 250cc title. KTM machines were always at the forefront of technical development and ideas developed on factory bikes were soon adopted on production models. This 1989 125cc model is typical. It features water cooling, upside-down forks, and rising-rate rear suspension. KTM also produced more mundane motorcycles and mopeds with two-stroke Sachs or Puch engines mounted in simple, tubular frames.

KTM *RC8R*

THROUGHOUT THE 1970s, '80s, and '90s, KTM were famous for their off-road bikes, winning several motocross world championships and exporting their bikes around the world. Their first sports bike RC8 was introduced in 2008. The V-twin engine was based on their existing power unit, but capacity was initially increased to 1148cc, and later raised to 1195cc. As a newcomer to the market, KTM produced a distinctive machine with a steel trellis frame, stacked headlights, and sharp-edged styling. It also had an adjustable riding position, making it one of the most comfortable bikes in this class. The higher specification R model was introduced in 2010.

SPECIFICATIONS

MODEL KTM RC8R
CAPACITY 1195cc
POWER OUTPUT 173 bhp @ 10,000 rpm
WEIGHT 410 lb (186 kg)
TOP SPEED 170 mph (274 km/h)
COUNTRY OF ORIGIN Austria

Tubular steel trellis frame

Linkage and bell crank transfer leverage from swingarm to shock absorber

10-spoke Marchesini alloy wheels fitted to R model

Cast alloy seat subframe

SPECIFICATIONS
Model KTM 990 Adventure
Capacity 999cc
Power Output 95 bhp @ 8750 rpm
Weight 456 lb (207 kg)
Top Speed 133 mph (214 km/h)
Country of Origin Austria

Fuel tank is split, with two filler caps. Total capacity is 5-gallon (19.5-liter)

Vertically arranged headlights allow a narrower fairing

The brakes' anti-lock system can be deactivated for off-road use

The engine is a 75° liquid-cooled V-twin with four valves per cylinder

KTM *990 Adventure*

KTM's first twin-cylinder production bike was the 950 Adventure, launched in 2003 following the success of the Austrian factory's rally bikes in the Dakar Rally earlier that year. The styling was distinctive, and inspired by the bikes that had given KTM such success in off-road competition. High-quality components were used, making this a purposeful, go-anywhere adventure tourer that soon gained an enthusiastic following. Capacity was increased to create the 990 model in 2006. The range was quickly expanded to include several other models using the same engine. The 2011 model is shown here.

LAVERDA *Jota*

LAVERDA INTRODUCED A NEW 980cc d.o.h.c. three-cylinder machine in 1973. The Jota, which was to become Laverda's most famous model, was a souped-up version of the original triple first produced for the British market in 1976. Limited engine tuning turned the already quick 3CL model into the fastest production bike of its day. Finesse was not a feature of the design; the entire machine is overengineered. Metal components are high quality and heavy. The controls are leaden and the handling suspect, yet construction quality was better than on many Italian bikes of the period. This is a 1982 model.

SPECIFICATIONS

MODEL Laverda Jota
CAPACITY 980cc
POWER OUTPUT 90 bhp @ 8000 rpm
WEIGHT 520 lb (236 kg)
TOP SPEED 139 mph (224 km/h)
COUNTRY OF ORIGIN Italy

QUIRKY RIDE

Top-heavy weight distribution made the Jota's handling quirky, and yet road tests of the period raved about the machine—which says more about its rivals than it does about the Laverda.

The bike's brutal frontal aspect was partly hidden behind a half fairing on later models

Handlebars adjustable in four places

Orange became the favorite color for Laverdas

Tubular cradle frame

Twin Brembo disc brakes

MAICO *MC350*

MAICO STARTED MAKING motorcycles in the mid-1930s, switched to aircraft parts shortly before World War II, and resumed motorcycle production in 1947. In the late 1950s, the company started producing motocross machines, which went on to earn Maico the manufacturers' World Championship. Maico was also runner-up three times in the individual 500cc title. The MC350 is typical of the oversized 250s that ousted the big four-strokes in 500cc motocross during the 1960s, being adequately powerful and easier to handle. The bike shown here is a 1969 model.

Four-speed engine/gearbox unit

Sturdy duplex frame

SPECIFICATIONS

MODEL Maico MC350
CAPACITY 352cc
POWER OUTPUT 28 bhp @ 6500 rpm
WEIGHT 250 lb (113 kg)
TOP SPEED Depends on gearing
COUNTRY OF ORIGIN Germany

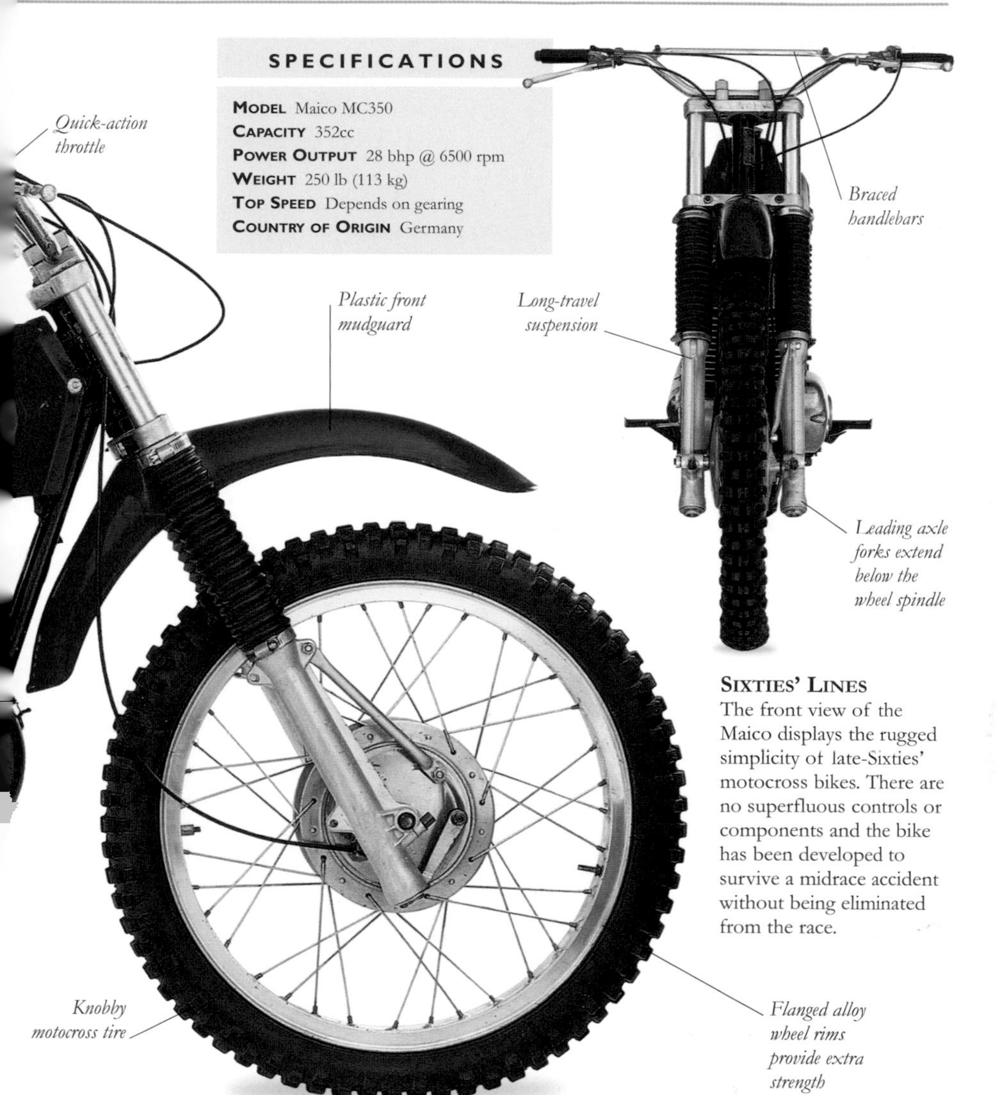

SIXTIES' LINES

The front view of the Maico displays the rugged simplicity of late-Sixties' motocross bikes. There are no superfluous controls or components and the bike has been developed to survive a midrace accident without being eliminated from the race.

MATCHLESS *Model B Silver Hawk*

FOUNDED IN 1899, MATCHLESS was one of the first British motorcycle manufacturers. The company unveiled its Silver Hawk at the 1931 Motorcycle Show, with the bike's engine essentially two Silver Arrow units side by side, surmounted by a single overhead camshaft. To minimize the effect of fuel surge, the single Amal carburetor was often equipped with twin float chambers. The handshift gearbox was driven by a duplex primary chain with an automatic tensioner allowing the unit to remain fixed. Under pressure the Silver Hawk became noisy and tended to develop cylinder head joint leaks. Designed as a luxury touring machine to rival Ariel's Square Four *(see pp.30–31)*, only 500 examples were built. Production ceased in 1935 because the Silver Hawk was too expensive and too problematic for a depressed market.

SPECIFICATIONS

MODEL Matchless Model B Silver Hawk
CAPACITY 592cc
POWER OUTPUT 26 bhp
WEIGHT 380 lb (172 kg)
TOP SPEED 80 mph (129 km/h)
COUNTRY OF ORIGIN UK

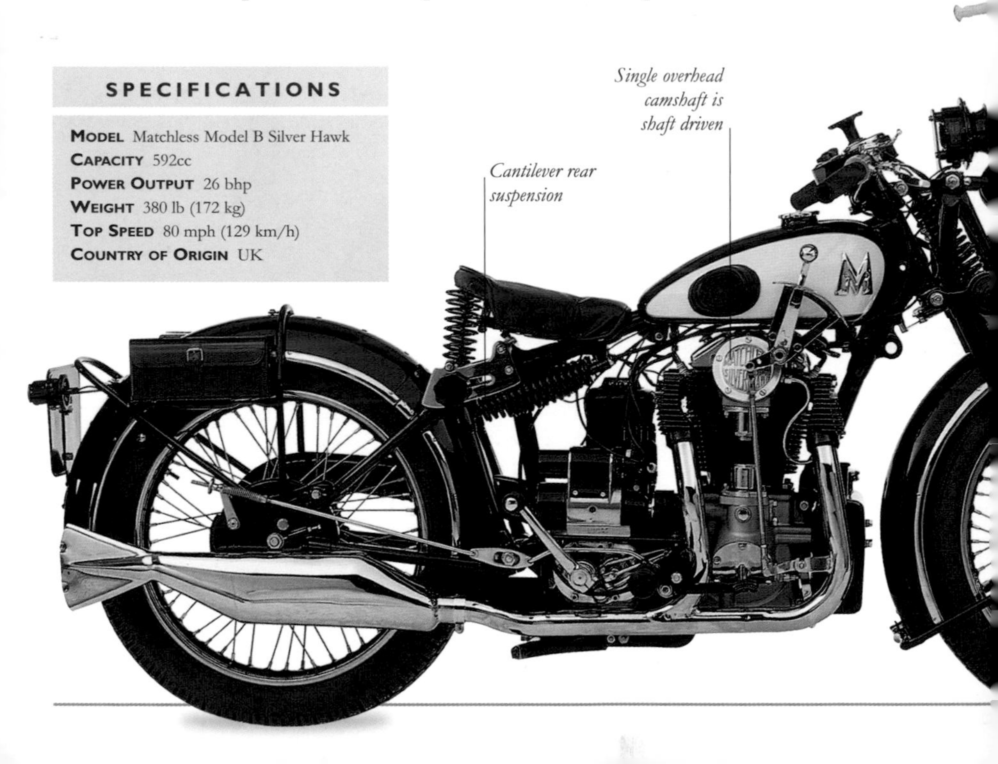

Pannier rack

Wartime blackout lighting

Girder front forks

The G3L was one of the first British production models to feature telescopic forks

SPECIFICATIONS

MODEL Matchless 41/G3L
CAPACITY 347cc
POWER OUTPUT 16.6 bhp
WEIGHT 392 lb (134 kg)
TOP SPEED 70 mphh (113 km/) (est.)
COUNTRY OF ORIGIN UK

MATCHLESS *41/G3L*

DURING WORLD WAR II, Matchless supplied over 80,000 41/G3Ls to the British War Department. The origins of the o.h.v. G3L can be traced to the G3 Clubman launched in 1935, but the rigid simplex cradle frame planned for the 1940 season was used, as well as new "Teledraulic" forks. The Lucas magneto and the dynamo were chain-driven. The only criticism of the 1930s and '40s G3 series was the inaccessibility of the dynamo tucked under the magneto platform.

MATCHLESS *G45*

DEVELOPED BY THE AJS RACE SHOP for the 1952 season, the G45 was based on the 500cc parallel-twin that came fourth in the 1951 Manx Grand Prix. Twin Amal Grand Prix carburetors, a Lucas racing magneto, and a tachometer drive were standard along with the Burman racing gearbox. The engine included a one-piece forged steel crank, alloy barrels, and triple valve springs. Seventeen G45s entered the 1955 Senior race and only 10 finished. Although the G45 was not successful compared to other models, it allowed AJS—which Matchless had taken over in 1931—riders to compete in Senior and Junior Clubman TTs. The bike shown dates from 1957—the year that production of the G45 was halted.

SPECIFICATIONS

MODEL Matchless G45
CAPACITY 498cc
POWER OUTPUT 48 bhp @ 7200 rpm
WEIGHT 320 lb (145 kg)
TOP SPEED 120 mph (193 km/h) (est.)
COUNTRY OF ORIGIN UK

The G45 is equipped with a 2.6 x 2.86-in (66 x 72.8-mm) engine

Steering shock absorber

Twin Amal Monobloc carburetors are used, along with half-race camshafts

The G11 CSR has a small fuel tank

SPECIFICATIONS

MODEL Matchless G11 CSR
CAPACITY 593cc
POWER OUTPUT 40 bhp @ 6000 rpm
WEIGHT 380 lb (172 kg)
TOP SPEED Not known
COUNTRY OF ORIGIN UK

The engine has separate iron barrels and alloy cylinder heads

Telescopic forks

MATCHLESS *G11 CSR*

INTRODUCED FOR 1958, the G11 CSR was an early example of the factory custom bike. Developed initially for the US market, it adopted the street-scrambler style. Early CSRs used lightened versions of the bolt-up, duplex roadster frame. A Lucas alternator was mounted on the driveside crankshaft but magneto ignition was retained. A standard four-speed gearbox was driven by a simplex primary chain through a multiplate clutch in a new cast-alloy primary chaincase.

MATCHLESS *G50 CSR*

THIS WAS THE ULTIMATE EARLY SIXTIES factory-built hot rod. The G50 CSR was made in very limited numbers to homologate the potent overhead camshaft Matchless G50 racing engine for US competition, in particular the prestigious Daytona 200-mile (322-km) race. The genuine G50 was a pure racing machine but the rules stated that a road bike had to be available for the public. So the Matchless parent company, Associated Motorcycles, placed the race engine into the chassis of the existing G12 CSR models which were usually powered by a 650cc parallel-twin cylinder engine. A belt-driven dynamo was clamped to the front of the crankcase to power the lights. The racing carburetor and exhaust which were needed to achieve the engine's full potential were listed as optional extras. Because most machines were immediately stripped and turned into racers, original bikes are very rare.

SPECIFICATIONS

MODEL Matchless G50 CSR
CAPACITY 498cc
POWER OUTPUT 46 bhp @ 7200 rpm
WEIGHT 320 lb (145 kg)
TOP SPEED 115 mph (185 km/h) (est.)
COUNTRY OF ORIGIN UK

COMPONENTS
Fed by a 1½-in (38-mm) Amal Grand Prix carburetor, the inlet valve was a massive 2 in (51 mm) in diameter. All steel and tin running parts were finished in either red or blue, with white pinstriping on the scrambles fuel tank.

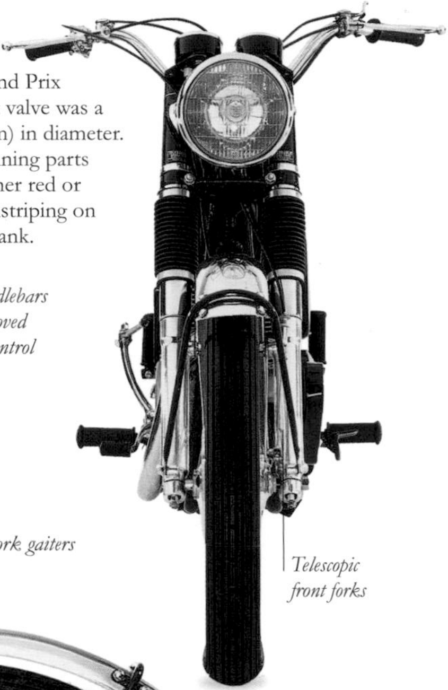

Wide handlebars offer improved off-road control

Rubber fork gaiters

Telescopic front forks

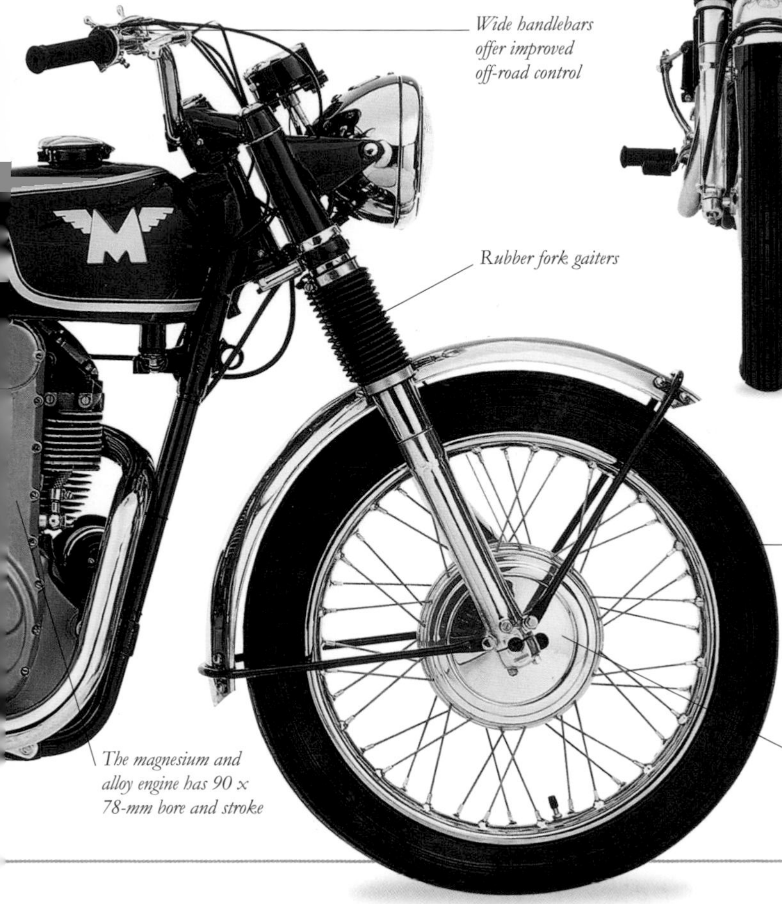

Grooved tire

The magnesium and alloy engine has 90 × 78-mm bore and stroke

Standard road-bike drum brake

MEGOLA *Racing Model*

THE MEGOLA PROBABLY ranks as the world's most unconventional motorcycle ever. Introduced in 1921, approximately 2,000 were built before the factory closed in 1924. This machine dates from 1923. Obviously inspired by the rotary-airplane engines of World War I, designer Fritz Cockerell came up with a five-cylinder side-valve radial engine mounted within the front wheel, which it drove via epicyclic gearing. During each forward rotation of the wheel the engine rotated six times in the opposite direction. There was no clutch or gearbox, but different-sized wheels were available to suit the intended use: a track-racing model fitted with a 29-in (74-cm) front wheel was clocked at 92 mph (148 km/h).

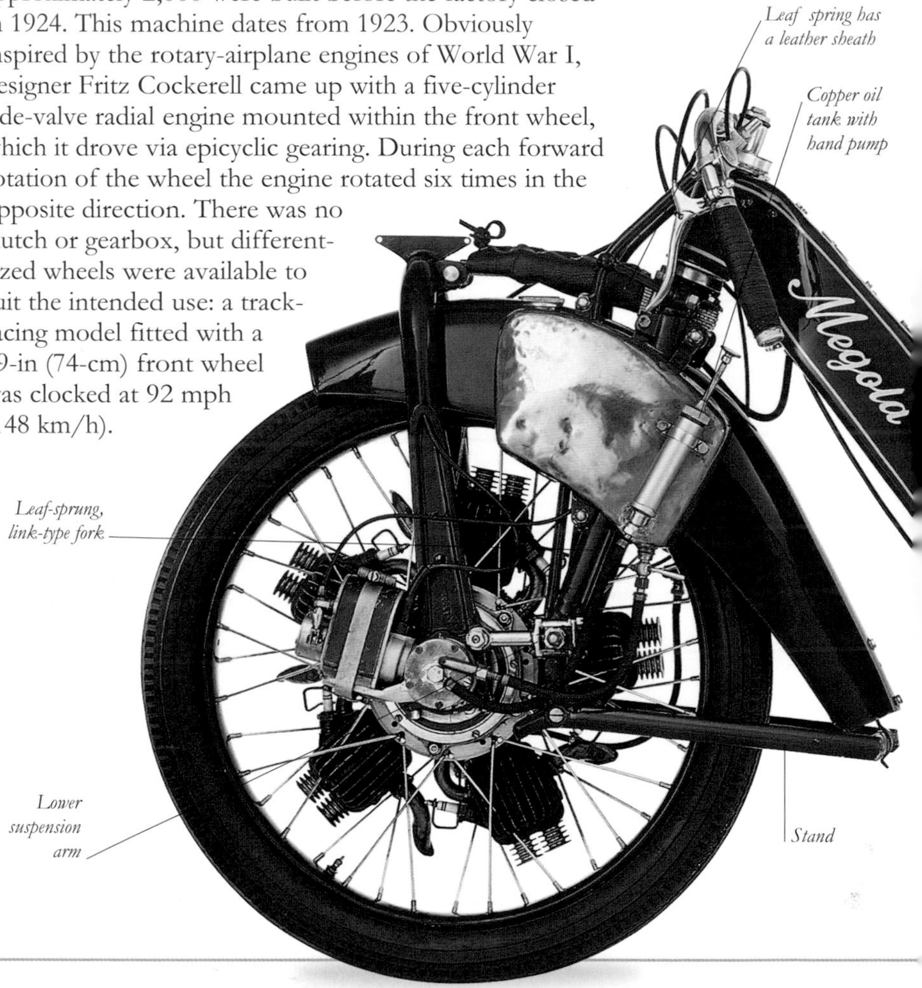

Leaf spring has a leather sheath

Copper oil tank with hand pump

Leaf-sprung, link-type fork

Lower suspension arm

Stand

SPECIFICATIONS

MODEL Megola Racing Model
CAPACITY 640cc
POWER OUTPUT 14 bhp @ 3600 rpm
WEIGHT Not known
TOP SPEED Not known
COUNTRY OF ORIGIN Germany

Racing handlebars

Magneto

Main fuel tank is built into box-section frame

NOVEL FRAME

The beam-type frame incorporated the fuel tank, which was pressurized by a hand pump feeding fuel to a header tank above the carburetor.

Unconventional rear fairing

Racing tire

Rear wheel has contracting band and drum brakes

MORBIDELLI *V8*

CONCEIVED BY Giancarlo Morbidelli, a millionaire industrialist and motorcycle enthusiast, the V8 was intended for very limited production. Styled by the car designer Pininfarina, the engine was suspended from the tubular space frame. If prospective buyers were not scared off by its ugly lines and complexity, the excessive price tag was highly restrictive, despite including the cost of airfreighting the machine back to the Italian factory for servicing. This model dates from 1994.

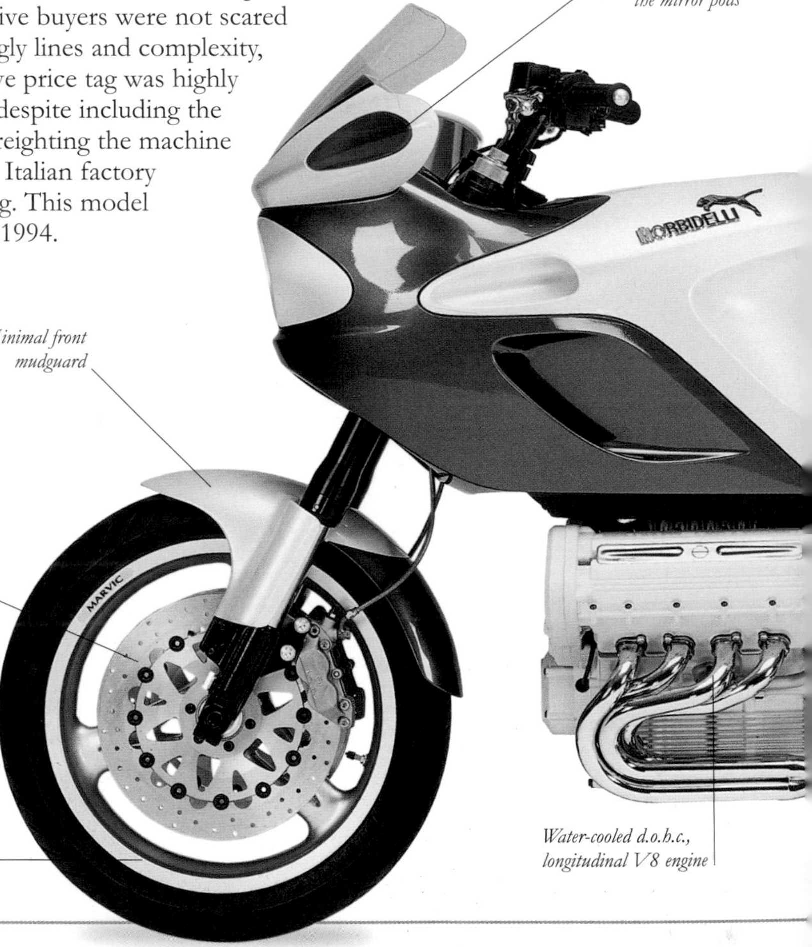
Indicators are incorporated in the mirror pods
Minimal front mudguard
Twin front disc brakes with four-piston Brembo calipers
Marvic three-spoke alloy wheel
Water-cooled d.o.h.c., longitudinal V8 engine
MORBIDELLI
MARVIC

SPECIFICATIONS

MODEL Morbidelli V8
CAPACITY 847cc
POWER OUTPUT Not known
WEIGHT 441 lb (200 kg)
TOP SPEED 150 mph (241 km/h)
COUNTRY OF ORIGIN Italy

OTHER V8S

The Morbidelli was not the first V8 motorcycle to be built. American pioneer Glenn Curtiss built a V8 in 1912, and Moto Guzzi made a 500cc Grand Prix bike in the late 1950s.

Bug-eyed twin headlights set in fairing/ bodywork

Removable passenger seat cover

Bodywork conceals a tubular trellis frame

Road tire

Rear brake caliper

Massive exhaust muffler

Gearbox drives the rear wheel via a shaft

MORINI *3½ Sport*

INTRODUCED IN 1972, the 344cc Morini was one of the most innovative machines of the period. Designed for efficiency of performance and manufacturing, the 72° V-twin engine used a "Heron" cylinder head design that had parallel valves, a flat cylinder head, and the combustion chamber machined into the crown of the piston. Electronic ignition, tachometer, and fuel tap were supplied. Two models were produced: a Sport version (a 1974 model is shown here), with a more powerful engine and extreme riding position, and the more mundane Strada.

Tubular steel cradle frame

Sports seat

Muffler

Flanged alloy wheel rims

Six-speed gearbox

SPECIFICATIONS

MODEL Morini 3½ Sport
CAPACITY 344cc
POWER OUTPUT 39 bhp @ 8500 rpm
WEIGHT 340 lb (154 kg)
TOP SPEED 100 mph (161 km/h)
COUNTRY OF ORIGIN Italy

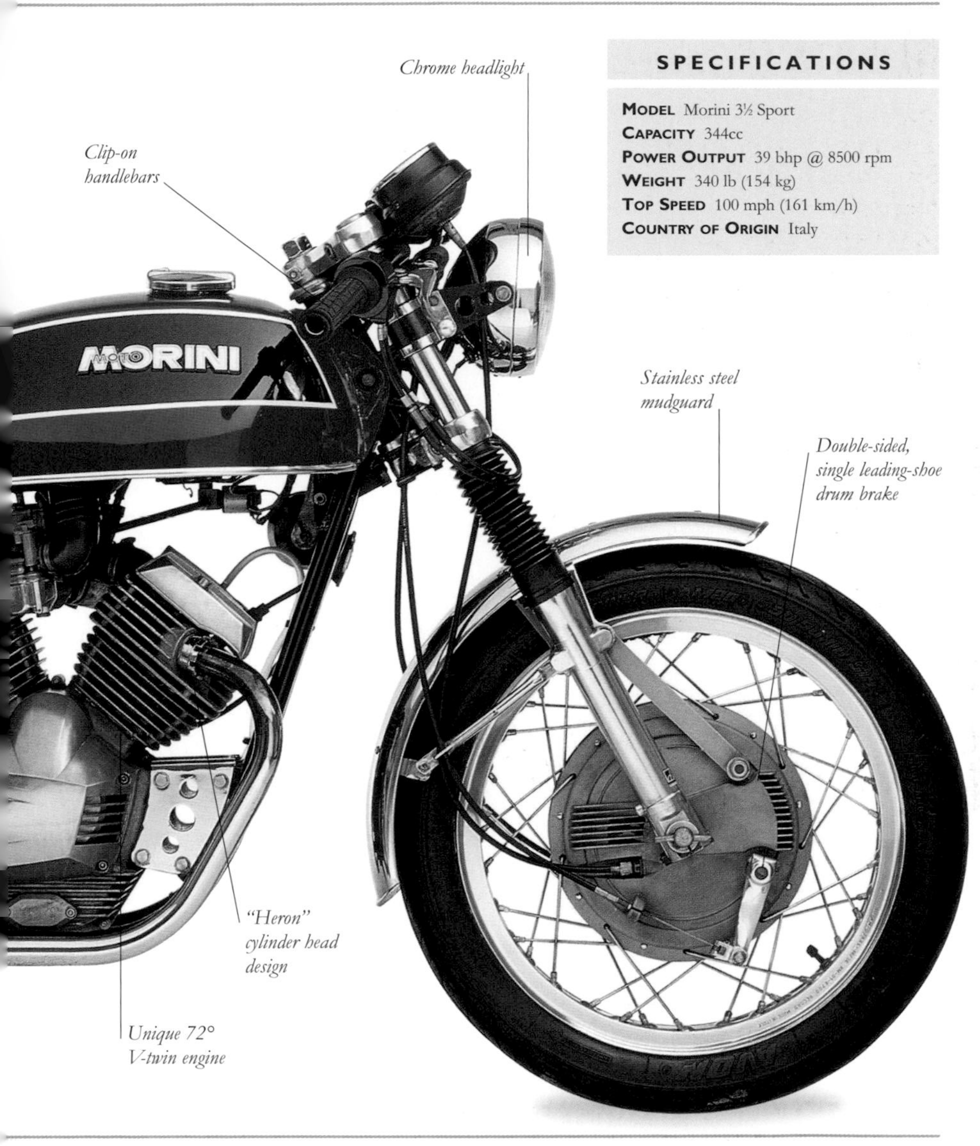

Morini *Camel*

Morini's V-twin engine was produced in three sizes. As well as the original 344cc version, 239cc and 478cc models were also made. Originally, the engine was available only in conventionally styled road bikes, but the fashion for four-stroke trail bikes in the early 1980s encouraged Morini to produce an off-road version. The combination of a flexible and compact engine into a sturdy frame equipped with good-quality suspension produced a surprisingly good off-road machine.

35-in (89-cm) seat height is the by-product of massive suspension travel

Air filters are hidden under the fuel tank

SPECIFICATIONS

MODEL Morini Camel
CAPACITY 478cc
POWER OUTPUT 42 bhp @ 7400 rpm
WEIGHT 380 lb (140 kg)
TOP SPEED 95 mph (153 km/h)
COUNTRY OF ORIGIN Italy

Moto Guzzi *500S*

After World War I, former Italian Air Force pilot Giorgio Parodi set up the Moto Guzzi company with Carlo Guzzi, his former mechanic and driver. Guzzi designed its first prototype in 1920. The design of the single-cylinder machine was so advanced that the company's last horizontal single, built in 1976, had the same bore and stroke dimensions as the original. The 500's four-stroke motor had an interesting feature for those worried about breaking down. A retaining fork attached to the combustion chamber made it impossible for the exhaust valve to drop into the engine. Guzzi continued this feature until the 1940s. The "bacon slicer" external flywheel appeared on all models until the late 1960s. The S model, with its hand gear-change and rigid rear end, was one of the most basic versions. The machine shown here dates from 1928.

SPECIFICATIONS

MODEL Moto Guzzi 500S
CAPACITY 498cc
POWER OUTPUT 18 bhp @ 4000 rpm
WEIGHT 287 lb (130 kg)
TOP SPEED 62 mph (100 km/h)
COUNTRY OF ORIGIN Italy

FRONT VIEW

The front view of the 500S reveals the pleasing symmetry of layout. The girder forks with central springs are a prominent feature, as is the deeply grooved front tire.

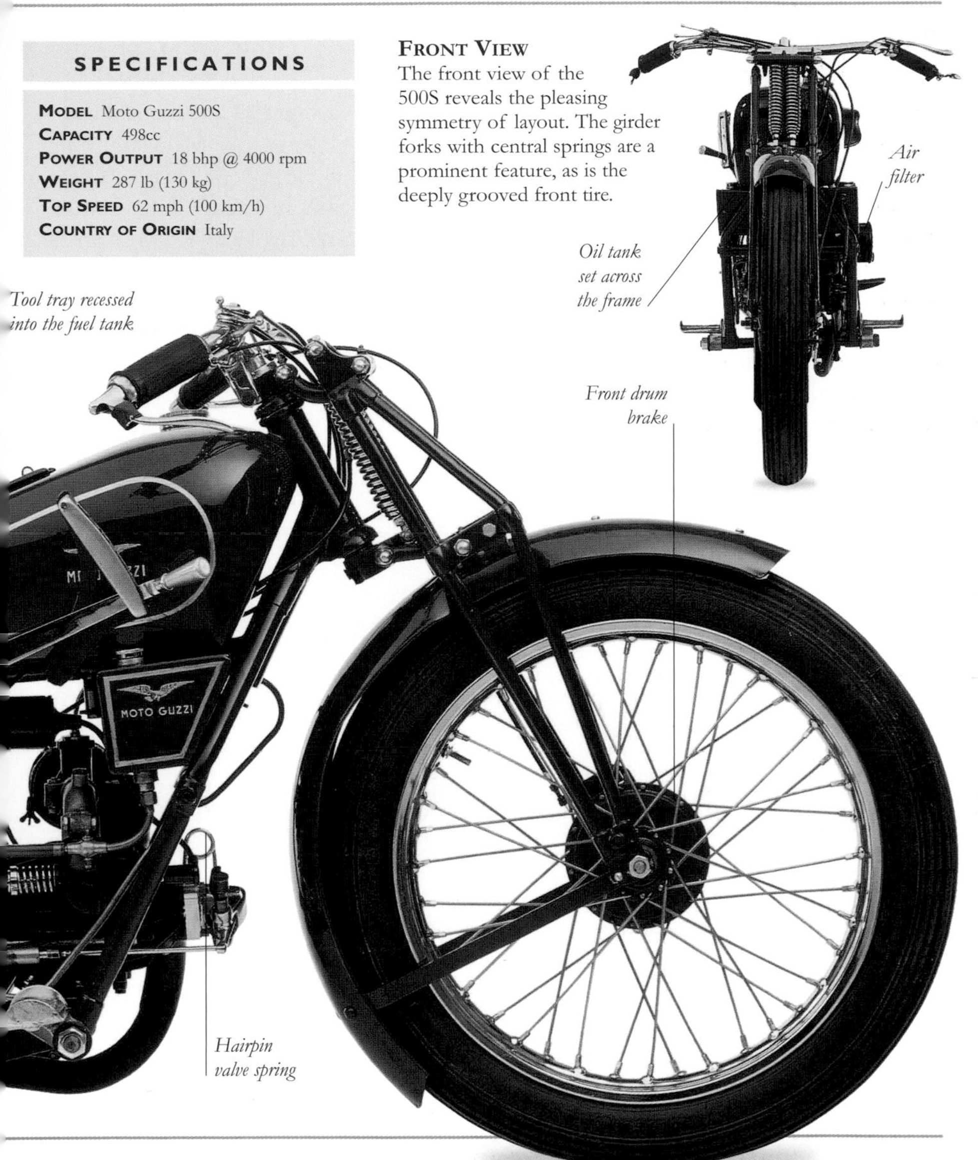

Moto Guzzi *V7 Special*

Although Moto Guzzi first developed its V-twin motor in the 1950s, motorcycle interest in Italy was at a low ebb and the motor was used in a three-wheeled military vehicle. The V-twin motorcycle was built to fulfill an order from the Italian police. Overseas interest created a need for a civilian version. The 703cc V7 came out in 1967. In 1969 the bore was enlarged by 3 mm—increasing the capacity to 757cc—and called the V7 Special. It ran until 1971 (the year of this model), had many features seen on modern Guzzi V-twins, and was the forerunner of models such as the V7 Sport, S3, and Le Mans *(see pp.298–99)*.

SPECIFICATIONS

Model Moto Guzzi V7 Special
Capacity 757cc
Power Output 40 bhp @ 5000 rpm
Weight 516 lb (234 kg)
Top Speed 112 mph (180 km/h)
Country of Origin Italy

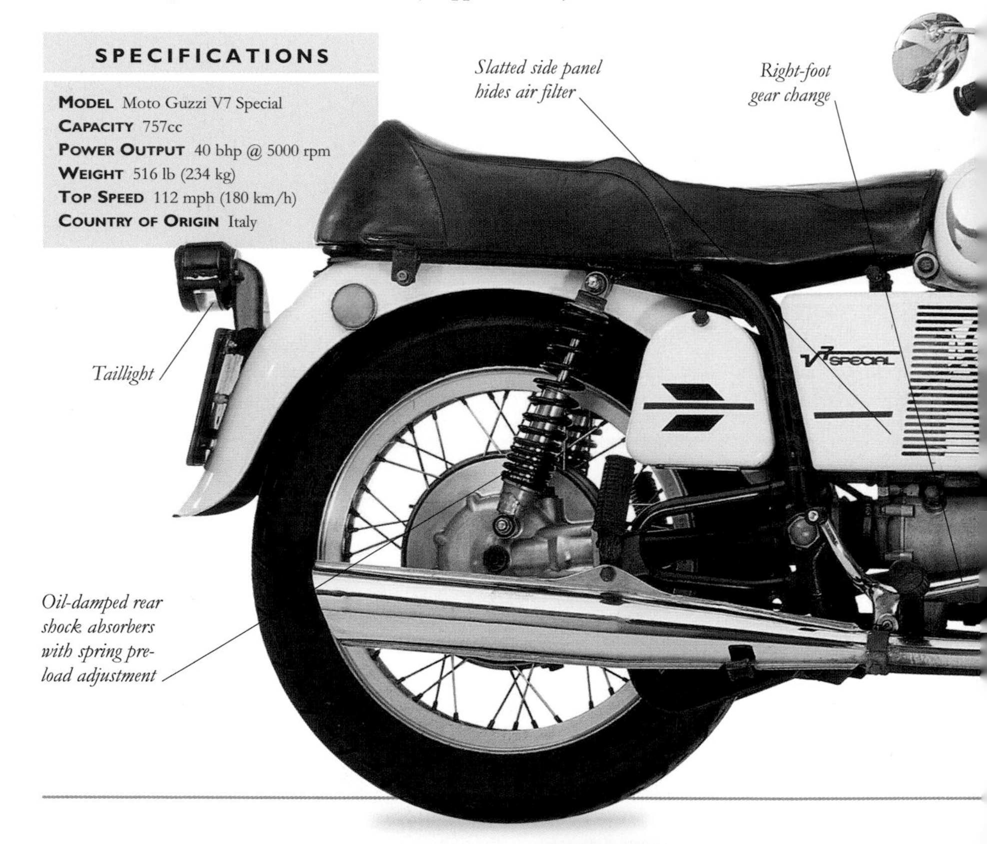

Slatted side panel hides air filter

Right-foot gear change

Taillight

Oil-damped rear shock absorbers with spring pre-load adjustment

US NAMES

The Special was aimed at the US market, where it sold under various names, such as the Ambassador, Eldorado, and California. It formed the basis of many of the company's successful sports models.

The windshield is a US option

Cylinder protectors

Crash bars

Valanced mudguard

Moto Guzzi *850 Le Mans 1*

The Le Mans 1 was one of the fastest of the Moto Guzzi bikes. Low, aggressive, and very stylish, it was always the center of attention. Its massive, V-twin engine was designed by Giulio Carcano and was developed by Lino Tonti into the shaft-drive model that has since served so well. Much of this 850's weight was high in the frame, making it a very responsive bike when cornering. The left-hand front brake disc and the rear disc were pedal-activated, while the right-hand front disc was operated by hand lever.

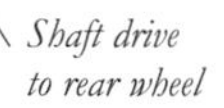

Shaft drive to rear wheel

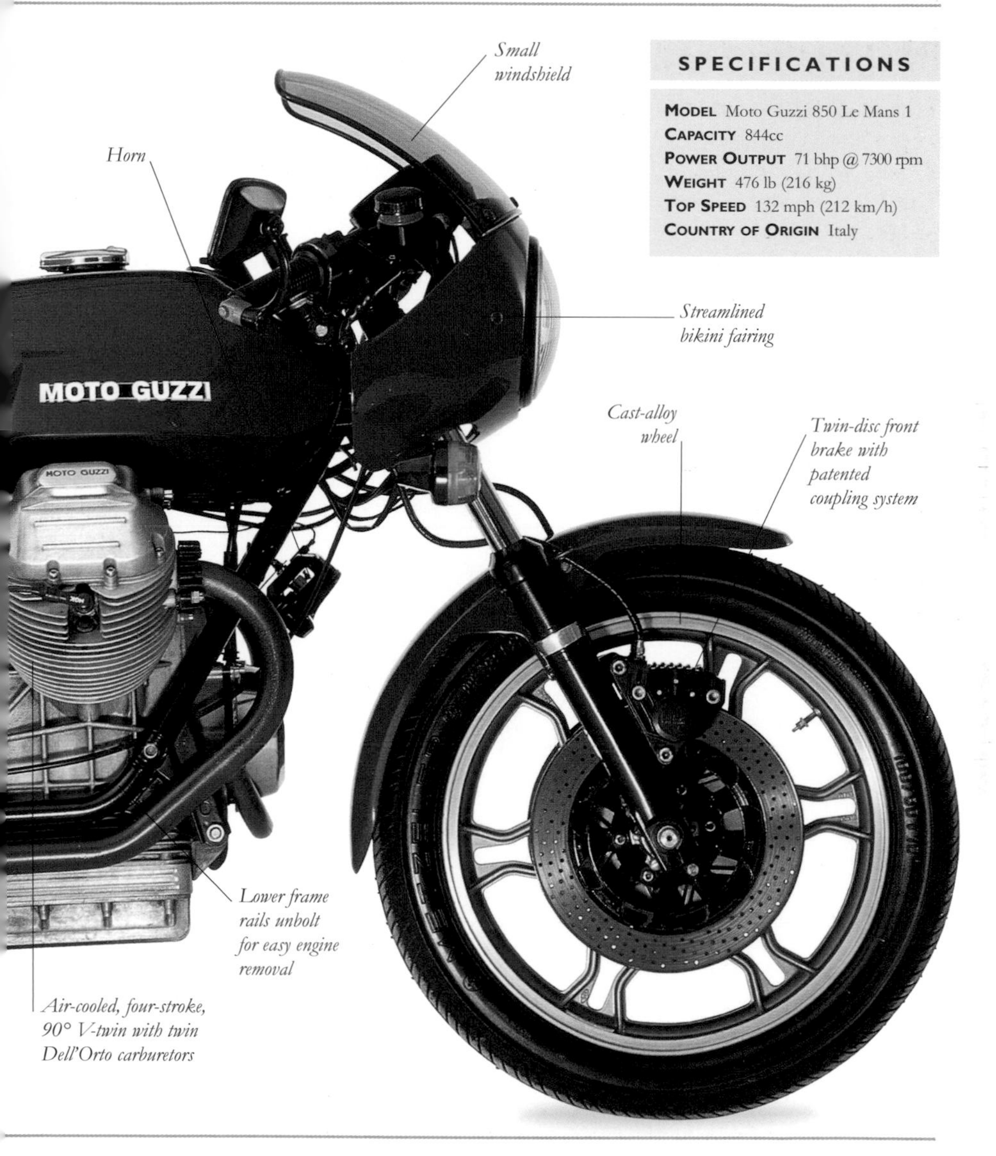

SPECIFICATIONS

MODEL Moto Guzzi 850 Le Mans 1
CAPACITY 844cc
POWER OUTPUT 71 bhp @ 7300 rpm
WEIGHT 476 lb (216 kg)
TOP SPEED 132 mph (212 km/h)
COUNTRY OF ORIGIN Italy

Motosacoche *Jubile 424*

The name Motosacoche roughly translates as "the motor in a bag" and refers to the Swiss company's early products. These were simple, lightweight, four-stroke engines that could be bolted into a bicycle frame. Motosacoche continued to supply its engines to other manufacturers under the name MAG even when the products became more sophisticated. By the 1930s, the line that emerged from its Geneva factory included everything from lightweight singles to 1000cc V-twins with various valve arrangements. This 1932 model has a single cylinder side-valve engine. The design is very conventional, and although the company made its own engines, parts like carburetors, gearboxes, and forks were bought from various suppliers throughout Europe. Motosacoche production finally ended in 1956 after more than half a century.

SPECIFICATIONS

MODEL Motosacoche Jubile 424
CAPACITY 498cc
POWER OUTPUT Not known
WEIGHT Not known
TOP SPEED 68 mph (109 km/h) (est.)
COUNTRY OF ORIGIN Switzerland

EARLY INNOVATION

By the time of the Jubile 424's manufacture, battery-and-coil ignition had been proven in the car world. Its use on the 424 marked an early application in a motorcycle.

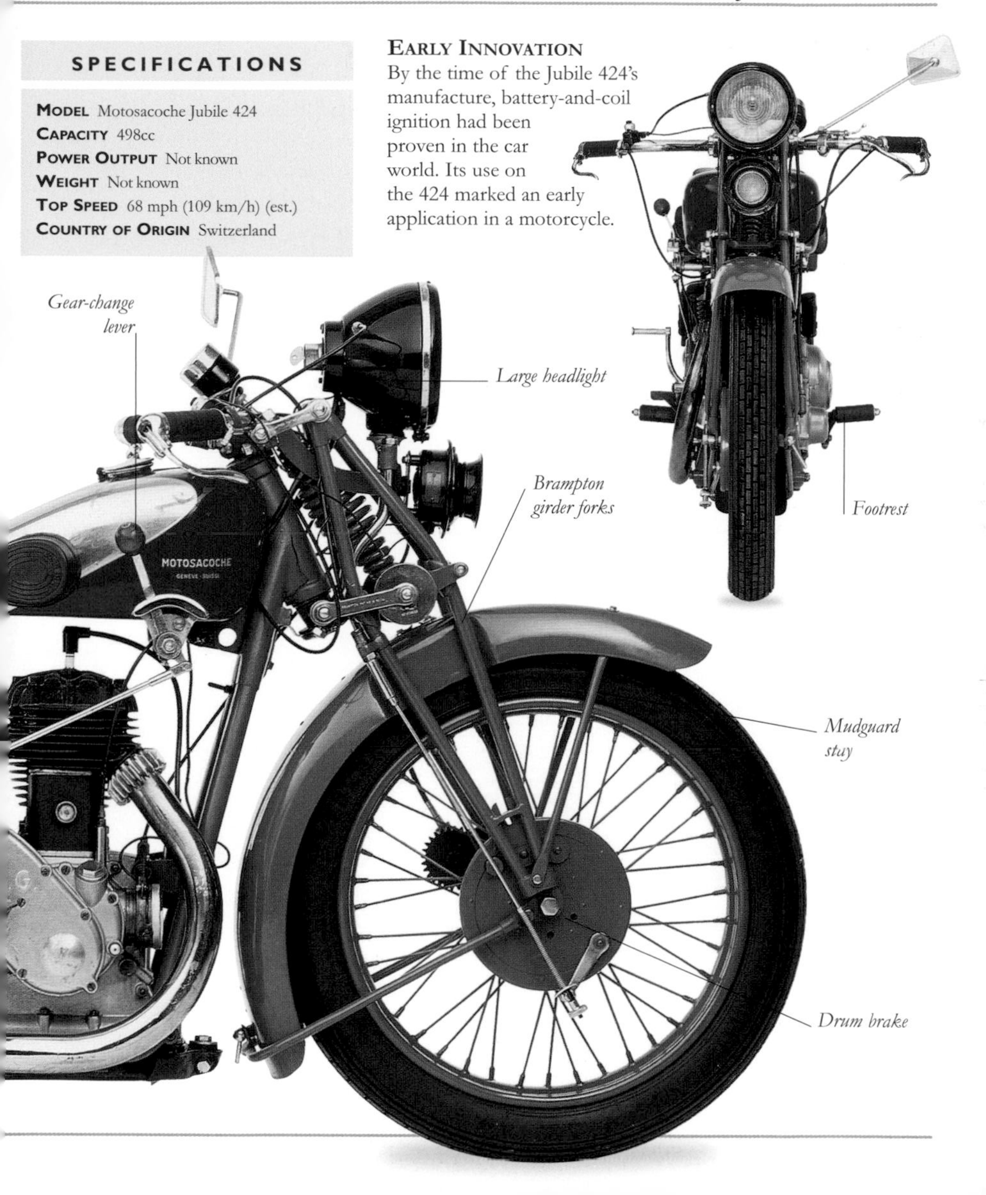

MÜNCH *Mammoth*

TAKING AS ITS BASE AN ENGINE from an NSU Prinz car, former Horex engineer Fried Münch created a remarkable motorcycle in the Mammoth. Only produced in limited numbers from 1966 and sold at a price that could have bought three of Norton's heralded Commandos *(see pp.326–29)*, the four-cylinder Mammoth was capable of performance levels that were unequaled at the time. Although its weight made it awkward to maneuver at low speed, it was a useful high-speed tourer capable of cruising all day at 110 mph (177 km/h) and reaching a top speed of 137 mph (220 km/h). Münch produced many prototypes, but the Mammoth is his most famous machine.

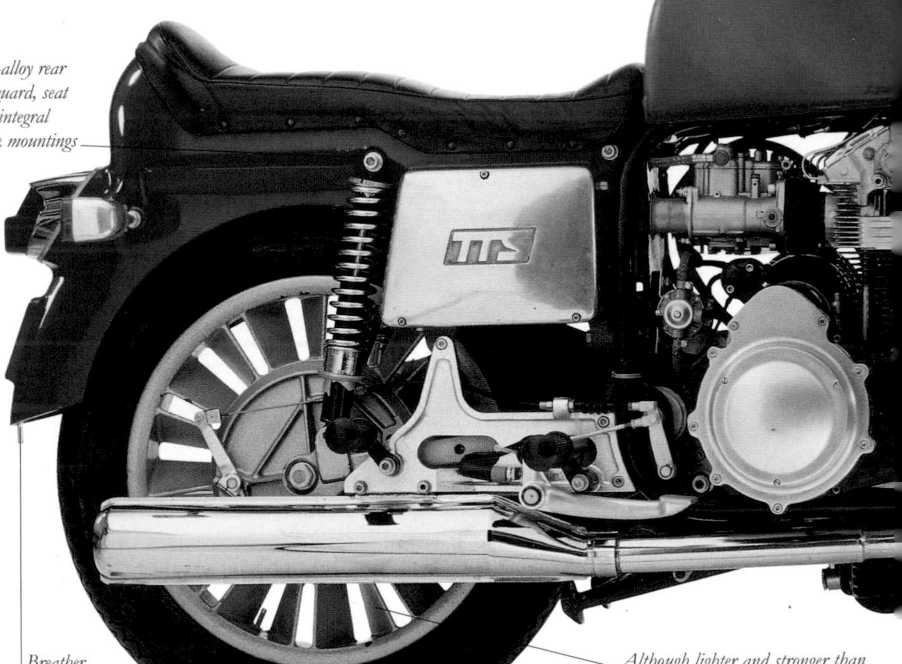

Cast-alloy rear mudguard, seat with integral shock mountings

Breather pipe

Although lighter and stronger than steel, elektron-cast alloys are notoriously prone to corrosion

Large-capacity fuel tank

SPECIFICATIONS

MODEL Münch Mammoth
CAPACITY 1177cc
POWER OUTPUT 88 bhp @ 6000 rpm
WEIGHT 656 lb (298 kg)
TOP SPEED 137 mph (220 km/h)
COUNTRY OF ORIGIN Germany

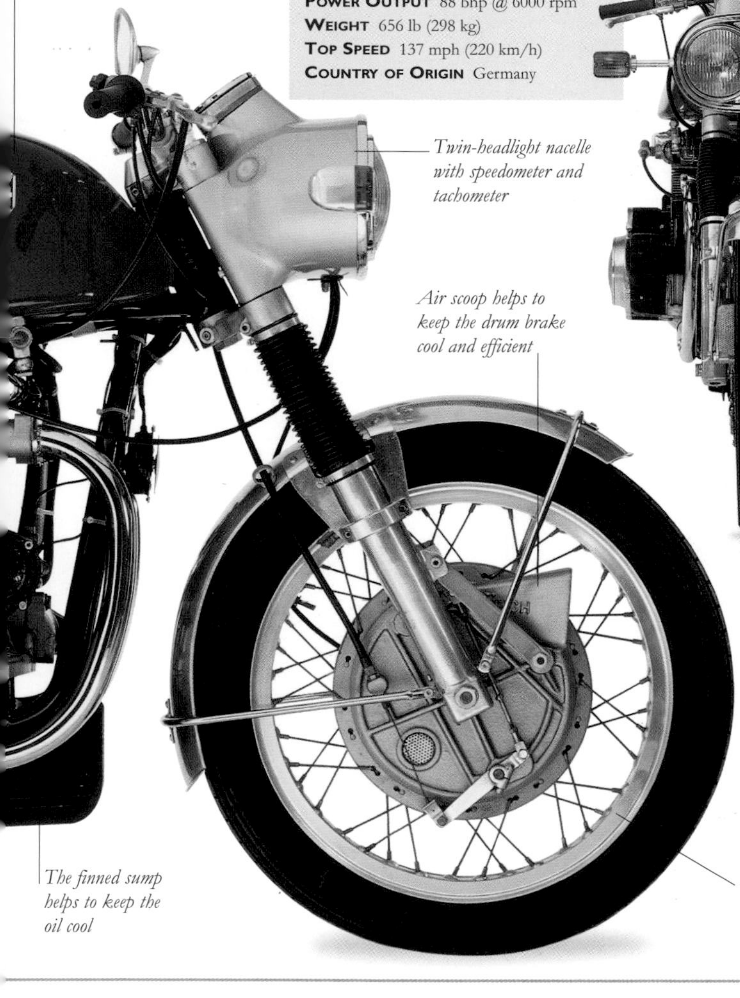

Twin car headlights provide a high degree of nighttime visibility

Twin-headlight nacelle with speedometer and tachometer

Air scoop helps to keep the drum brake cool and efficient

Camshaft drive-chain cover

FRONT VIEW
The massive proportions of the Münch Mammoth's frame, front forks, fuel tank, and powerful twin headlights are clear in this front shot of the bike.

The finned sump helps to keep the oil cool

Alloy wheel rims are used for lighter weight

MV Agusta *350GP*

The legend of MV Agusta is based on phenomenal racing success. From its debut, in 1950, to 1976, when the challenge from Japanese two-strokes proved to be too powerful, MV Agusta won 37 World Championships and 273 Grand Prix. In 1958, 1959, and 1960, the bikes won every solo championship. MV Agusta won its last world title in 1974, and its last Grand Prix in Germany in the 1976 season. MV's first Grand Prix bikes were single cylinder 125cc and 500cc fours. The 350cc version of the four was built in 1953, but did not achieve a notable success until 1958, when it won the first of four successive World Championships.

Engine Longevity
Factory racing bikes evolve year by year. The MV 350's frame (seen from the front, above) was first used in the 1960 season, but the engine was used from 1954 to 1961.

SPECIFICATIONS

MODEL MV Agusta 350GP
CAPACITY 347cc
POWER OUTPUT 42 bhp @ 11,000 rpm
WEIGHT 320 lb (145 kg)
TOP SPEED 130 mph (210 km/h)
COUNTRY OF ORIGIN Italy

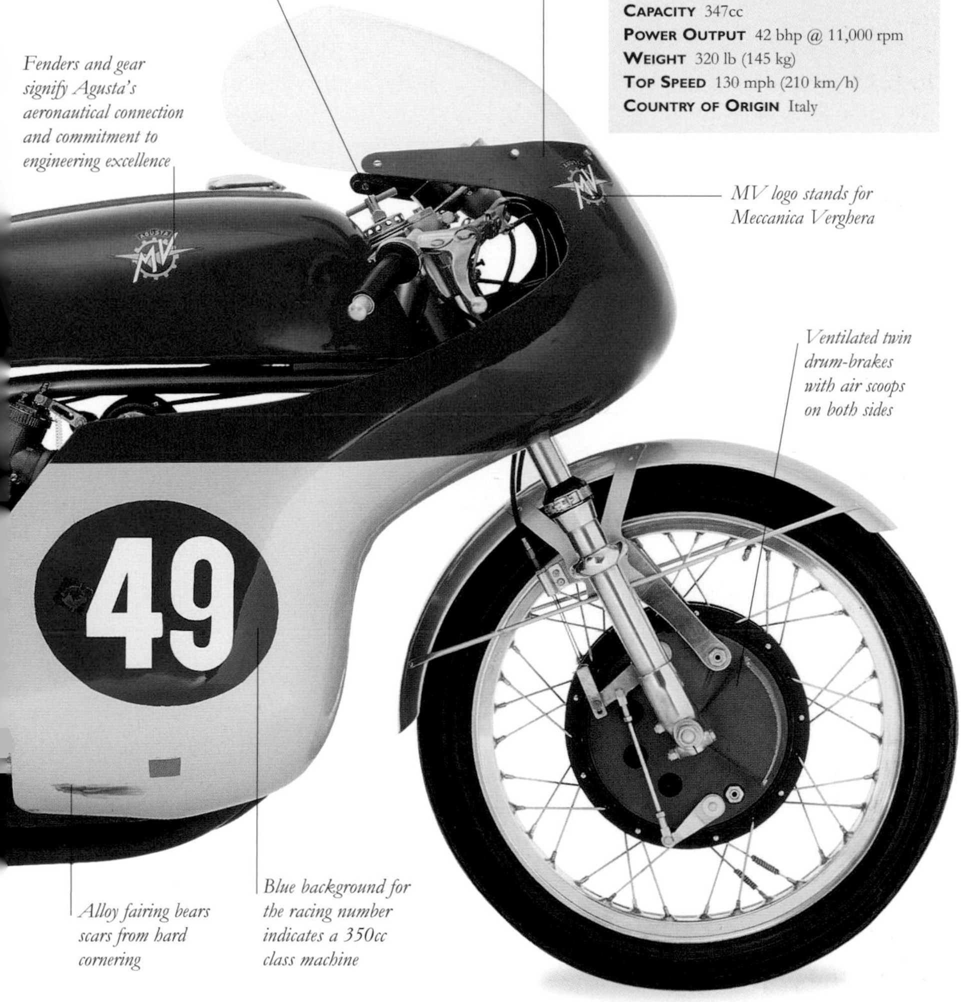

Steering shock absorber

"Dolphin" fairing

Fenders and gear signify Agusta's aeronautical connection and commitment to engineering excellence

MV logo stands for Meccanica Verghera

Ventilated twin drum-brakes with air scoops on both sides

Alloy fairing bears scars from hard cornering

Blue background for the racing number indicates a 350cc class machine

MV Agusta *America*

In 1971 MV produced its first sporty four-cylinder road bike. It retained the shaft drive of the touring 600 but the capacity and compression ratio were increased. Also supplied with four carburetors, the performance was significantly improved. The new machines also looked better with sporty styling and significant splashes of red. The 1975 model shown here was part of a special batch of machines; intended to impress the American market, they failed to make any impact and many of the bikes were eventually sold in other countries.

Upmarket specification includes a suede seat

Mufflers for the US market are quiet but ugly

SPECIFICATIONS

MODEL MV Agusta America
CAPACITY 789cc
POWER OUTPUT 75 bhp @ 8500 rpm
WEIGHT 529 lb (240 kg)
TOP SPEED 130 mph (210 km/h)
COUNTRY OF ORIGIN Italy

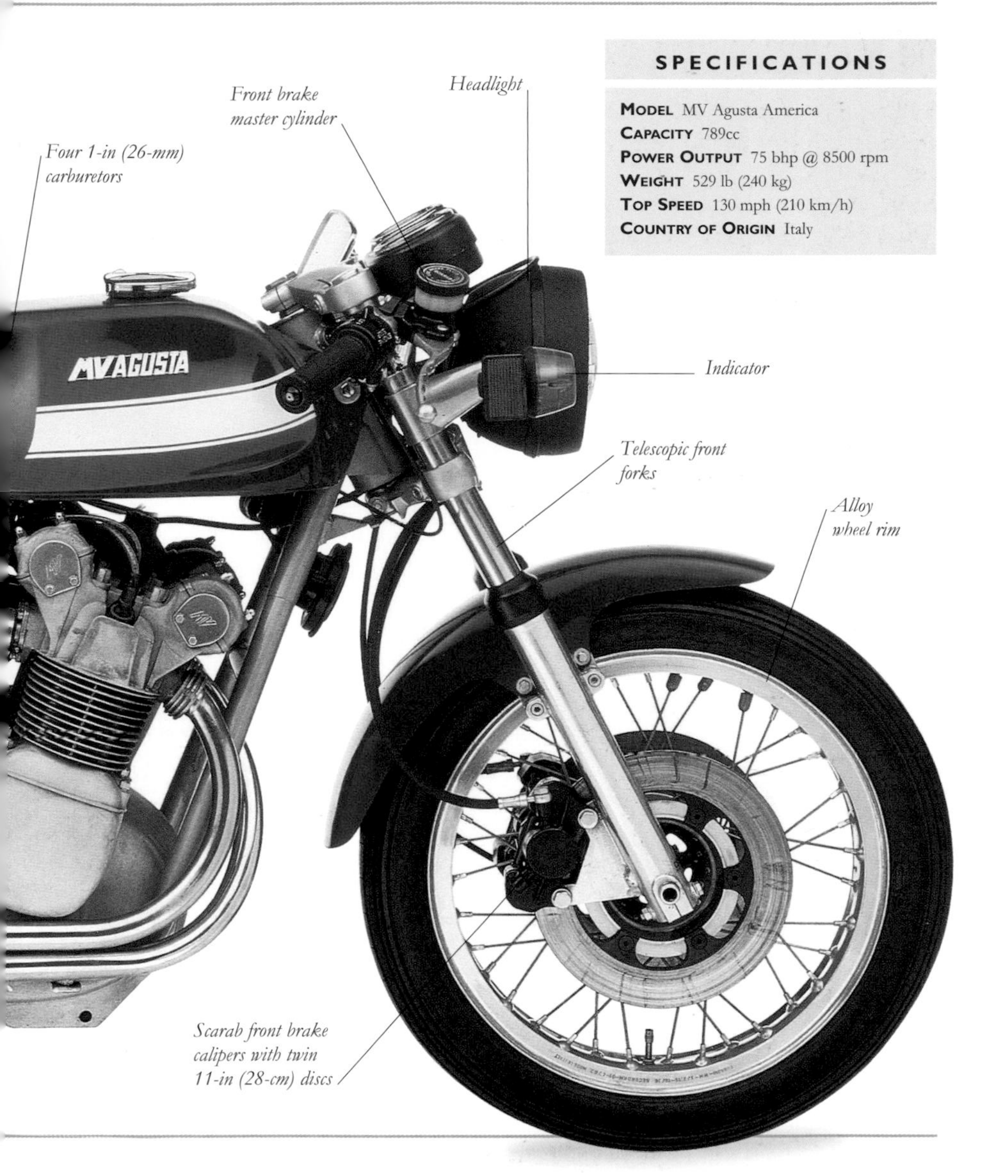

MV Agusta *F4S*

MV Agusta is a name with a huge motorcycling reputation—75 rider and manufacturer World Championships and 270 Grand Prix wins from the early Fifties to the mid-Seventies. MV disappeared from the motorcycle scene in 1980, but in the late Nineties it was ready for a comeback. The new MV had to live up to the reputation, and it did. The F4 is one of the most beautiful motorcycles ever built. Designer Massimo Tamburini, who is also responsible for the earlier Ducati 916 *(see pp.110–11)*, created a bike that was technically innovative, with excellent performance and stunning looks.

SPECIFICATIONS

Model MV Agusta F4S
Capacity 749cc
Power Output 106 bhp
Weight 419 lb (190 kg)
Top Speed 168 mph (270 km/h)
Country of Origin Italy

Indicators are mounted in the mirror pods

Limited-vision windshield

Fairing cutouts and ducts are an important part of the MV design

Distinctive stacked headlights

CRC logo refers to the Cagiva Research Center, where the bike was designed

Radiator

Front View
Because the twin headlights of the MV are stacked, rather than side-by-side, the frontal area can be reduced and the distinctive shape is possible.

Five-spoke alloy wheel

MV Agusta *910S Brutale*

When the MV Agusta marque was re-launched in 1999, its single model was a 750cc four-cylinder sports bike. In 2002, the Brutale model was created by removing the fairing but retaining the attitude and technology of the sports bike. The Brutale was a naked machine with aggressive attitude, though like all MVs it was attractively styled with jewel-like detailing. In 2005, capacity was increased to 920cc and it later grew again to a full 1000cc. Increasing capacity gave the bike more engine flexibility, making it an easier ride but without compromising its sports bike edginess. The bike shown here is a 2007 model.

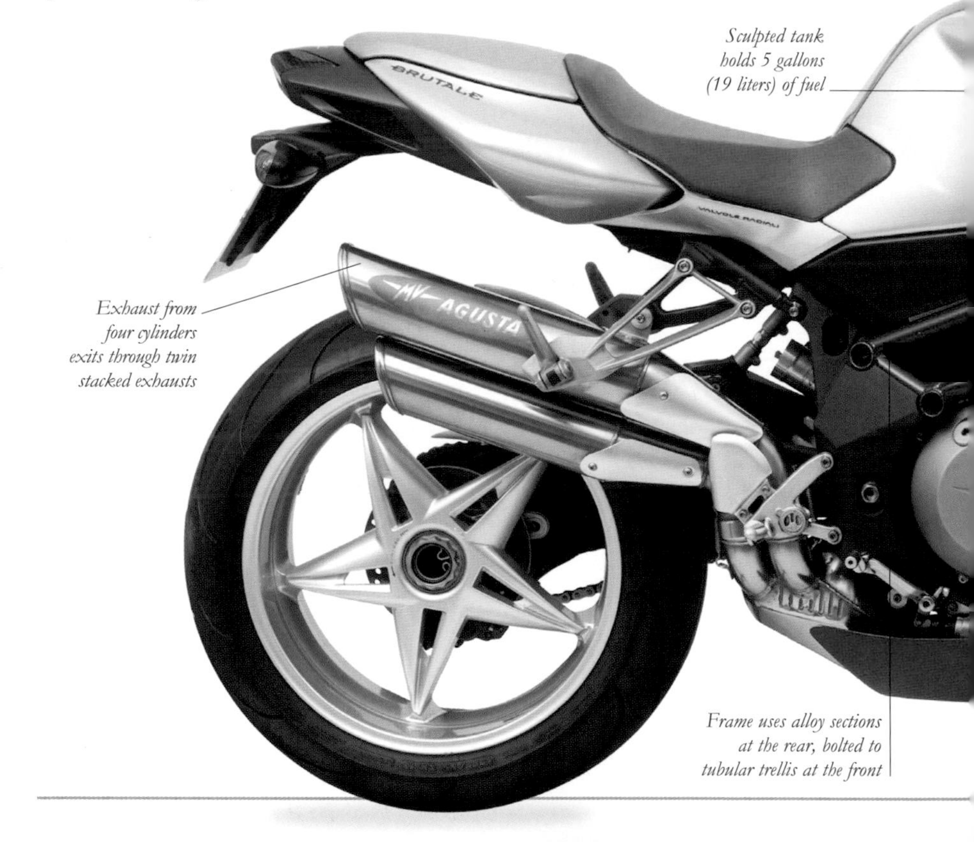

Sculpted tank holds 5 gallons (19 liters) of fuel

Exhaust from four cylinders exits through twin stacked exhausts

Frame uses alloy sections at the rear, bolted to tubular trellis at the front

SPECIFICATIONS

Model MV Agusta 910S Brutale
Capacity 909cc
Power Output 136 bhp @ 11,000 rpm
Weight 408 lb (185 kg)
Top Speed 155 mph (250 km/h)
Country of Origin Italy

Instrument binnacle houses a white-faced analog rev-counter, digital speedometer, and other crucial displays

Headlight has a distinctive shape, as if it has melted

Front brakes use Nissin six-piston callipers to grip the discs. 50-mm Marzocchi forks have quick-release axle retainers

MZ *RE125*

PRODUCTION MZs WERE RUGGED, utilitarian machines, yet much of the success of Japanese two-stroke racing machines can be attributed to MZ and its RE125. Chief designer for MZ during the 1960s was Walter Kaaden, a brilliant engineer who pioneered many of the advances made in two-stroke technology since World War II. The MZ racing team did not enjoy the resources of their Japanese competitors in the 125cc and 250cc World Championships, yet they could still produce some remarkable feats of engineering, not least the development of a single-cylinder 125cc engine capable of producing more than 200 bhp per liter.

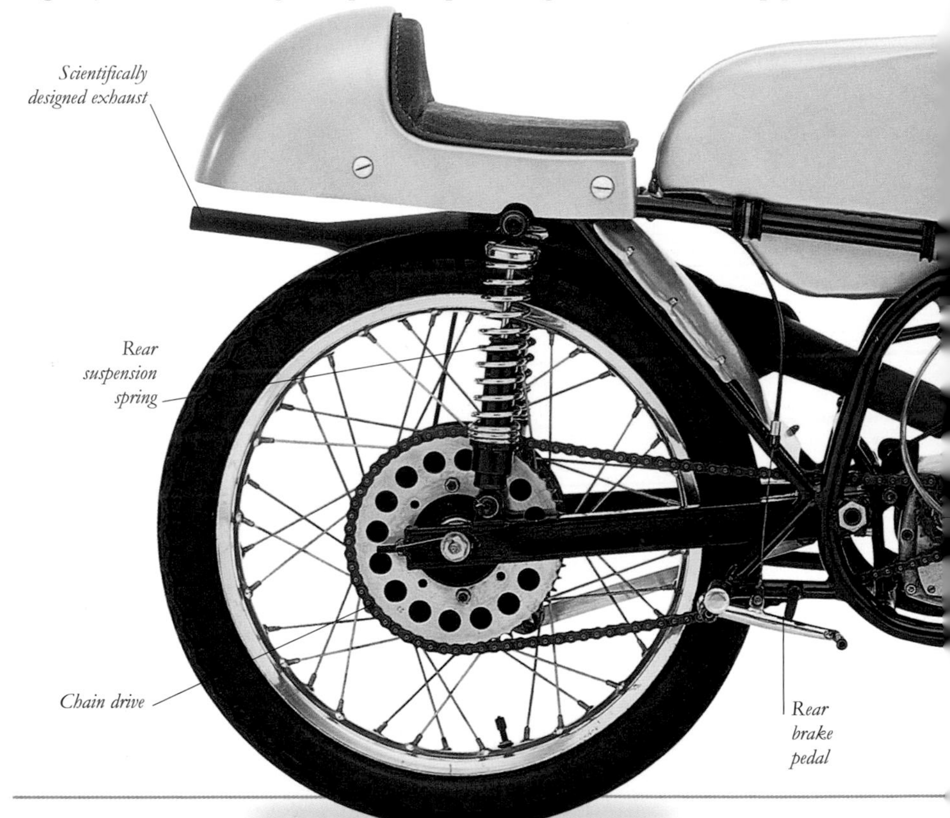

SPECIFICATIONS

Model MZ RE125
Capacity 124cc
Power Output 25 bhp
Weight 200 lb (91 kg)
Top Speed 131 mph (210 km/h)
Country of Origin Germany

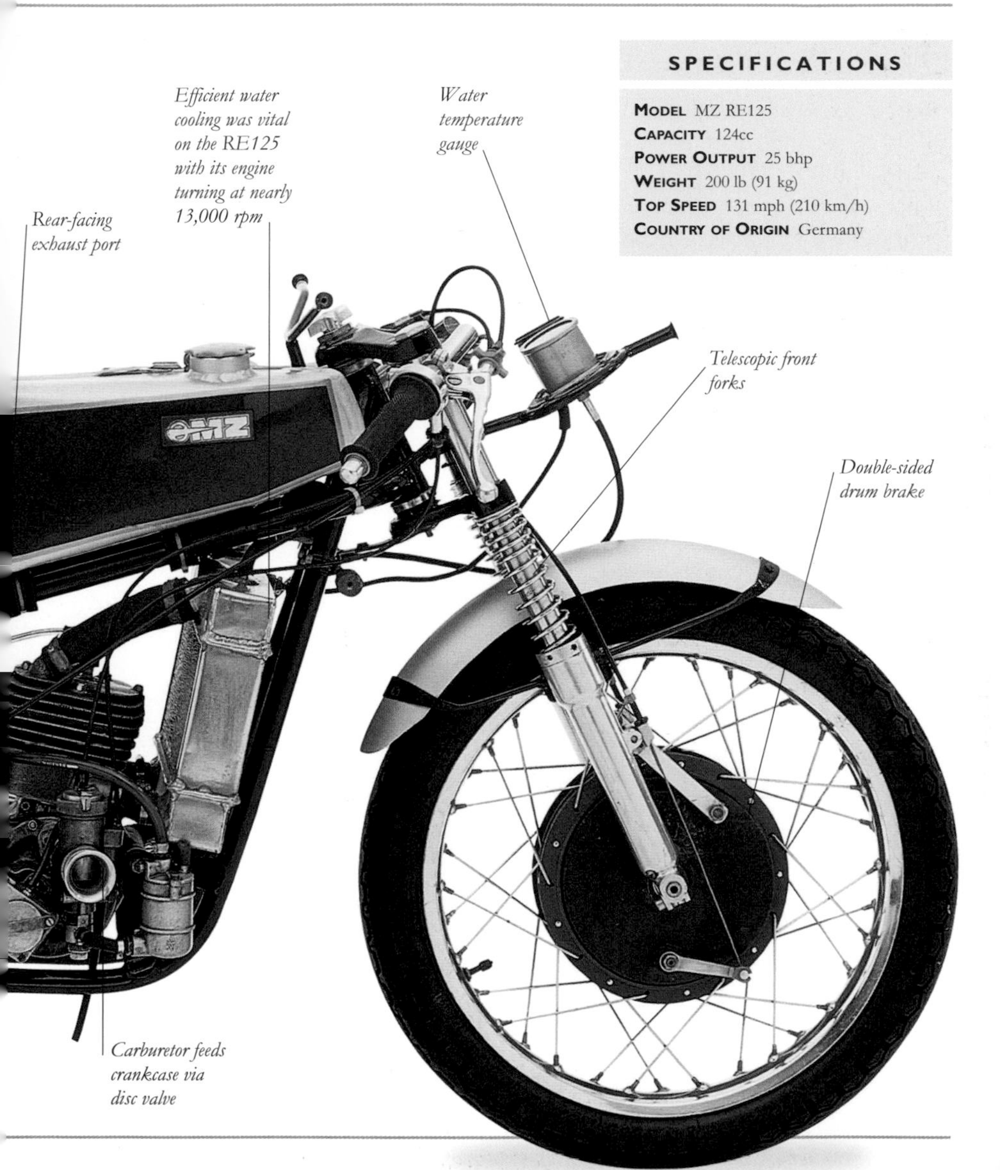

NEW IMPERIAL

ALTHOUGH DESIGNED AS a racing machine, it was as a record breaker that the 500cc, V-twin New Imperial achieved lasting fame. The engine was virtually two pushrod, o.h.v. 250s on a common crankcase, driven via a four-speed, Sturmey-Archer gearbox. As a racing machine, it was beset with handling problems. Even so, ridden by Ginger Woods, it raised the Brooklands 500cc lap record to 115.82 mph (186 km/h) in 1935. In 1928, *The Motor Cycle* offered a prize for the first British 500cc multicylinder machine to cover 100 miles (161 km) in one hour on a British track. Despite the efforts of both Triumph and Ariel with supercharged machines, it was not until the New Imperial twin appeared in 1934 that the prize was won, with 102.27 miles (165 km) covered.

SPECIFICATIONS

MODEL New Imperial
CAPACITY 491cc
POWER OUTPUT 40 bhp @ 7500 rpm
WEIGHT Not known
TOP SPEED 110 mph (177 km/h)
COUNTRY OF ORIGIN UK

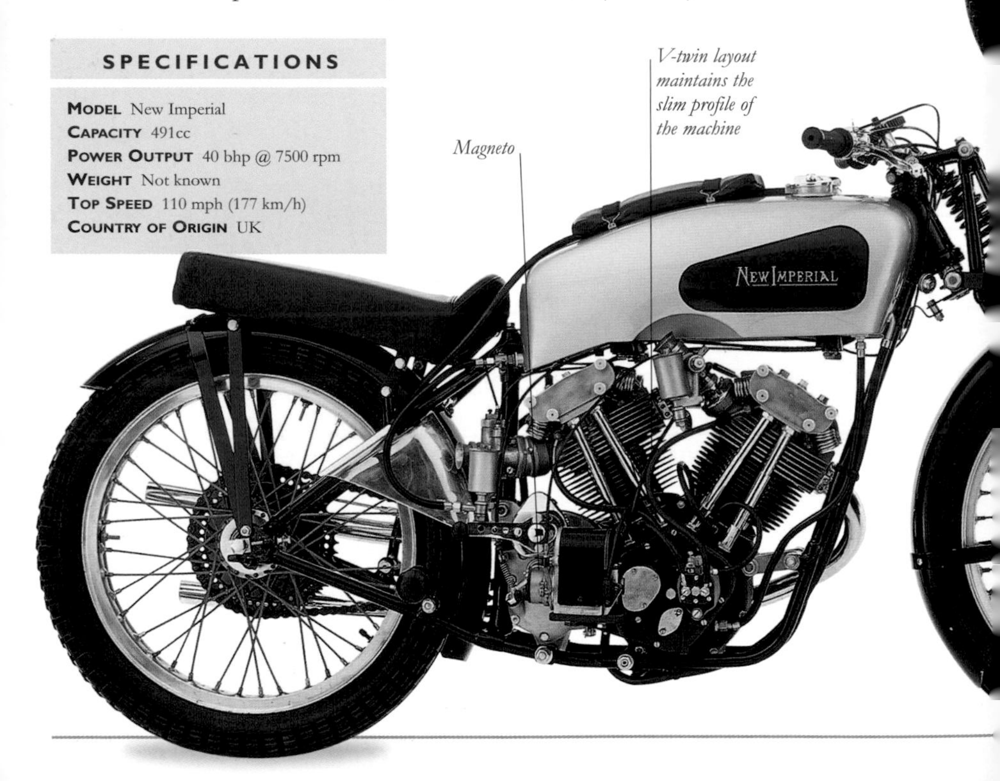

Magneto

V-twin layout maintains the slim profile of the machine

Metal-plate frame

Pressed-steel handlebars

Leading-link front forks

Nimbus bikes always had good telescopic forks

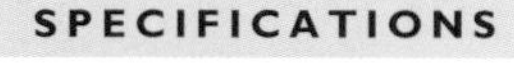

SPECIFICATIONS

MODEL Nimbus MkII
CAPACITY 746cc
POWER OUTPUT 22 bhp @ 4000 rpm
WEIGHT 380 lb (172 kg)
TOP SPEED 75 mph (121 km/h)
COUNTRY OF ORIGIN Denmark

NIMBUS *MkII*

PRODUCED IN DENMARK, the Nimbus was hardly changed from 1934 to its demise in 1959. Its sophisticated engine was housed in a crude frame made of steel plates and strip. A dash panel incorporating the handlebars carried the instruments and electrical switchgear, while the machine was completed by a fishtail straight-through exhaust system. The Nimbus was rarely sold outside its country of origin.

NORTON *Side Valve*

JAMES LANSDOWNE NORTON started building motorcycles in Birmingham, England, shortly after the turn of the 20th century. The first bikes were powered by an assortment of proprietary engines. A Peugeot power unit was used to propel the Norton that won the first ever TT race at the Isle of Man and which really began Norton's racing reputation. Norton introduced its own engine the following year, and this side-valve design continued the company's reputation for quality products and racing success with excellent results at the Brooklands racetrack. It soon became available in 633cc and 490cc versions. The 490cc engine was used in this 1920 racing model which also had a three-speed gearbox and chain drive. Earlier machines were single-speed, belt-drive devices. Improved versions of this power unit were made all the way through until 1954.

SPECIFICATIONS

MODEL Norton Side Valve
CAPACITY 490cc
POWER OUTPUT 12 bhp @ 3000 rpm
WEIGHT 252 lb (114 kg)
TOP SPEED 75 mph (121 km/h)
COUNTRY OF ORIGIN UK

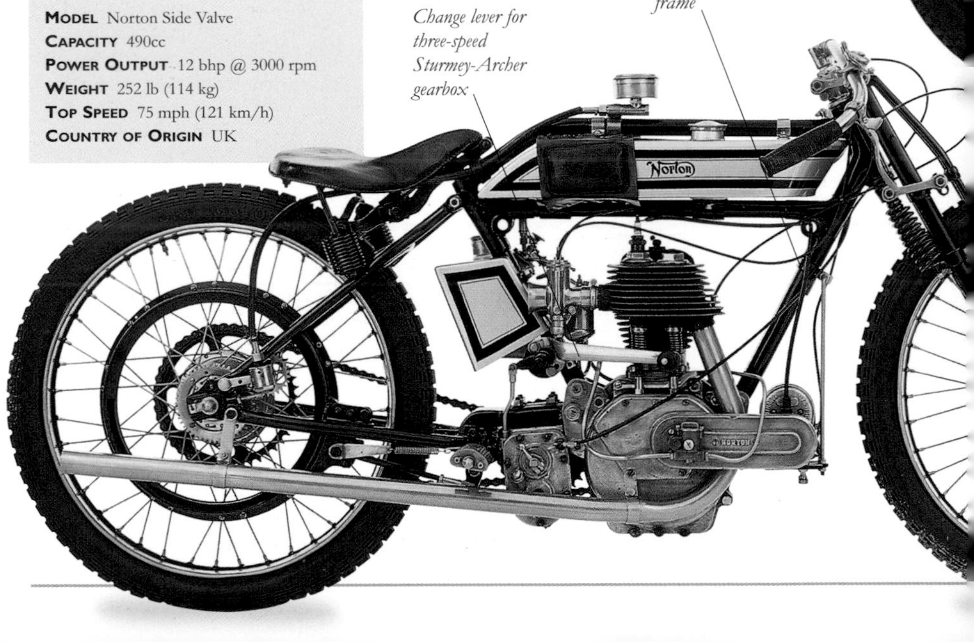

Change lever for three-speed Sturmey-Archer gearbox

Diamond-pattern frame

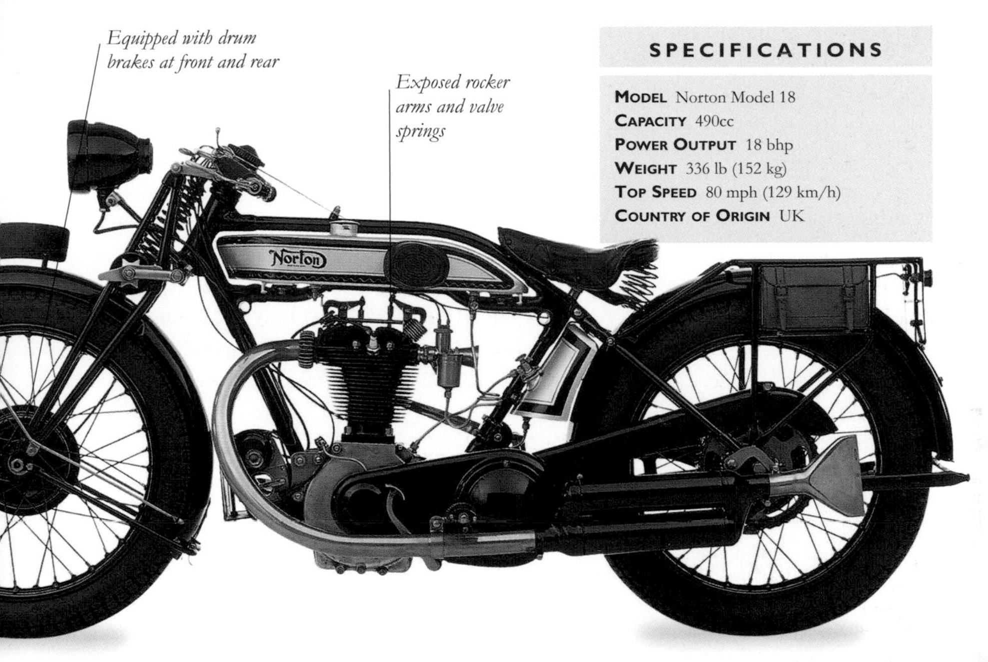

SPECIFICATIONS

MODEL Norton Model 18
CAPACITY 490cc
POWER OUTPUT 18 bhp
WEIGHT 336 lb (152 kg)
TOP SPEED 80 mph (129 km/h)
COUNTRY OF ORIGIN UK

NORTON *Model 18*

THE MODEL 18 first appeared in 1922, along with a number of Norton's new overhead-valve singles. The Model 18 remained available, with some modifications to the suspension and the addition of a four-speed gearbox, until 1954. The bottom half of the engine was virtually identical to that of the Model H side valve, but instead of side valves, the cams operated overhead valves by means of pushrods and rockers. The chain drive to the three-speed Sturmey-Archer gearbox ran in a new pressed-tin "oilbath" case that survived until the 1960s on Norton roadsters. The bike shown here was produced in 1924.

NORTON *CS1*

THE CS1—MEANING CAMSHAFT ONE—was the first of Norton's hugely successful overhead camshaft, single cylinder bikes which the company continued to make until 1962. The CS1 was developed in response to the success of Velocette's overhead-cam racer and was unveiled in May 1927. It was designed by Walter Moore, who later went to work for German manufacturer NSU, for whom he produced the very similar 500SS design *(see p.335)*. The CS1 had a new cradle frame as well as the new engine and the latest Webb girder forks. The CS1 achieved a maiden victory in the 1927 Senior TT which was won by factory rider Alec Bennett. Redsigned for 1930, the overhead camshaft Norton went on to win seven out of ten Senior TT races in the 1930s.

Rear seat pad allowed riders to adopt a racing crouch

Two pairs of bevel gears drive the camshaft

SPECIFICATIONS

MODEL Norton CS1
CAPACITY 490cc
POWER OUTPUT Not known
WEIGHT 330 lb (150 kg)
TOP SPEED 71 mph (114 km/h)
COUNTRY OF ORIGIN UK

REVISED REAR

The "cricket bat" CS1, so called because of the shape of its bevel gear housings, gained a more efficient fishtail muffler in 1928. Production of the CS1 ended in 1929.

Girder front forks

Footrest

A muffler would have been attached for racing at Brooklands

Grooved tire

Semidry sump lubrication

Brakes are big 8-in (21-cm) drums, front and back; contemporary testers found the front brake almost useless

NORTON *500*

NORTON INTRODUCED PLUNGER rear suspension on its factory-standard racing machines in 1936. Production versions of the chassis, nicknamed the "garden gate," soon followed. This 1938 racing machine is not equipped with the telescopic forks or twin-camshaft engine that appeared on the team bikes that year. Although increasingly uncompetitive against more powerful foreign opposition, the "Manx Specification" o.h.c. Norton with alloy cylinder head was still almost obligatory equipment in the 500 class for amateur racers. The "Manx" name was not adopted until after World War II.

"Garden gate" frame with plunger rear suspension

Sturmey-Archer-based gearbox

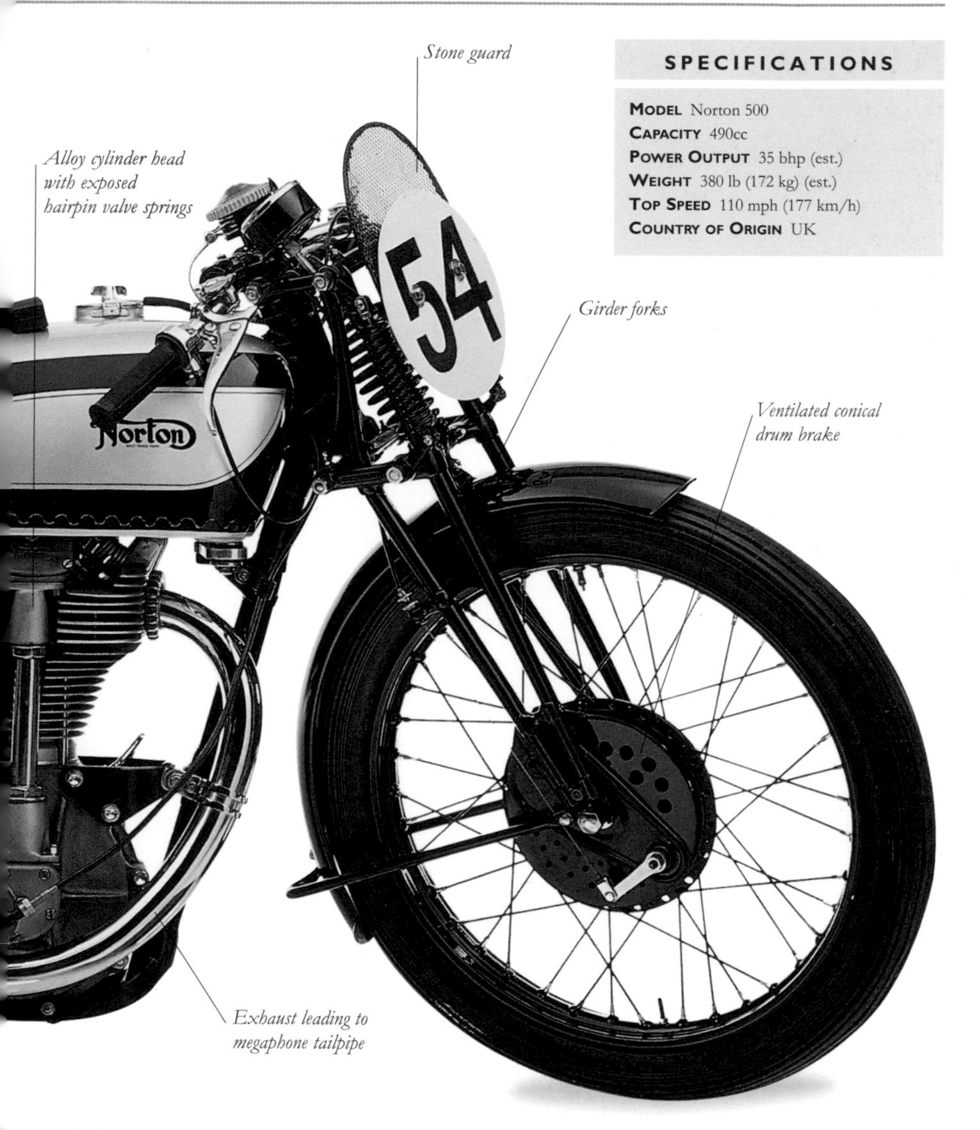

SPECIFICATIONS

MODEL Norton 500
CAPACITY 490cc
POWER OUTPUT 35 bhp (est.)
WEIGHT 380 lb (172 kg) (est.)
TOP SPEED 110 mph (177 km/h)
COUNTRY OF ORIGIN UK

NORTON *Model 7 Dominator*

LIKE OTHER FACTORIES, Norton was eager to emulate the success of Triumph's Speed Twin *(see pp.408–09)*, but World War II caused a delay until November 1948. The Model 7 Dominator comprised a new 500cc o.h.v. parallel-twin engine with a revised gearbox to fit the shape of the engine in the plunger-framed, rolling chassis from the ES2 single. The new engine had a single camshaft mounted in front of the cylinders, and pushrods passed through a tunnel in the iron barrel. It was designed by Bert Hopwood. The Norton was a viable alternative to the Triumph and other British twins, especially after 1951 when the famous "featherbed" frame was used, giving Norton a clear handling advantage. Despite this, it was never as popular as the Triumph either at home or in the US. Illustrated is a 1949 machine.

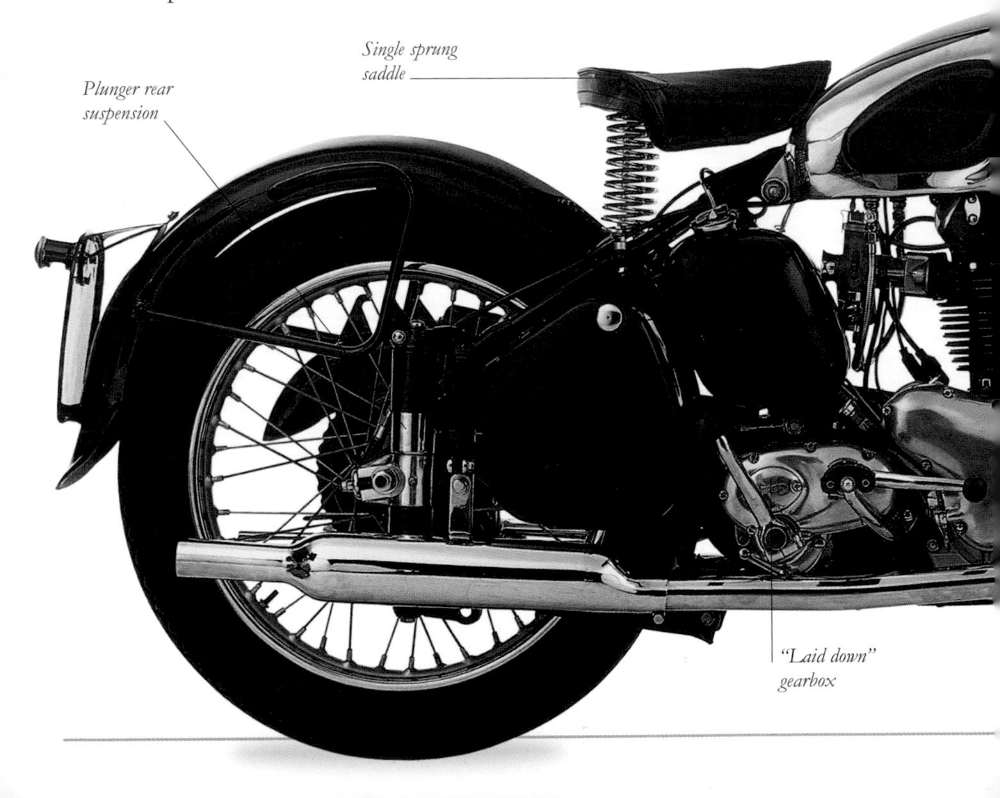

SPECIFICATIONS

MODEL Norton Model 7 Dominator
CAPACITY 497cc
POWER OUTPUT 29 bhp @ 6000 rpm
WEIGHT 420 lb (190 kg)
TOP SPEED 88 mph (142 km/h)
COUNTRY OF ORIGIN UK

THE ORIGINAL Although the plunger-framed Model 7 was eclipsed by featherbed-framed versions from 1952 onward, it was the predecessor to a line of twins that culminated in the Commando *(see pp.326–29)*.

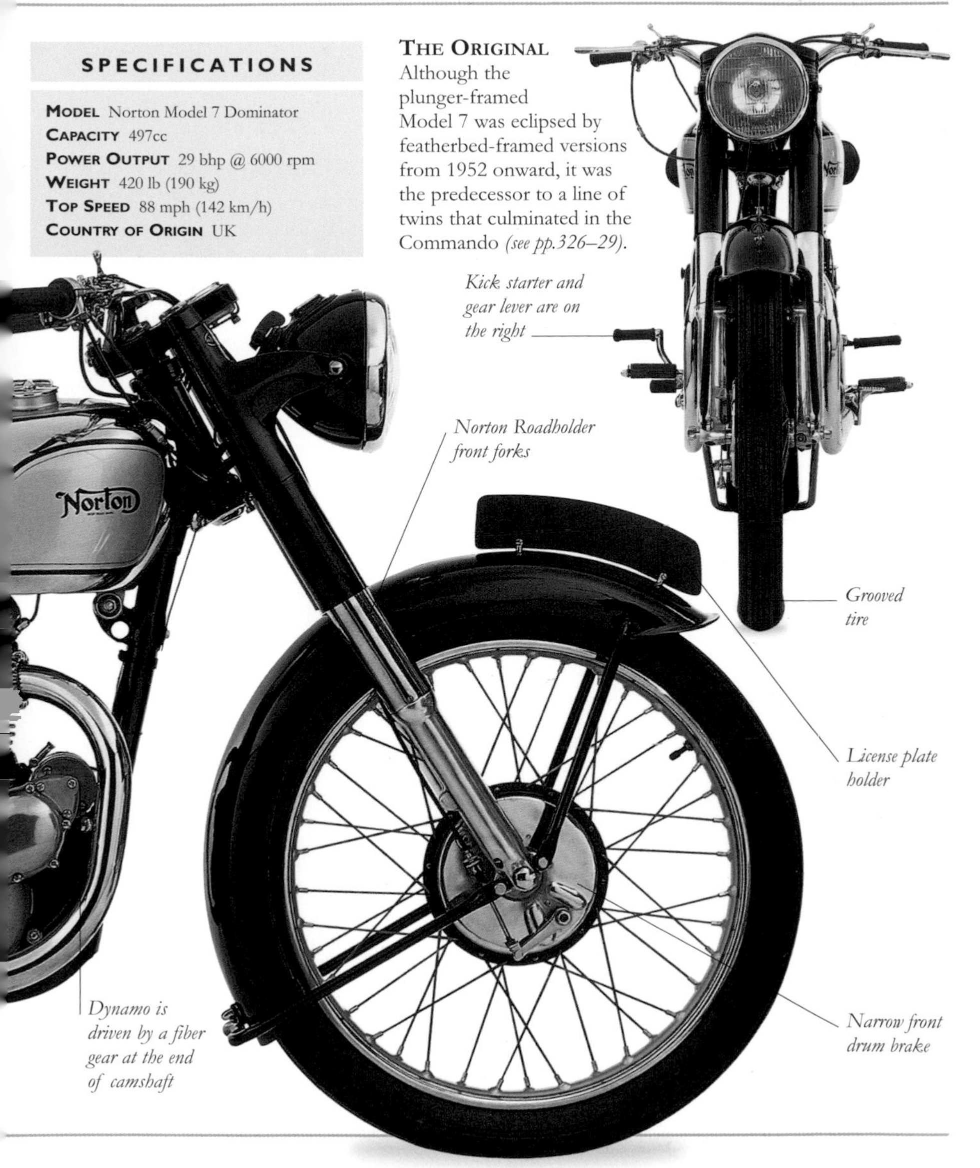

NORTON *Manx*

NORTON'S OVERHEAD-CAMSHAFT racing singles had received 35 years of development by the time the final machines emerged from the firm's factory at Bracebridge Street, Birmingham, England, in 1962. Direct factory involvement was reduced from 1955 but the Manx remained competitive until much later. The most significant postwar improvement to the machine was the use of the famous "featherbed" frame, which endowed the Manx with faultless handling. The twin camshaft design, which the factory bikes used in 1938, was adopted on postwar production models, and a revised short-stroke design was used from 1954. The Manx was available with a 350cc or 500cc engine.

SPECIFICATIONS

MODEL Norton Manx
CAPACITY 498cc
POWER OUTPUT 47 bhp @ 6500 rpm
WEIGHT 309 lb (140 kg)
TOP SPEED 140 mph (225 km/h)
COUNTRY OF ORIGIN UK

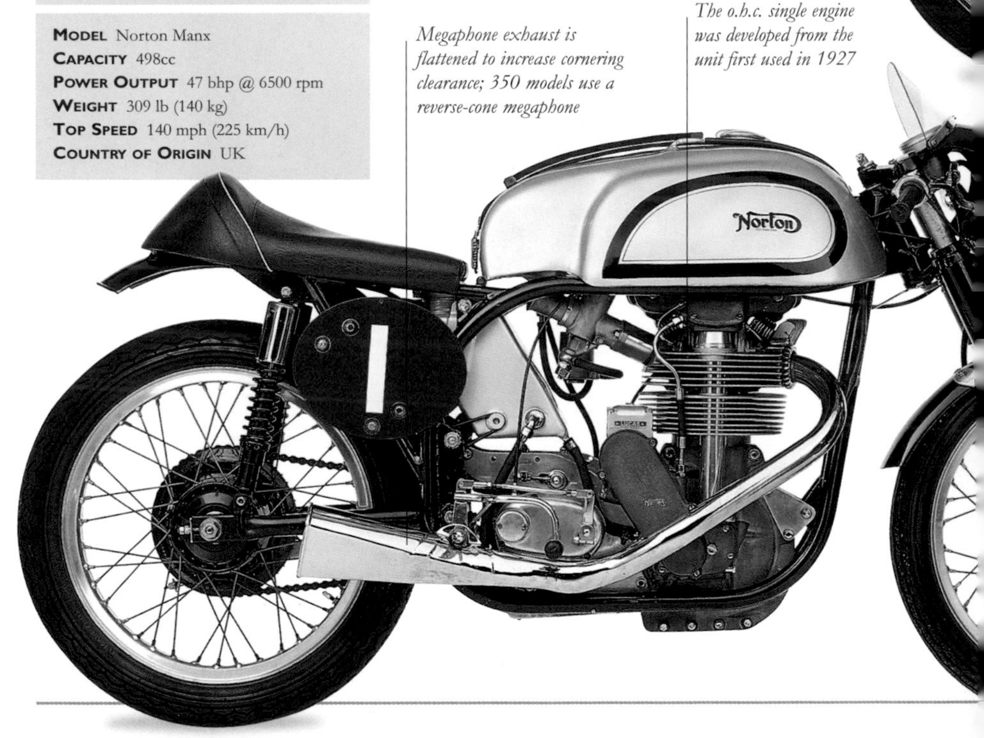

Megaphone exhaust is flattened to increase cornering clearance; 350 models use a reverse-cone megaphone

The o.h.c. single engine was developed from the unit first used in 1927

Post-1960 featherbed frames were known as "slimline"

SPECIFICATIONS

MODEL Norton 650SS Dominator
CAPACITY 646cc
POWER OUTPUT 49 bhp @ 6800 rpm
WEIGHT 434 lb (197 kg)
TOP SPEED 115 mph (185 km/h)
COUNTRY OF ORIGIN UK

Norton's famous Roadholder telescopic forks

NORTON *650SS Dominator*

MANY CONSIDERED THE 650SS to be the best of the featherbed Dominators. The engine was enlarged to 646cc and equipped with steeply angled twin Amal carburetors. The lighting system was fed by a 12-volt alternator, but magneto ignition was retained long after other makers had changed to coil. The gearbox was AMC, following Norton's 1958 takeover by Associated Motor Cycles. The revised, "slimline" frame had waisted top rails to accommodate pressed-steel paneling and was attached to some models from 1960 to 1963. This bike dates from 1962.

NORTON *Fastback Commando*

LAUNCHED IN 1967, the radically styled Commando gained the name Fastback in 1969 (the year of this model), to distinguish it from the more conventional 1968 version. The engine, gearbox, swingarm, exhaust system, and rear wheel were mounted together as a single assembly, held onto the massive tubular spine frame by three "Isolastic" rubber insulators. This reduced the vibration from the 750cc engine to acceptable levels without compromising handling. The Atlas engine was slanted forward and the four-speed AMC gearbox was driven by an uprated triplex chain. The front wheel gained an efficient twin, while the leading-shoe brake and the Roadholder forks were retained.

SPECIFICATIONS

MODEL Norton Fastback Commando
CAPACITY 745cc
POWER OUTPUT 56 bhp @ 6500 rpm
WEIGHT 398 lb (180 kg)
TOP SPEED 115 mph (185 km/h)
COUNTRY OF ORIGIN UK

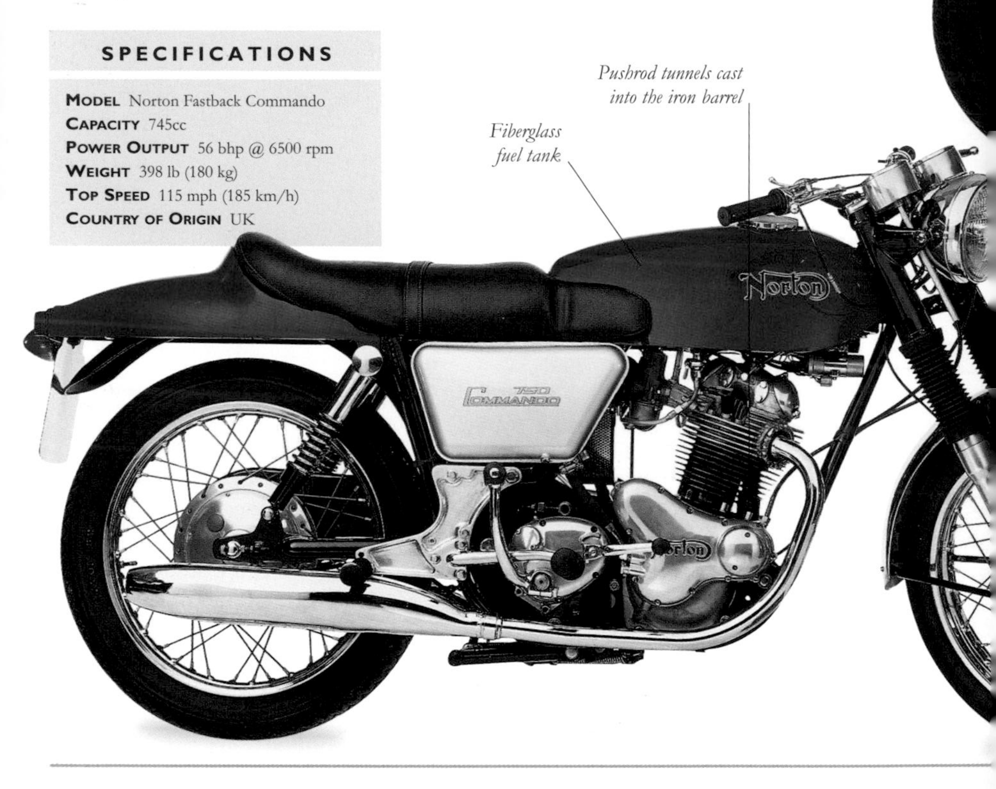

Fiberglass seat

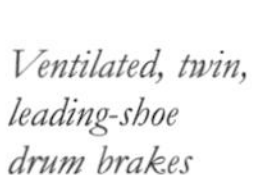

Ventilated, twin, leading-shoe drum brakes

SPECIFICATIONS	
MODEL	Norton Commando Formula 750
CAPACITY	750cc
POWER OUTPUT	Not known
WEIGHT	Not known
TOP SPEED	Not known
COUNTRY OF ORIGIN	UK

Angled exhaust allows steeper cornering

NORTON *Commando Formula 750*

LACK OF POWER IN THE Commando's engine meant that the rolling chassis of Norton's 1972 team racers had to be lightened to make them competitive. Team rider Peter Williams developed a lightweight sheet-steel, monocoque frame. A fuel pump is operated by movement of the rear swingarm, and this 1973 version uses a steel monocoque in which the fuel is housed in pannier fuel tanks. Despite a 25 bhp disadvantage, Peter Williams won the 1973 Isle of Man Formula 750 TT on a Commando team racer, confounding critics who thought Nortons were outdated and had no chance against the Japanese.

NORTON *Commando Interstate MkIII 850ES*

NORTON WAS ON THE VERGE OF bankruptcy by 1975 when this bike—the final version of the Commando—was produced. Based on 1973's MkI 850, the MkIII was equipped with an electric starter in addition to the kick starter. It had souped-up mufflers, Lockheed disc brakes front and rear, left-hand gearchange, halogen lighting, and a full range of accessories. Nothing could conceal the age of the design, and compared to the increasingly sophisticated Japanese opposition it was very dated, though it had some good points. The four-speed gearbox and twin-cylinder o.h.v. engine might have been underpowered and of dubious reliability, but its light weight and fine handling were to be commended. Many riders around the world mourned its passing in 1977.

Passenger handrail

Annular discharge mufflers were souped up to give a deep engine note

The Commando's preunit gearbox was outdated by 1975

SPECIFICATIONS

MODEL Norton Commando Interstate MkIII
CAPACITY 829cc
POWER OUTPUT 60 bhp @ 6000 rpm
WEIGHT 430 lb (195 kg)
TOP SPEED 115 mph (185 km/h)
COUNTRY OF ORIGIN UK

US ADJUSTMENTS

For the 850ES to comply with strict environmental legislation in the US, 140 improvements were required. It was sold in black-and-gold Roadster form, or as the Interstate with Manx-style silver and black as seen here.

Chrome headlight and mounting brackets

Avon Roadrunner tires were among the best of the period

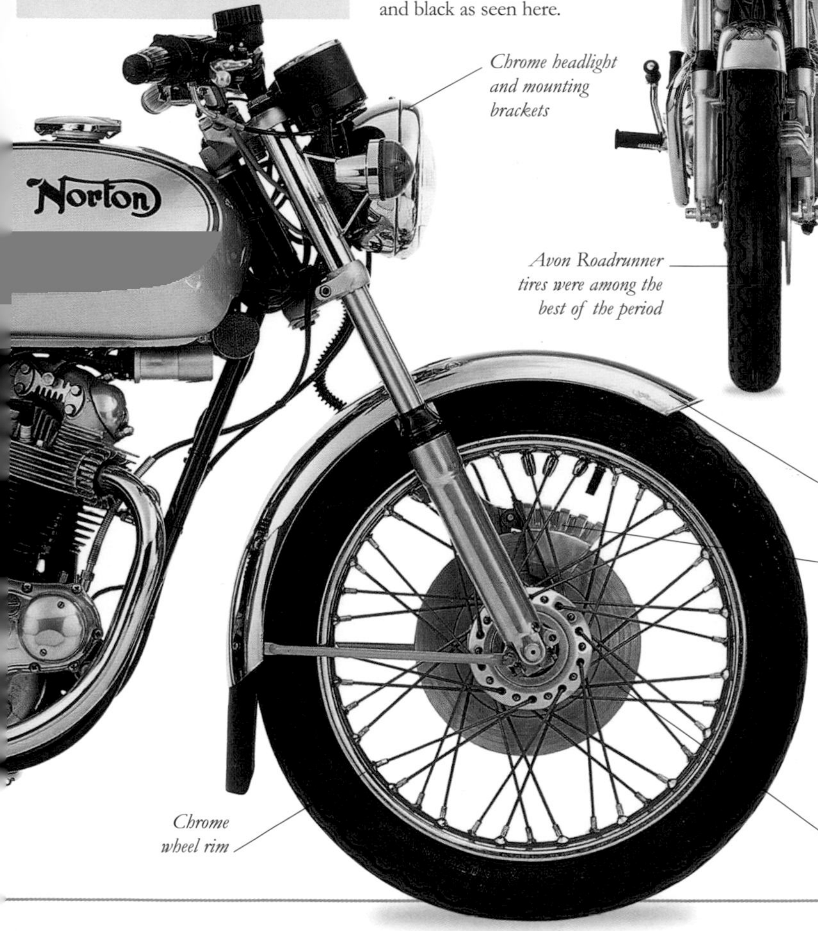

Chrome mudguard

Lockheed brake calliper

Chrome wheel rim

Disc brake

NORTON *NRS588 Rotary Racer*

TWO DECADES AFTER ITS LAST TT victory, a Norton Wankel ridden by Steve Hislop won 1992's Senior race. The bike was the culmination of 20 years of Wankel development begun by the BSA-Triumph group. The chassis had a twin-spar, alloy frame, with upside-down front forks and monoshock rear suspension. Despite the success of the racers, the Wankel-engined road bikes did not sell well.

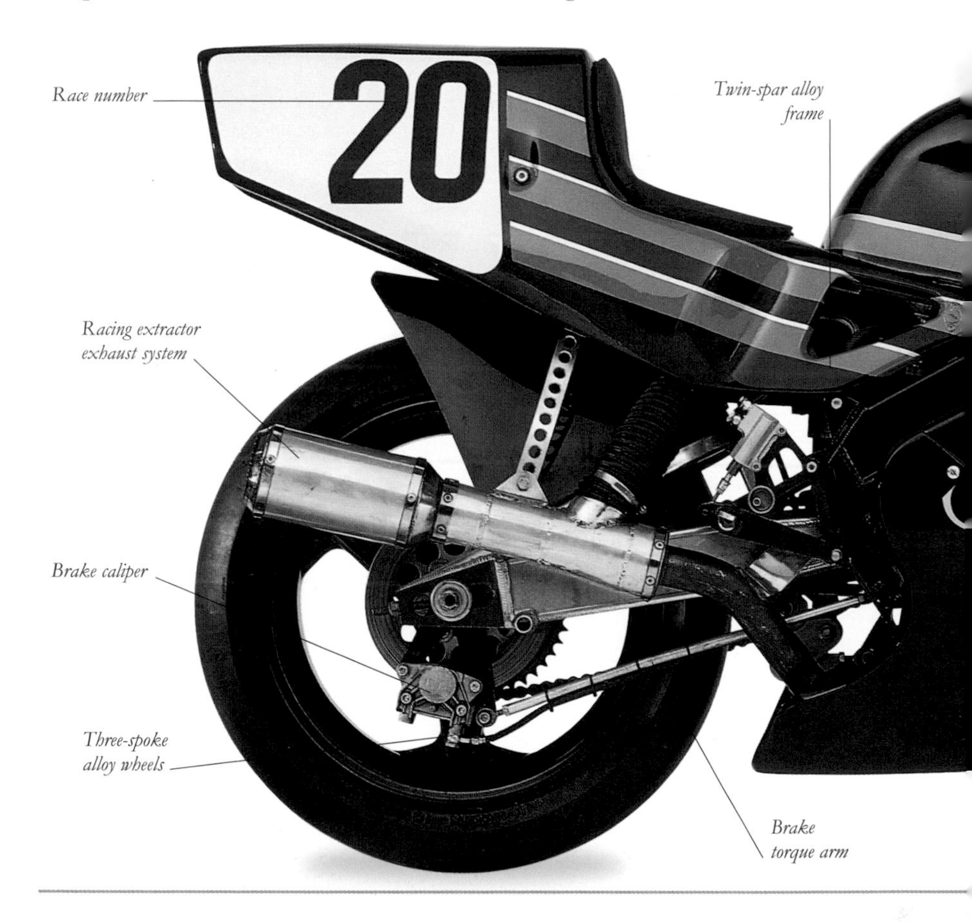

SPECIFICATIONS

MODEL Norton NRS588 Rotary Racer
CAPACITY 588cc (est.)
POWER OUTPUT 135 bhp (est.)
WEIGHT Not known
TOP SPEED Not known
COUNTRY OF ORIGIN UK

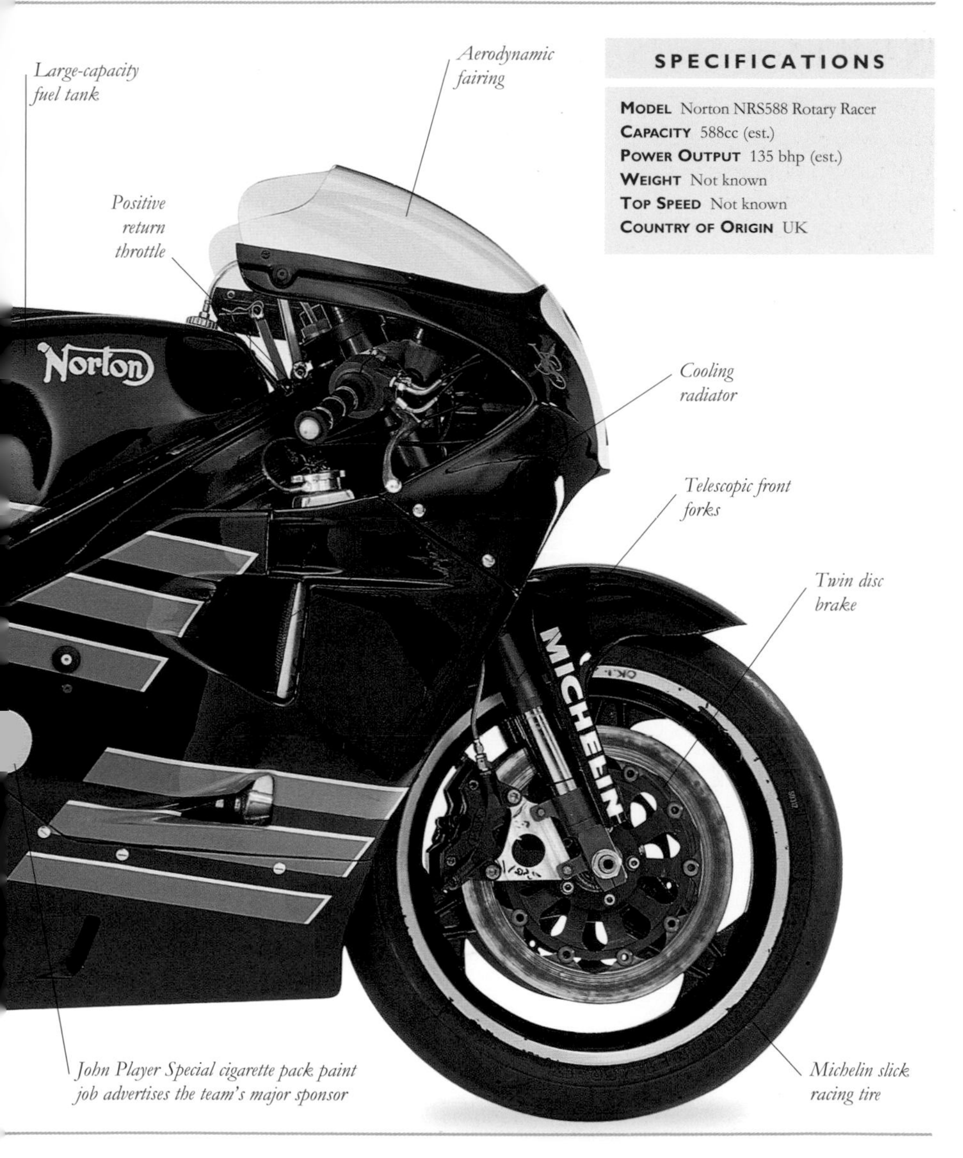

NORTON COMMANDO *961 Sport*

THE COMPLEX HISTORY OF THE NORTON MARQUE took another twist with the revival of the famous brand and the introduction of a new model in 2009. The new bike was inspired by, and named after, the classic Commando model of the 1970s but was a fresh design using modern technology, albeit in a retro package. Like its namesake, the new bike has a parallel twin-cylinder engine with two valves per cylinder operated by pushrods, and cooled by air. But the new Norton engine has a five-speed gearbox built-in unit with the engine, and is fitted with modern fuel injection. Three different versions of the bike were made.

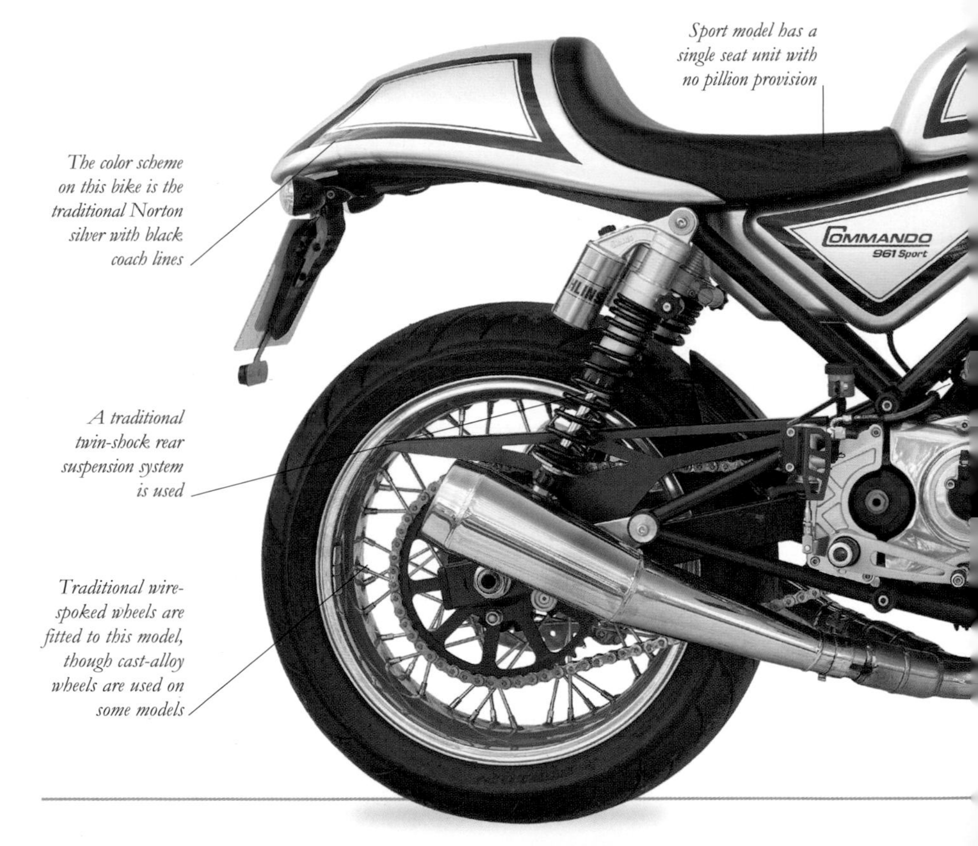

SPECIFICATIONS

MODEL Norton Commando 961 Sport
CAPACITY 961cc
POWER OUTPUT 79 bhp @ 6500 rpm
WEIGHT 414 lb (188 kg)
TOP SPEED 130 mph (209 km/h)
COUNTRY OF ORIGIN UK

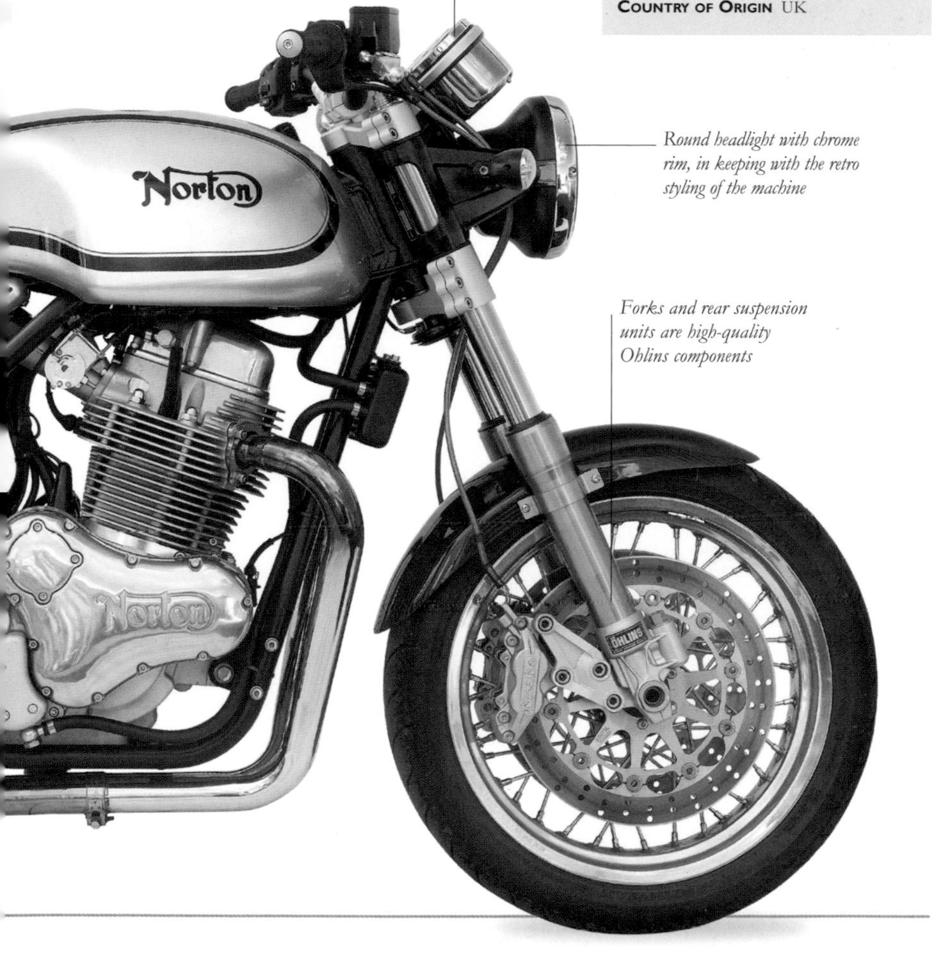
Traditional-looking analog speedometer and rev-counter have modern electronic internals
Round headlight with chrome rim, in keeping with the retro styling of the machine
Forks and rear suspension units are high-quality Ohlins components
Norton
Norton
ÖHLINS

NSU *350TT*

LIKE MANY PIONEER MOTORCYCLE manufacturers, NSU got into the business via bicycles and, before that, sewing machines. The advantages of the twin-cylinder engine over the single when more power was required were appreciated very early on, and, since the V-twin was the easiest form to accommodate in a motorcycle frame, it was that layout which most manufacturers followed. This 349cc NSU from 1912, with its mechanically operated, overhead inlet and side exhaust valves, had an up-to-date engine but the transmission was crude, with direct belt drive from the engine pulley to the rear wheel. The suspension was nonexistent. Even so, the machine achieved fourth and seventh places in the 1913 Isle of Man Junior TT.

SPECIFICATIONS

MODEL NSU 350TT
CAPACITY 349cc
POWER OUTPUT 7 bhp @ 2500 rpm
WEIGHT Not known
TOP SPEED Not known
COUNTRY OF ORIGIN Germany

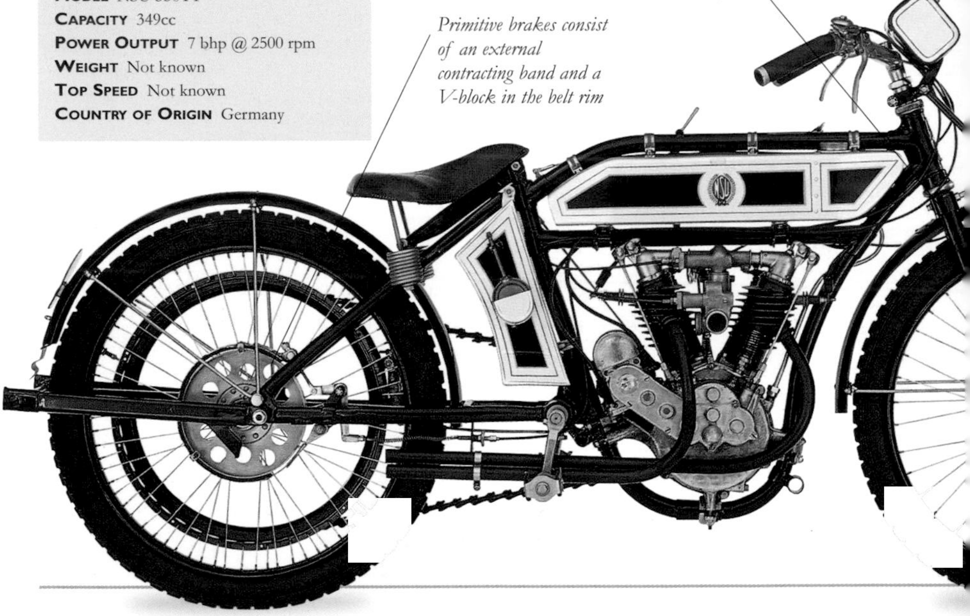

Girder forks

Primitive brakes consist of an external contracting band and a V-block in the belt rim

SPECIFICATIONS

MODEL NSU 500SS
CAPACITY 494cc
POWER OUTPUT 22 bhp @ 4400 rpm
WEIGHT 364 lb (165 kg)
TOP SPEED 92 mph (148 km/h)
COUNTRY OF ORIGIN Germany

Knobby tire

NSU *500SS*

SO SIMILAR WAS THE 500cc o.h.c. engine Walter Moore designed for NSU to his Norton design that it was said that NSU stood for "Norton Spares Used." It may be more true to say that, having produced one satisfactory design, he worked along similar lines in designing the NSU. Compare the NSU with the Norton CS1 *(pp.318–19)*. The bike shown dates from 1931, and that year NSU raced the model at the Isle of Man TT. Three bikes started the Senior race but only Ted Mellors finished in a respectable sixth place.

NSU *Kompressor*

Although the Kompressor was famous for its record-breaking activities after World War II, the supercharged NSU twin was an example of how not to make a racing motorcycle. At a time when the dominant 500cc machine was the 304-lb (138-kg) BMW Kompressor *(see p.49)*, a 441-lb (200-kg) 350 was clearly not the way to proceed. While there was much less opposition in the 350cc class, what there was had been developed over a number of years and the main threat to Norton and Velocette supremacy looked likely to come from DKW. Although it was clear that NSU's frames and suspension designs left much to be desired, the same could not be said of the company's engine, which, with double overhead camshafts, unit construction, and supercharging, represented the state-of-the-art at the time.

Plunger rear suspension

Supercharger

SPECIFICATIONS

MODEL NSU Kompressor
CAPACITY 347cc
POWER OUTPUT 46 bhp @ 8000 rpm
WEIGHT 441 lb (200 kg)
TOP SPEED 136 mph (219 km/h)
COUNTRY OF ORIGIN Germany

FRAME CHANGE

The NSU Kompressor appeared in 1938, at first in rigid-framed form, but by the following year the engine was mounted in a plunger-sprung frame, and still with girder forks.

Hand-formed, alloy fuel tank

Duplex, cradle frame has widely spaced frame tubes

Twin overhead camshafts are shaft driven

This bike dates from 1939

21-in (53.3-cm) front wheel

Front drum brake

In 1956, Kompressors raised the record for the "world's fastest" motorcycle to 211 mph (340 km/h)

NSU *Max*

REVOLUTIONARY WAS HARDLY a strong enough word to describe this 1952 Max on its introduction. The cycle parts were basically those of the 200cc Lux introduced the previous year, but the engine was entirely new. Its unique feature was the drive to the overhead camshaft via paired eccentrics and connecting rods. As the engine warmed up the tie-rod caused the camshaft housing to rotate slightly around its axis, which also passed through the points of contact between the rockers and the valves, thus maintaining valve clearances as the engine grew taller.

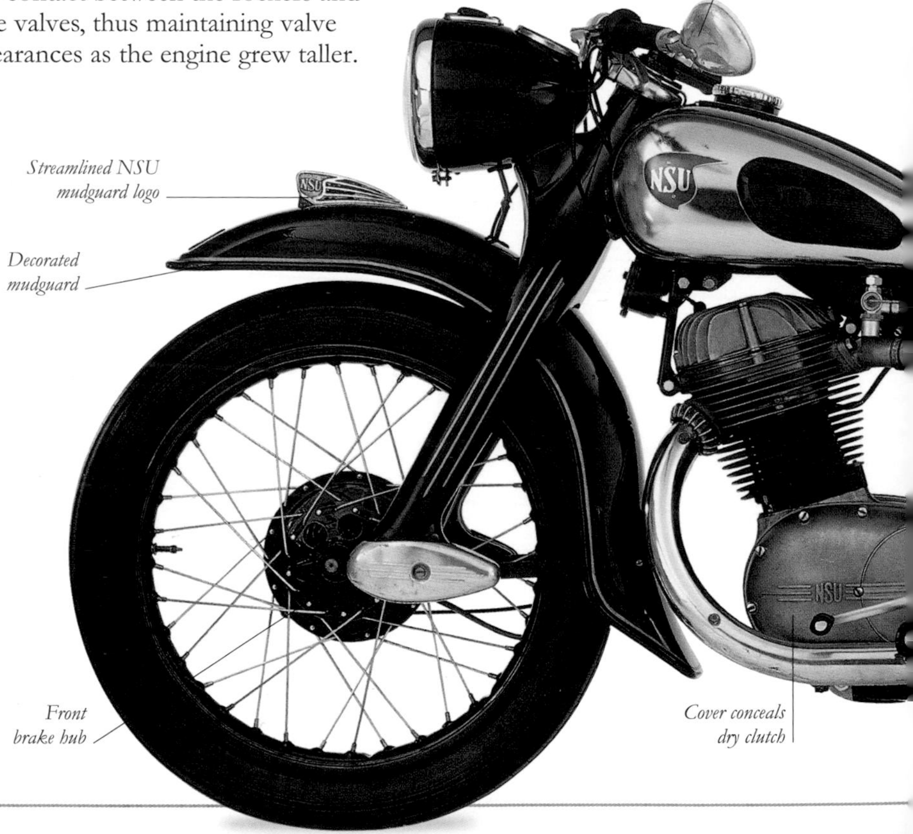

Handlebar end-mirror

Streamlined NSU mudguard logo

Decorated mudguard

Front brake hub

Cover conceals dry clutch

SPECIFICATIONS

Model NSU Max
Capacity 247cc
Power Output 15 bhp @ 5800 rpm
Weight 364 lb (165 kg)
Top Speed 72 mph (116 km/h)
Country of Origin Germany

Front View
NSU design was held in high regard by Japanese manufacturer Soichiro Honda. Compare the front fork and other details of the Max with early Hondas and the source of Honda's inspiration is obvious.

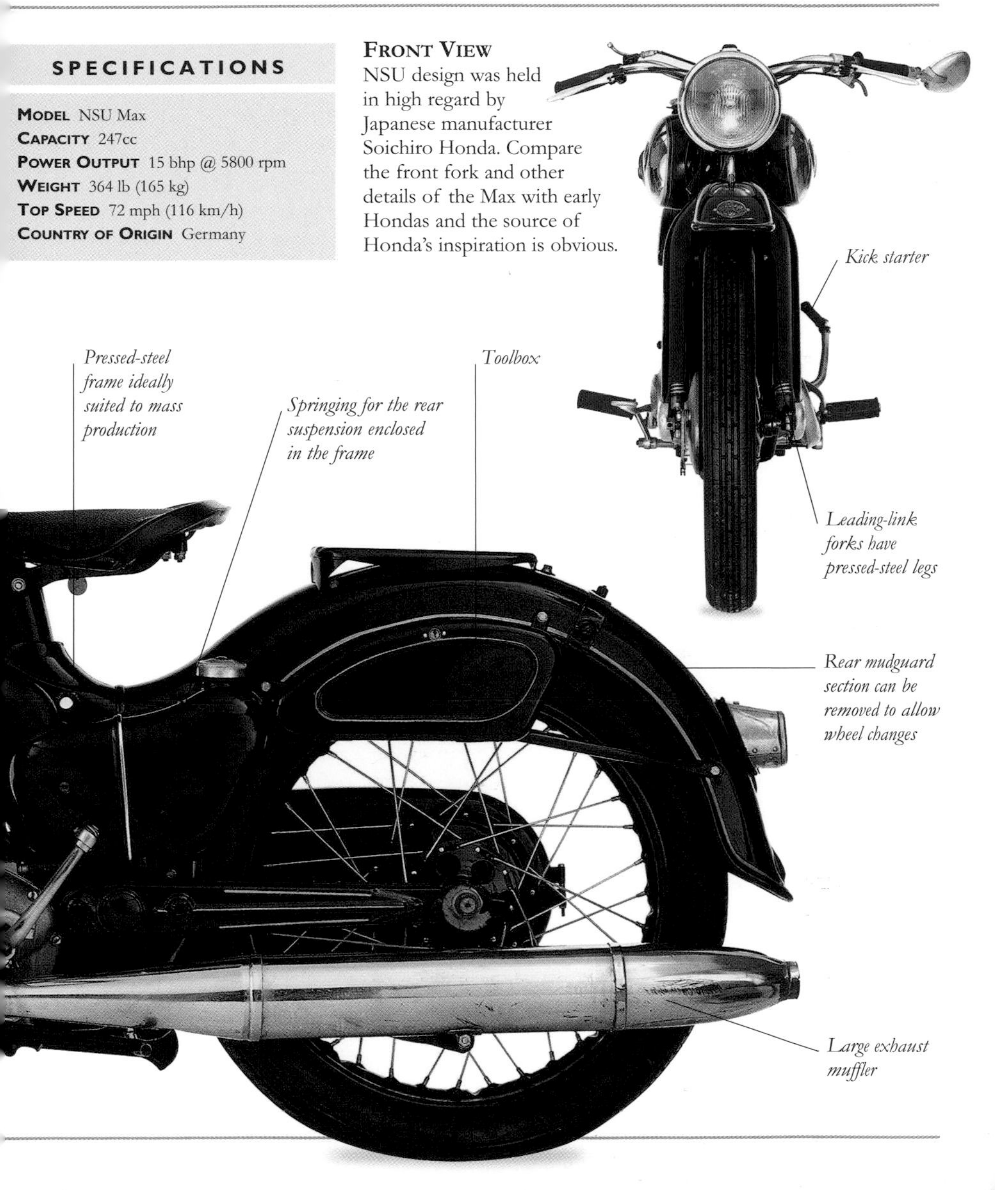

NSU *Rennmax*

The Rennmax was introduced in 1952 and showed promise from the start, but this 1953 twin-cylinder model was ready to take on the world. Ridden by Werner Haas and backed up by Reg Amstrong, the machine secured the two top placings in the World Championship table from strong Guzzi opposition. Haas repeated his placing the following year on a revised Rennmax. The original 1952 Rennmax used a conventional tubular frame, then, in 1953, a new pressed-steel chassis, similar to that used on Max roadsters, was introduced. In 1954, the design was updated again with a six-speed gearbox and improved camshaft drive. Though performance was equal to most machines in the 350cc class, NSU withdrew its factory team at the end of the year.

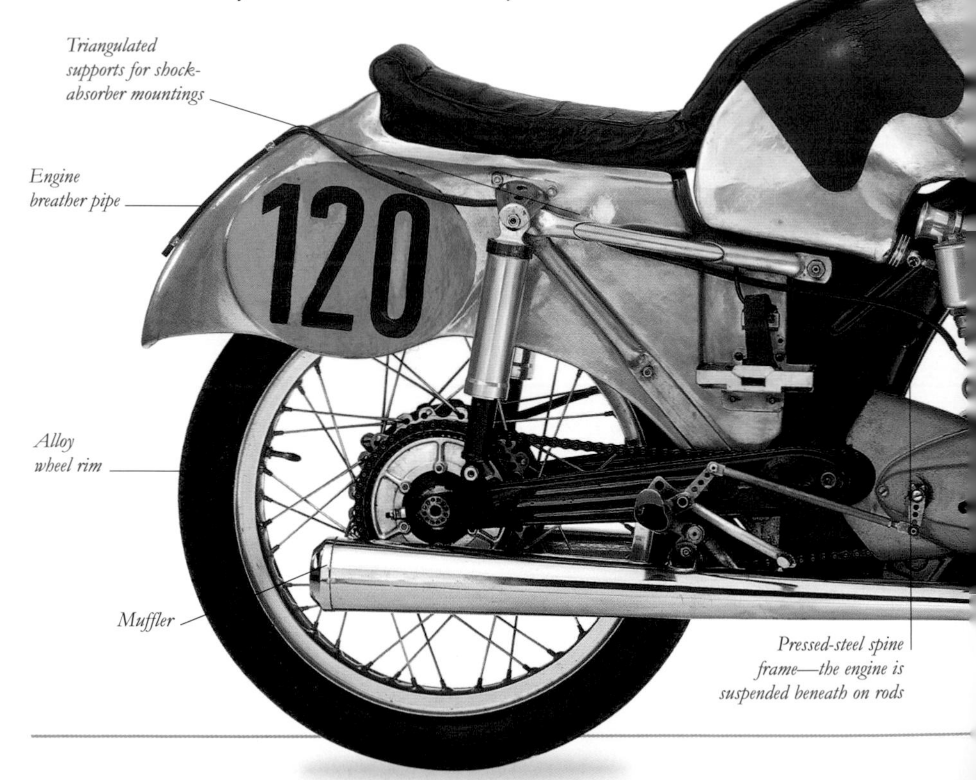

Triangulated supports for shock-absorber mountings

Engine breather pipe

Alloy wheel rim

Muffler

Pressed-steel spine frame—the engine is suspended beneath on rods

SPECIFICATIONS

MODEL NSU Rennmax
CAPACITY 247cc
POWER OUTPUT 15 bhp @ 5800 rpm
WEIGHT 364 lb (165 kg)
TOP SPEED 72 mph (116 km/h)
COUNTRY OF ORIGIN Germany

ENGINE DETAILS

The d.o.h.c. engine had shaft-driven camshafts and a four-speed gearbox. Revised camshaft drive and six-speed box were introduced in 1954. A rudimentary fairing was part of the hand-formed alloy fuel tank.

Hand-formed alloy bodywork is shaped to suit the rider

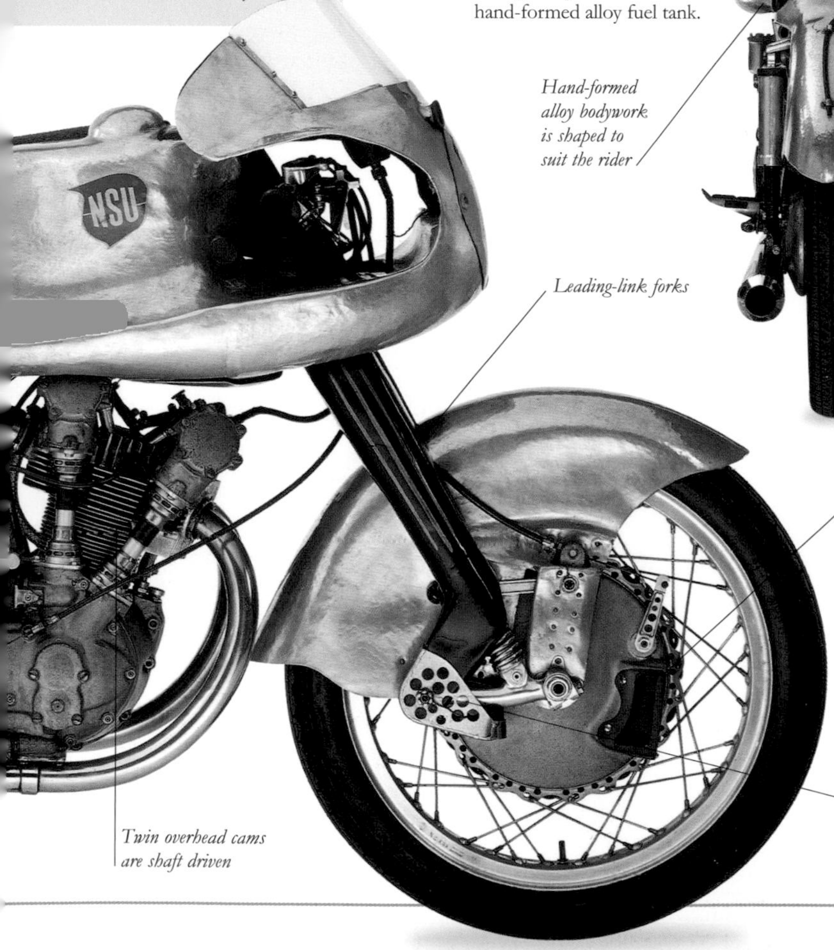

Leading-link forks

Air scoop keeps drum brakes cool

Twin overhead cams are shaft driven

Components have holes drilled to reduce weight

NSU *Sportmax*

In 1953 and 1954 NSU won the 250cc World Championship with its Rennmax racing machines, after which the factory withdrew from competition. The company continued to sell Sportmax racers based on its 250cc road bikes. Herman-Peter Müller became World Champion in 1955 riding a Sportmax. Dustbin fairings, as seen on this Sportmax, were banned from racing bikes in 1958. From the mid-to-late Fifties until the early Sixties, the Sportmax was an essential tool for any rider competing in the 250 class. Many stars of the future, including John Surtees and Mike Hailwood, competed on these well-engineered and competitive German machines.

SPECIFICATIONS

Model NSU Sportmax
Capacity 247cc
Power Output 28 bhp @ 9000 rpm
Weight 246 lb (112 kg)
Top Speed 125 mph (201 km/h)
Country of Origin Germany

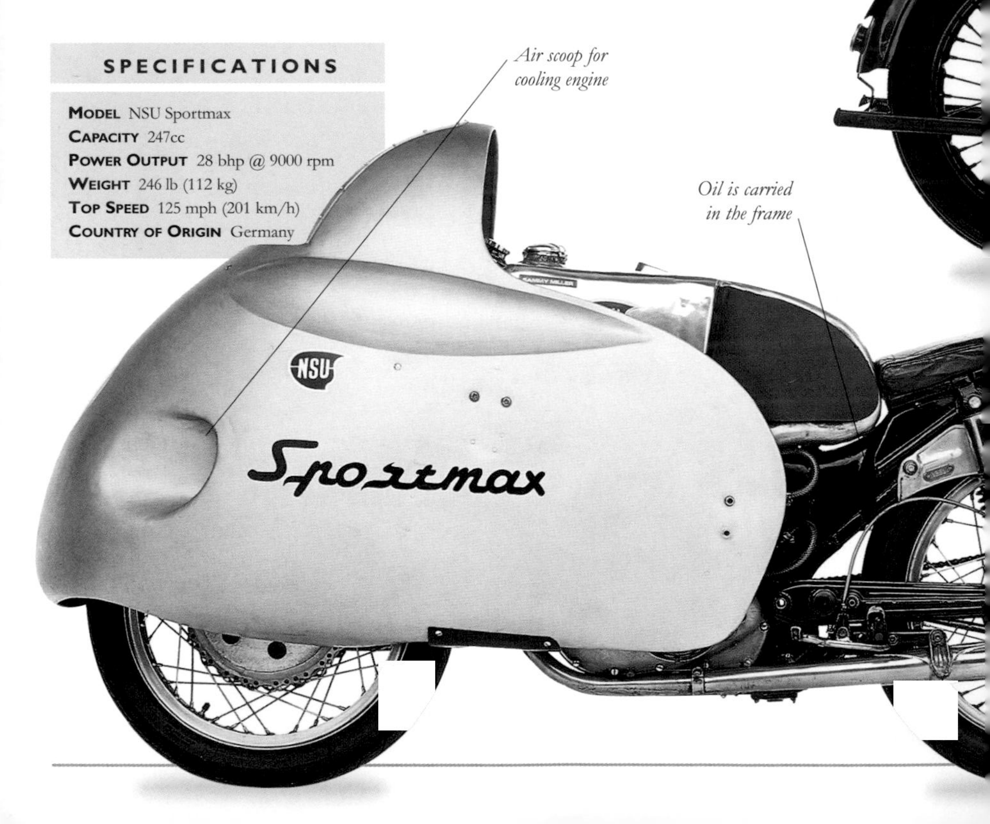

SPECIFICATIONS

MODEL OD TS50
CAPACITY 498cc
POWER OUTPUT 13 bhp @ 3500 rpm
WEIGHT 287 lb (130 kg)
TOP SPEED 71 mph (115 km/h)
COUNTRY OF ORIGIN Germany

19-in (48-cm) wheels with drum brakes

OD *TS50*

WILLY OSTNER IN DRESDEN (Ostner Dresden—OD) was typical of many small manufacturers in Germany. He built high-quality motorcycles with proprietary components between 1927 and 1935. This 1931 machine is a classic example. It uses an inclined single-cylinder, i.o.e. that was made by the Swiss company MAG, which also supplied larger capacity engines, including V-twins, to OD. Some lightweight models used Bark two-stroke engines. This particular bike has a three-speed Hurth gearbox, Amal carburetor, and Bosch magneto. The silver-and-blue paintwork was the usual color scheme for OD machines.

OEC *Commander*

THE OEC FACTORY WAS LOCATED on the south coast of England, far away from most of the rest of the industry in the Midlands. There was also some distance between OEC's ideas and those of other manufacturers. Some said that OEC stood for Odd Engineering Concepts rather than Osborne Engineering Company. This 1938 Commander model used an AJS/Matchless engine and Burman gearbox. Most models had optional rear suspension at a time when this was very rare in Britain. The extraordinary duplex front suspension was standard, though a conventional front end could be specified by the customer. Other models used engines supplied by JAP, Villiers, and Blackburne in every capacity from 148 to 1100cc.

SPECIFICATIONS

Model OEC Commander
Capacity 498cc
Power Output 20 bhp (est.)
Weight 433 lb (196 kg)
Top Speed 75 mph (121 km/h) (est.)
Country of Origin UK

Stable Ride

The famed duplex steering system, introduced in 1927, defies simple explanation but ultimately gave the machine exceptional stability at the expense of limited steering lock.

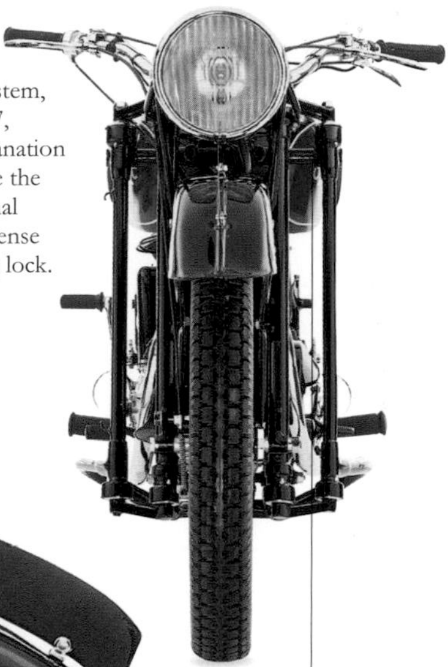

Equipped with a speedometer as standard

Strange steering components gave the OEC an unusual front end

Drum front brake

Wheel rims have painted center line

Opel *Motoclub*

The German carmaker Opel also built motorcycles intermittently. The Motoclub model was introduced in 1928 and survived until 1930, when Opel stopped motorcyc production entirely. This is a 1929 model. The confusing story of this machine also involves at least two other manufacturers. The bike was designed by Ernst Neander, and similar machines were also built by his Neander company. The bikes were actuall made for Opel by another motorcycle manufacturer, Diamant, which continued production until 1931. For further complication, some bikes carried an E.O. logo. All Opel-brand bikes were sold with Opel's own 499cc side valve or overhead valve engines. The other machines also used Küchen, MAG, or JAP engines. The bike had advanced pressed-steel frame construction and pivoting forks.

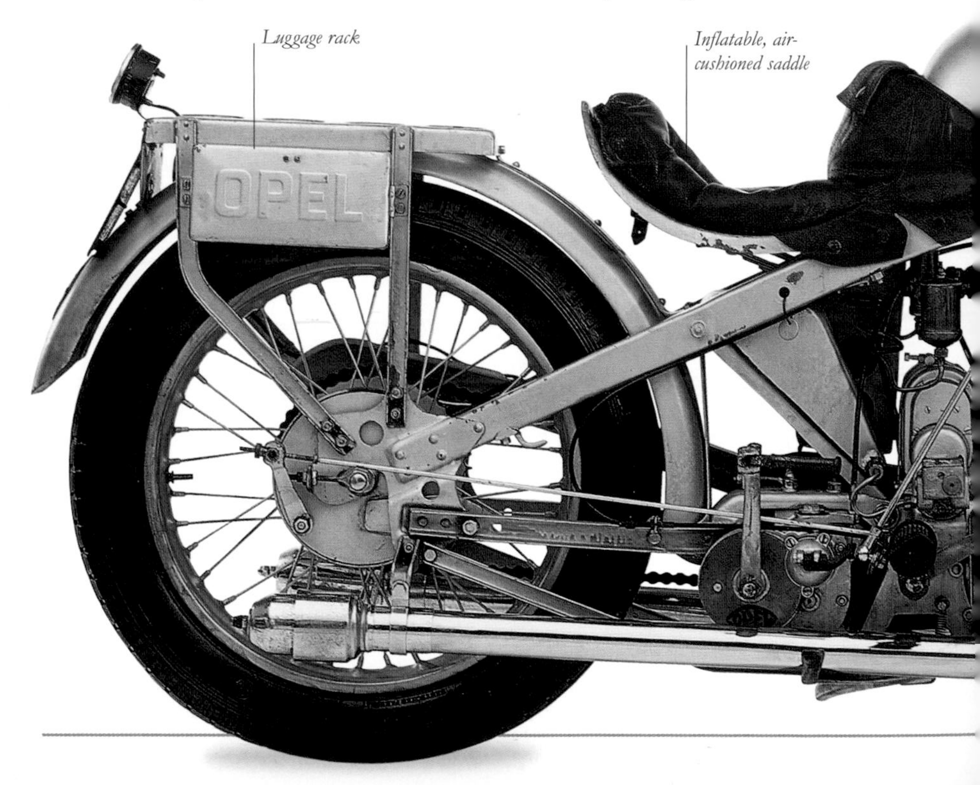

SPECIFICATIONS

MODEL Opel Motoclub
CAPACITY 496cc
POWER OUTPUT Not known
WEIGHT Not known
TOP SPEED 68 mph (109 km/h) (est.)
COUNTRY OF ORIGIN Germany

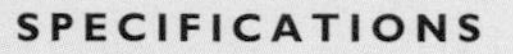

ENGINE CHOICE

The Motoclub used Opel's own 496cc o.h.v. twin-exhaust-port engine. When Opel production ended, a similar machine was briefly built using o.h.c. single-cylinder Küchen engines.

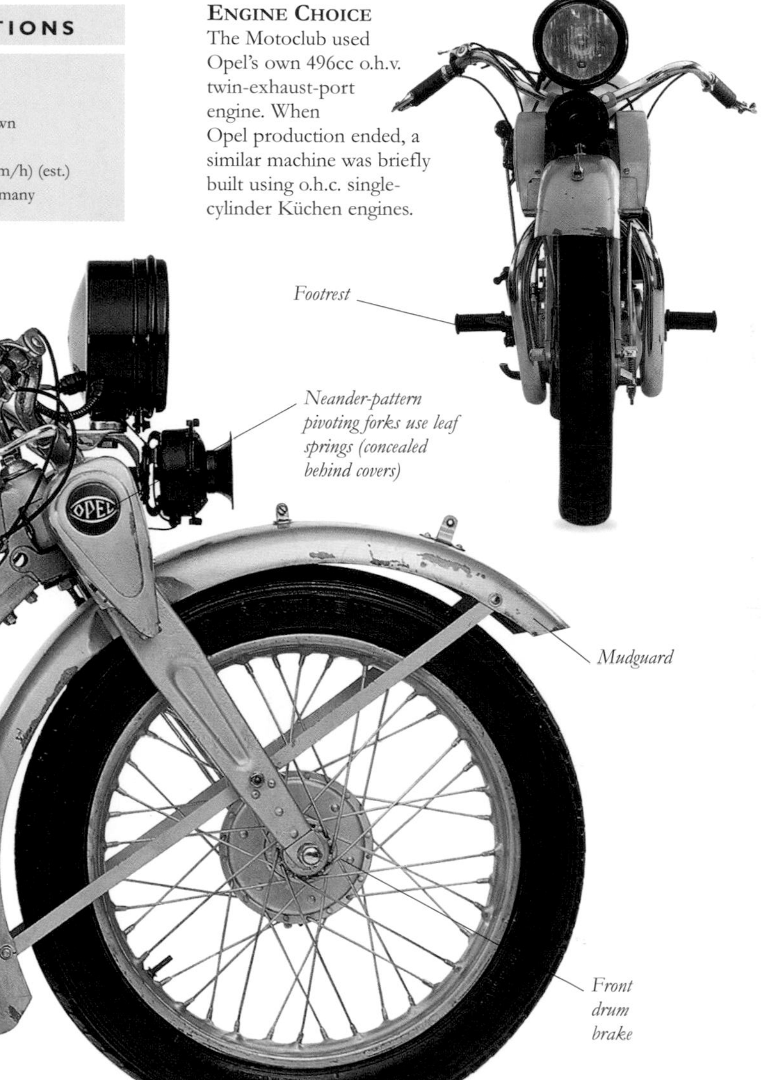

Fuel gauge and speedometer are set into the fuel tank

Footrest

Neander-pattern pivoting forks use leaf springs (concealed behind covers)

Mudguard

Front drum brake

PANTHER *Model 100*

JOAH PHELON MADE HIS FIRST motorcycle in 1900, and two of its features—all-chain drive and a sloping engine taking the place of the front downtube of the frame—were part of the specification of every large-capacity machine that the company produced before its demise in 1966. The Panther name was adopted in 1923 when a revised 500cc overhead-valve engine was introduced. This formed the basis for those that followed. By 1929 the capacity was up to 600cc, rising to 650cc in 1959. Panther bikes were not very powerful, but they were popular machines for pulling sidecars. This 600 is a 1935 model.

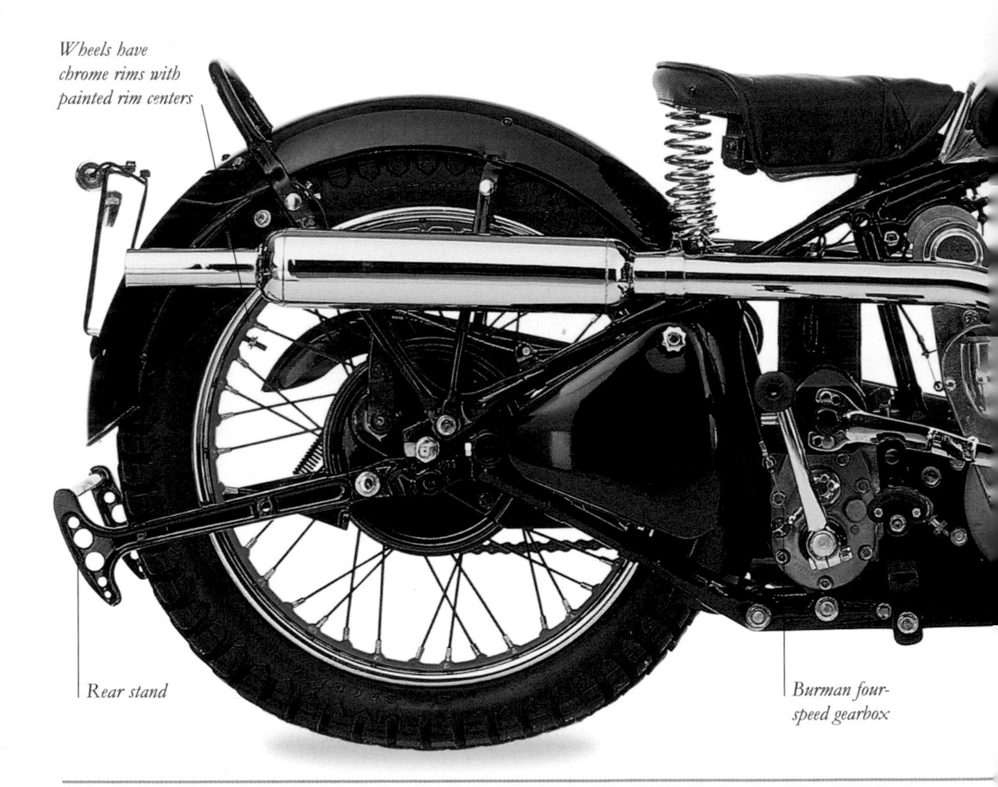

SPECIFICATIONS

MODEL Panther Model 100
CAPACITY 596cc
POWER OUTPUT 26 bhp @ 5000 rpm
WEIGHT 353 kg (160 kg)
TOP SPEED 85 mph (137 km/h)
COUNTRY OF ORIGIN UK

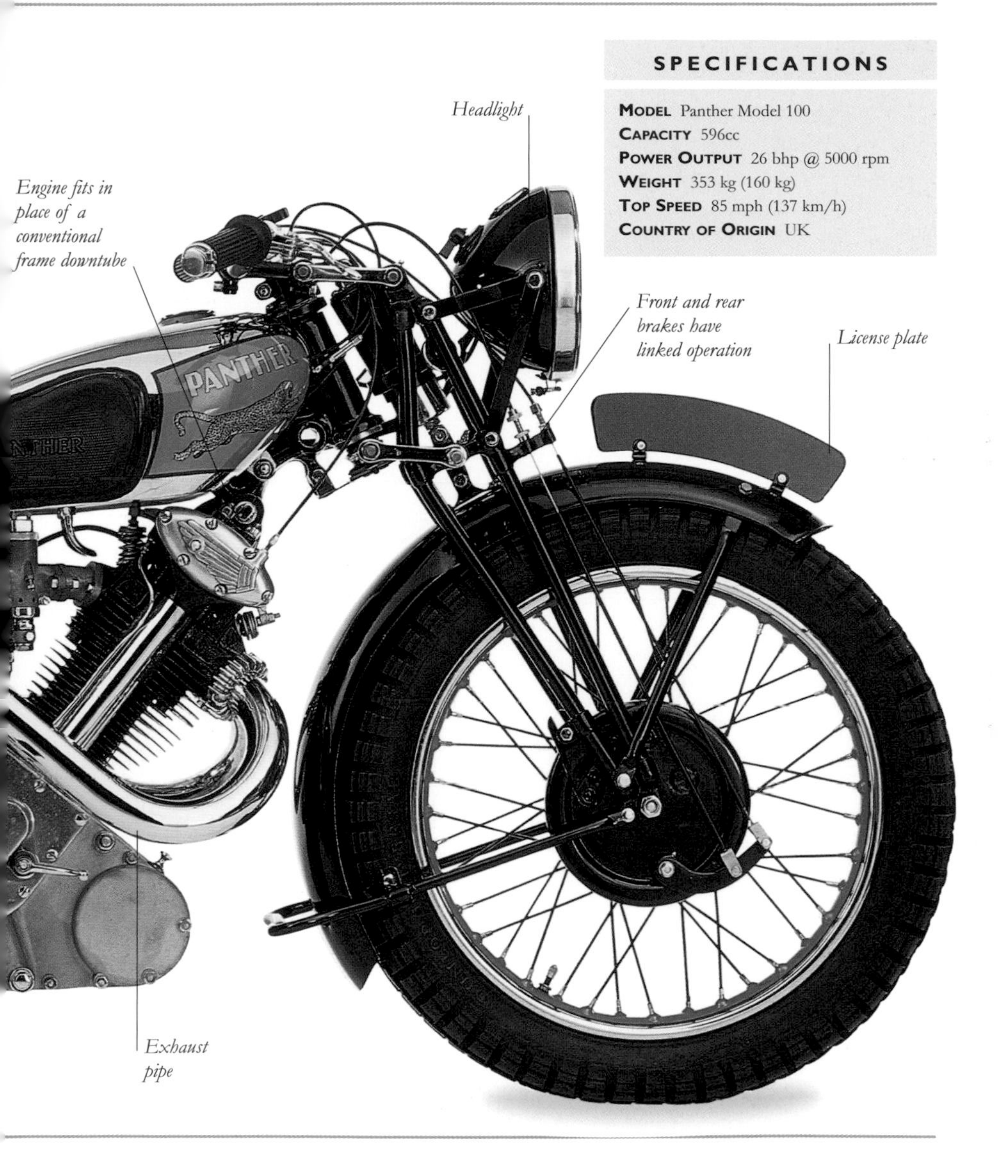

PARILLA *Wildcat Scrambler*

THE PARILLA WILDCAT features a distinctive engine in which the chain-driven camshaft is mounted high up in an alloy "sock" beside the cylinder. The valves are then operated by short pushrods. This system is meant to combine the advantages of both o.h.c. and o.h.v. layouts. The fact that the Wildcat could be revved to 10,500 rpm suggests that there is some truth in the idea. Parilla used this engine, also produced in 175 and 200cc versions, to power road, race, and off-road machines. The Wildcat was introduced in 1961.

High and wi handleba offer extra o road cont

Bashplate protects the underside of the engine

Alloy drum brake

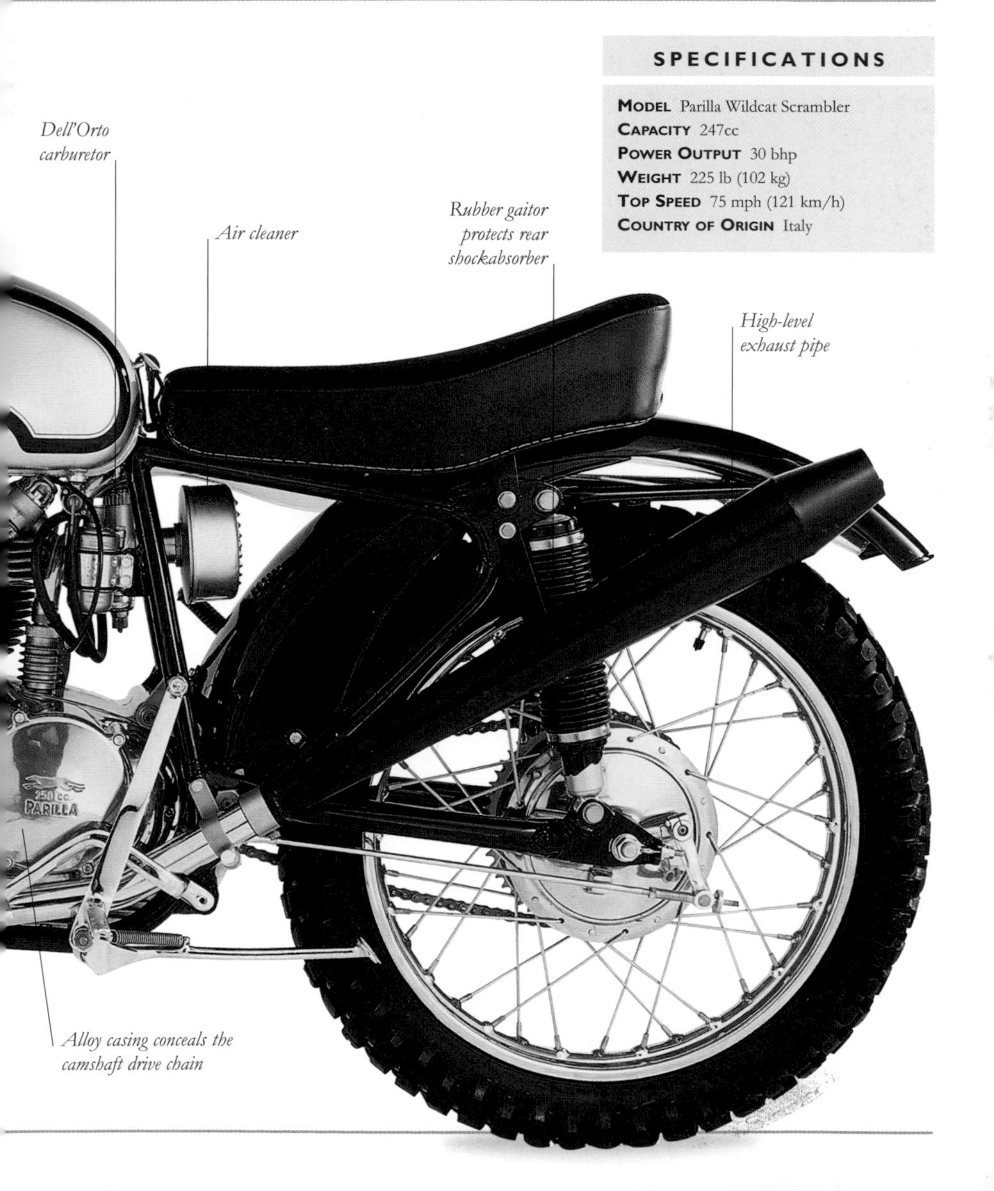

SPECIFICATIONS

MODEL Parilla Wildcat Scrambler
CAPACITY 247cc
POWER OUTPUT 30 bhp
WEIGHT 225 lb (102 kg)
TOP SPEED 75 mph (121 km/h)
COUNTRY OF ORIGIN Italy

REX-ACME *TT Sports*

REX-ACME ATTACHED A VARIETY of proprietary engines to its machines after it stopped making its own power units in 1922. Its most famous models of the 1920s were the Blackburne-engined TT Sports machines. The bike shown here was built in 1926, the year after Wal Handley won the Junior TT on a similar machine. The engine is the classic external flywheel single-cylinder, 2.8 x 3.5-in (71 x 88-mm) o.h.v. Blackburne. The frame is typical of the machines of the period, with the engine cases bolted into position between the front downtube and the seat post. The "saddle" gas tank was about to appear and give motorcycles a more modern appearance.

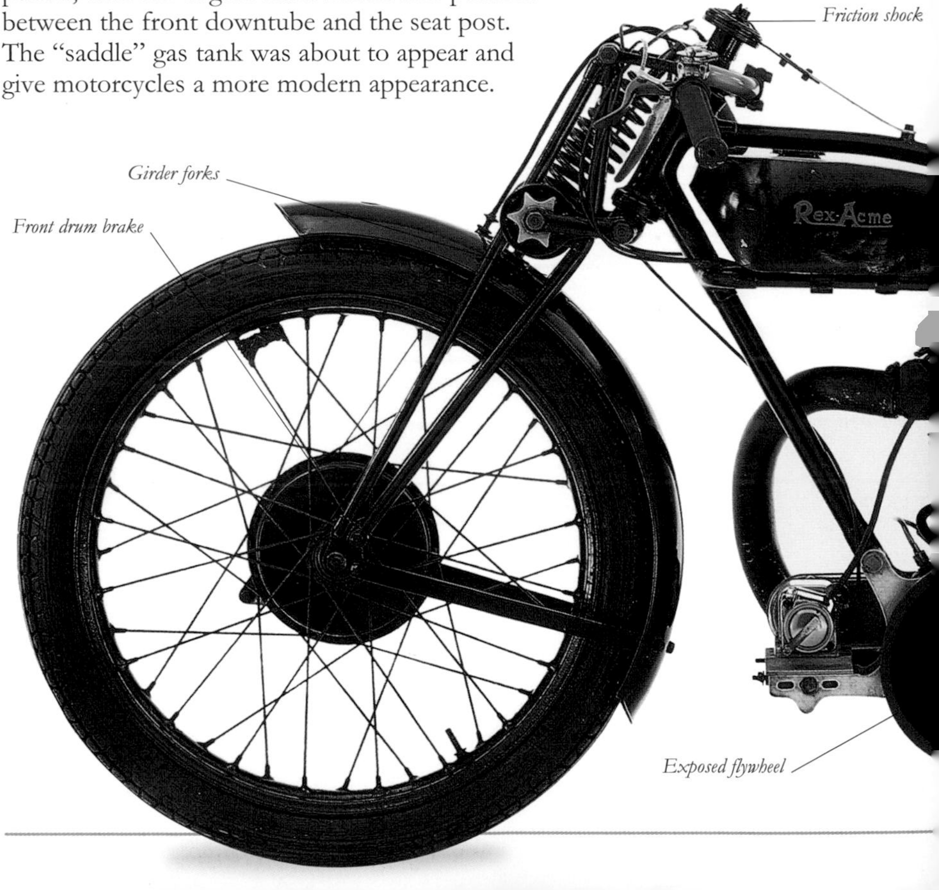

Friction shock

Girder forks

Front drum brake

Exposed flywheel

SPECIFICATIONS

Model Rex-Acme TT Sports
Capacity 348cc
Power Output Not known
Weight 237 lb (107 kg)
Top Speed 85 mph (137 km/h)
Country of Origin UK

Components
The gearbox is a three-speed Burman, and the carburetor was supplied by Amal. Despite the popularity of these machines among amateur competitors, Rex was in decline by the end of the 1920s.

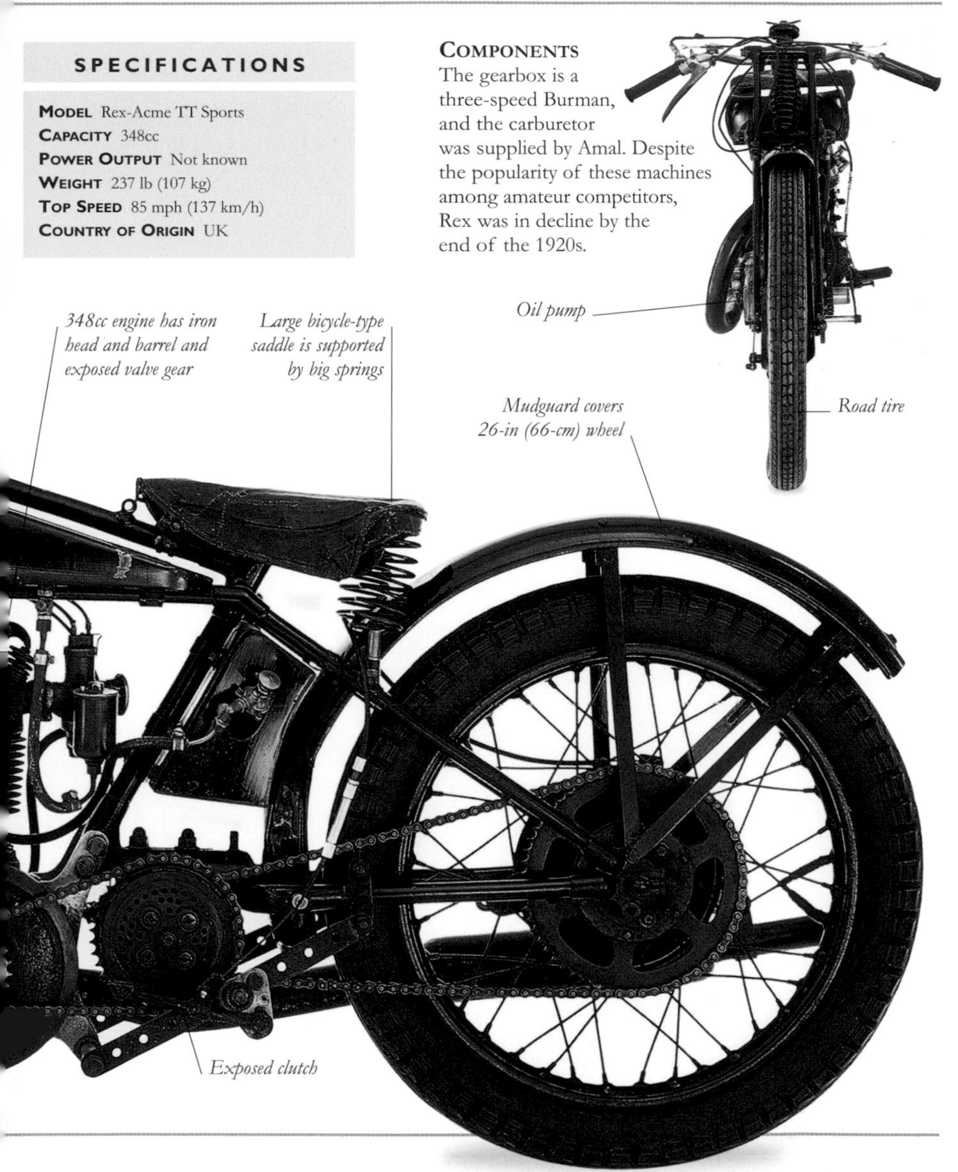

RICKMAN *Métisse Mk3 Scrambler*

THE RICKMAN BROTHERS began building specialty bikes after becoming disillusioned with production scramblers. Their first bikes were an amalgam of Triumph engine in BSA frame. The Métisse name (French for "mongrel") reflects their hybrid form. The Rickmans then began to build their own chassis, available in kit form or complete. This Métisse features a unit-construction Triumph TR6 SS Trophy engine. Equipped with 8.5:1 pistons and T120 camshafts, the engine was in essence a single carburetor version of the Bonneville unit. A single Amal Concentric carburetor was equipped with twin, high-level, open exhaust pipes. It is a replica of a 1965 model.

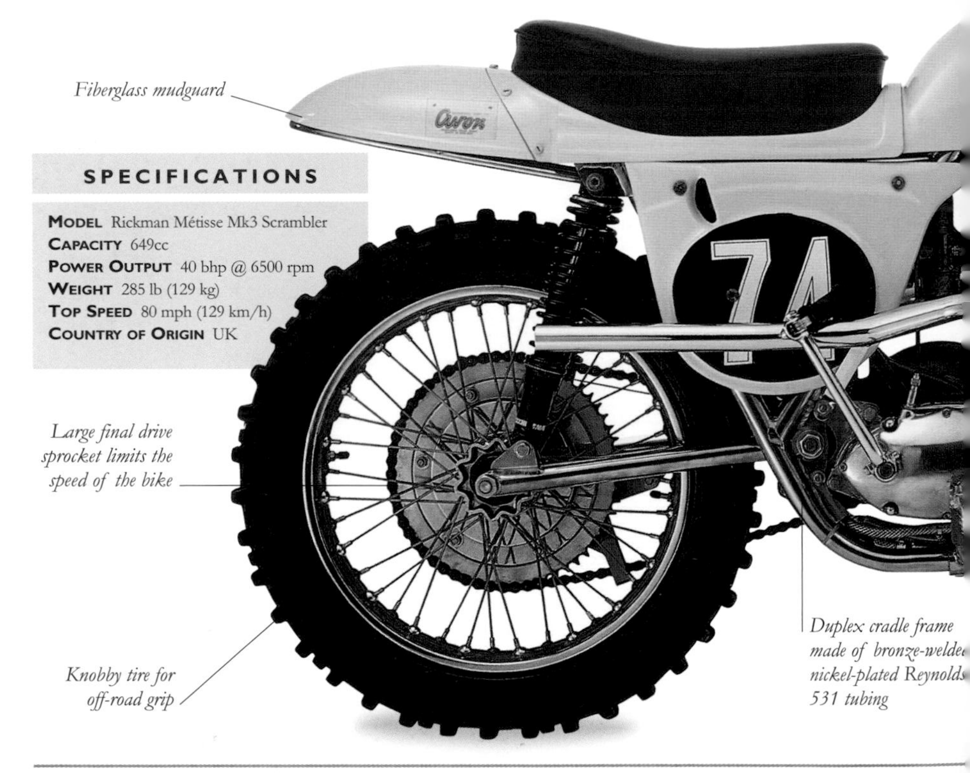

SPECIFICATIONS

MODEL Rickman Métisse Mk3 Scrambler
CAPACITY 649cc
POWER OUTPUT 40 bhp @ 6500 rpm
WEIGHT 285 lb (129 kg)
TOP SPEED 80 mph (129 km/h)
COUNTRY OF ORIGIN UK

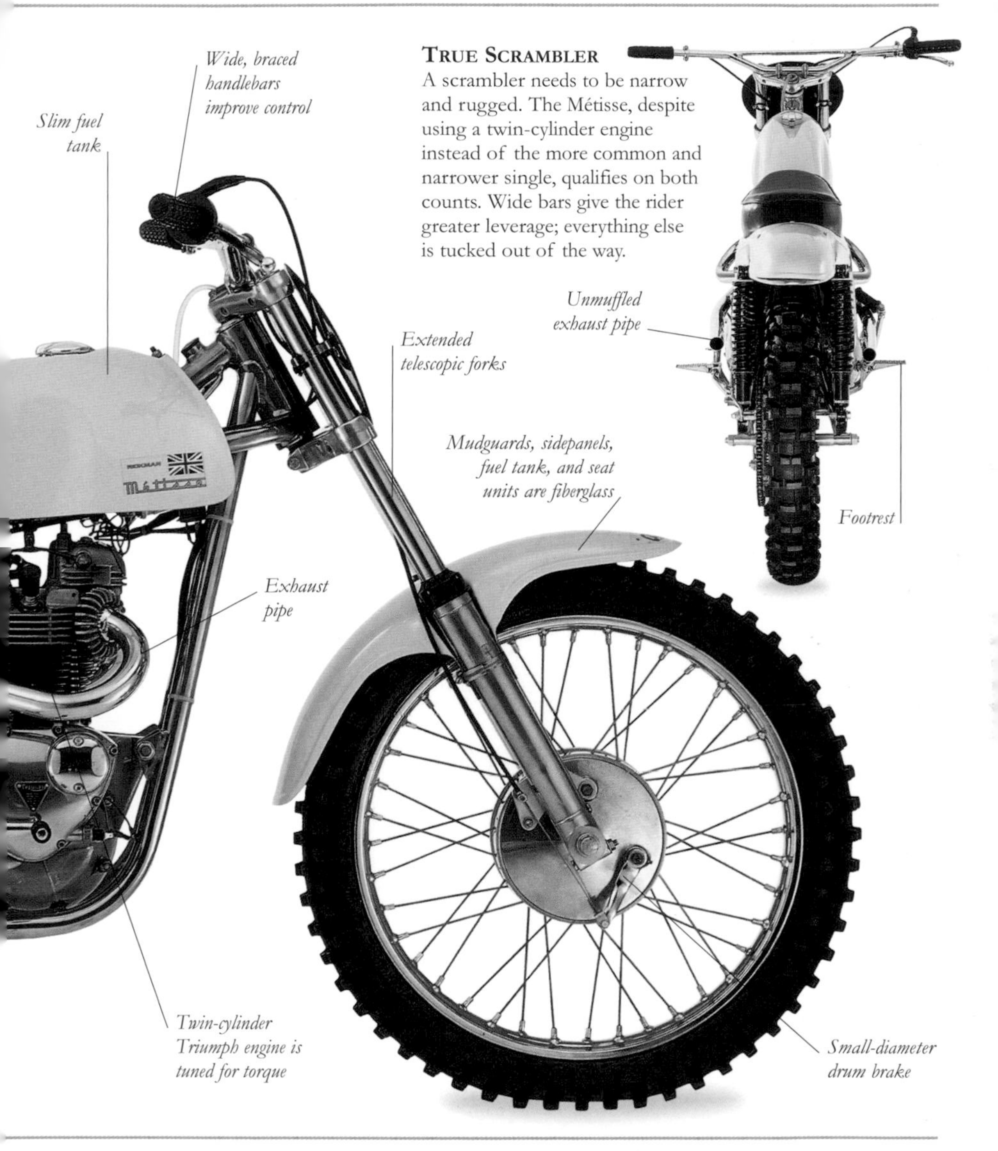

TRUE SCRAMBLER

A scrambler needs to be narrow and rugged. The Métisse, despite using a twin-cylinder engine instead of the more common and narrower single, qualifies on both counts. Wide bars give the rider greater leverage; everything else is tucked out of the way.

RICKMAN *Enfield Métisse*

WHEN THE ROYAL ENFIELD COMPANY went out of business in 1970, the company receiver was left with a batch of 200 736cc parallel-twin cylinder engines intended for Enfield's Mark 2 Interceptors. Unfortunately, there were no frames to put them in. So the final batch of Enfield engined machines was built at the Rickman brothers factory in Hampshire, England, using the Interceptor engines in a special Rickman frame. The first of these bikes appeared in 1970 although the bike shown here wasn't registered until four years later. The Rickman Enfield was a high-quality product with a nickel-plated frame, Rickman's own beefy forks, and disc brakes front and rear. Later in the Seventies, Rickman made good use of Kawasaki and Honda engines.

Footrests mount onto the exhaust pipes

Specifications

Model Rickman Enfield Métisse
Capacity 736cc
Power Output 52 bhp @ 6000 rpm
Weight 353 lb (160 kg)
Top Speed 110 mph (177 km/h)
Country of Origin UK

Light Bike
The slim Enfield-engined Rickman weighs just 353 lb (160 kg), 95 lb (43 kg) less than the Interceptor Mk2 for which the engine was intended. The lighter machine makes it a brisk performer.

ROYAL ENFIELD *V-Twin*

ROYAL ENFIELD WAS FOUNDED in the late 1890s and started making its first V-twins in 1910. Initially Enfield used MAG or JAP engines, but from 1914 onward, when this bike was made, Enfield used its own 425cc V-twin engine. The V-twin had a mechanical oil pump that supplied accurately pressurized oil to the engine. It was a great improvement over other bikes of the period, which were equipped with a hand pump that the rider had to remember to prime constantly. Valves on the bike were inlet-over-exhaust, and oil was pressure-fed to the big ends. It was equipped with a two-speed, all-chain transmission. Royal Enfield continued to produce V-twins until 1939.

SPECIFICATIONS

MODEL Royal Enfield V-Twin
CAPACITY 425cc
POWER OUTPUT 14 bhp
WEIGHT 312 lb (141 kg)
TOP SPEED 55 mph (88 km/h)
COUNTRY OF ORIGIN UK

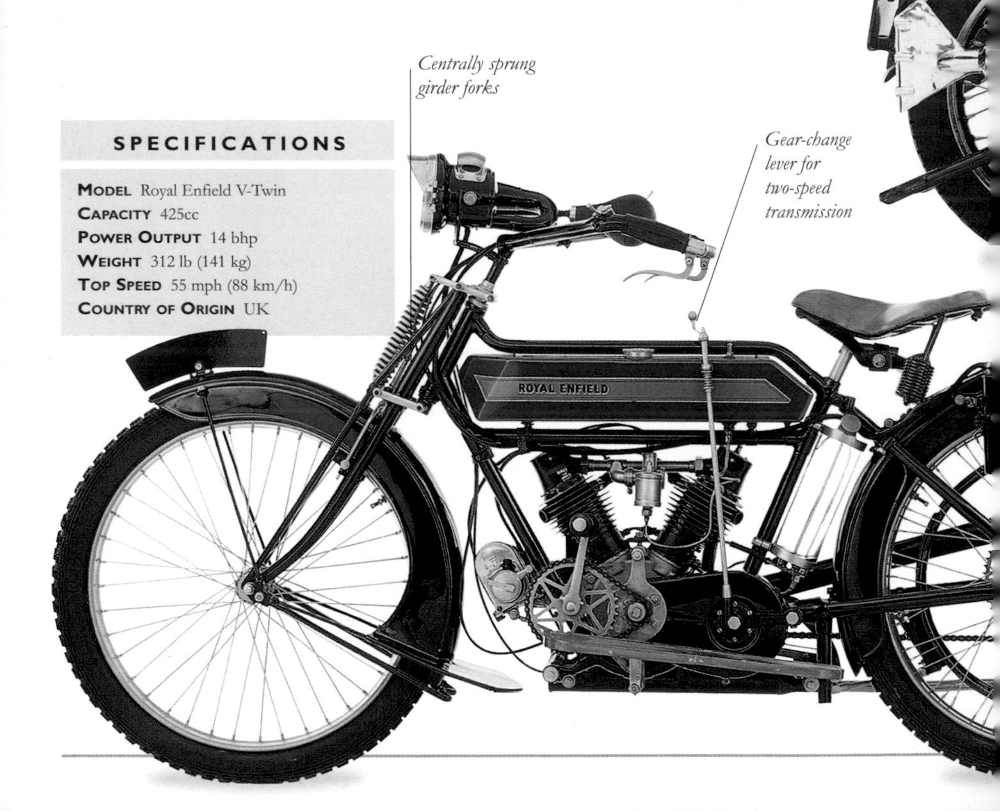

SPECIFICATIONS

Model Royal Enfield JF
Capacity 499cc
Power Output 19 bhp @ 5000 rpm
Weight 364 lb (165 kg)
Top Speed 80 mph (129 km/h)
Country of Origin UK

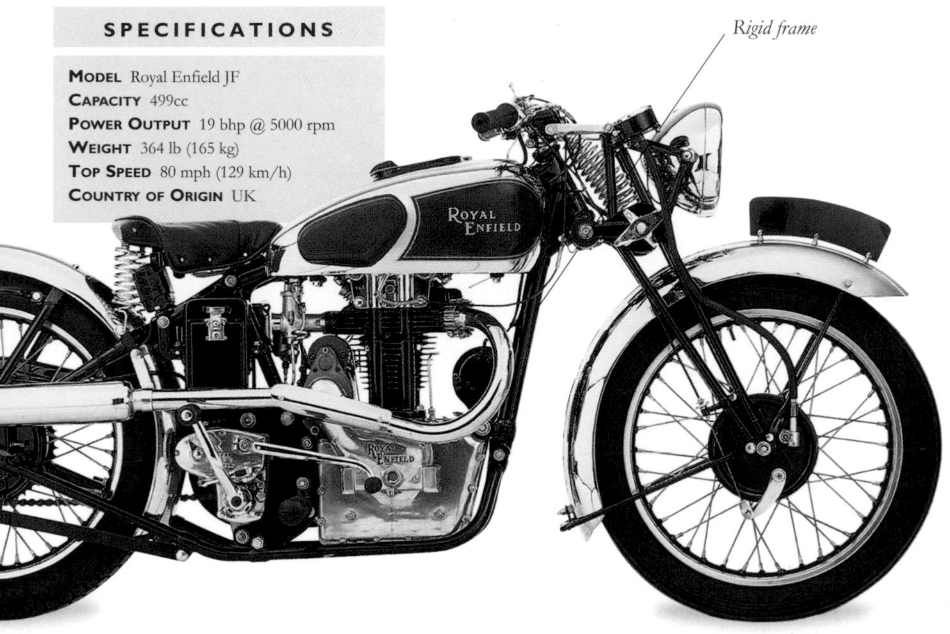

Royal Enfield *JF*

The JF was a four-valve single-cylinder machine. The idea of multiple-valve cylinders is commonly held to be a recent innovation in engine technology. The 1936 Royal Enfield on this page is proof that this is not the case, and in fact many brands deployed multivalve engines before World War II. The JF received unqualified praise on its release, not least for its extremely smooth and frugal engine that was able to push out 19 bhp at 5000 rpm. Unfortunately the four-valve model was short-lived: the complex cylinder head was the victim of cost-cutting, and was superseded by a two-valve version without the pace of its predecessor.

ROYAL ENFIELD *Trials Bullet*

PRODUCTION OF A NEW POSTWAR Bullet started in 1949. UK production ended in 1963, but the Bullet is still built in India. The 346cc, all-alloy, o.h.v. single-cylinder engine used an integral oil reservoir similar to the prewar machines, but it was moved behind the engine. A modern frame with swingarm rear suspension was used across the line. The success of the Trials Bullet was convincing publicity for the new suspension system. In addition to the trials machine, the Bullet was produced in scrambler and road versions, with different engine tuning, suspension, mudguards, and other components. A less successful 499cc version of the bike was built from 1953.

Pushrod inspection cover

SPECIFICATIONS

Model Royal Enfield Trials Bullet
Capacity 346cc
Power Output 17 bhp @ 5500 rpm
Weight 310 lb (141 kg)
Top Speed 65 mph (105 km/h)
Country of Origin UK

Trials Wins

Of all the versions of the Bullet, the trials machines were especially successful, winning an impressive series of competitions. The bike shown on these pages is a 1950 model.

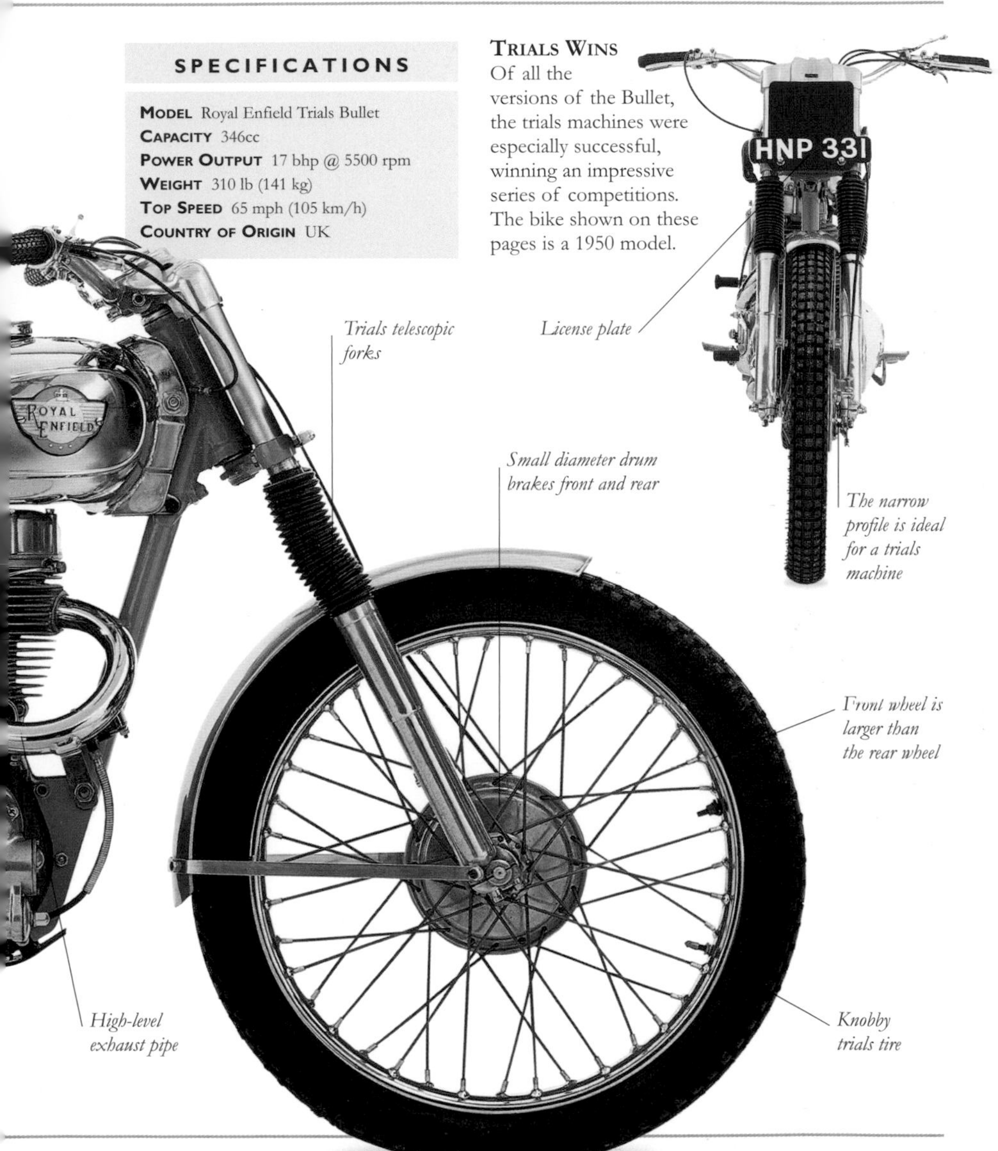

Royal Enfield *500*

Following the success of the Triumph Speed Twin *(see pp.408–09)*, other British manufacturers would have been quick to copy the Triumph if World War II hadn't interrupted development. Royal Enfield introduced its own 500 twin in 1948. It was a clean and elegant design in which the oil tank was incorporated into the engine to which the gearbox was bolted. The rear suspension setup on the 500 was copied from the Trials Bullet *(see pp.360–61)* and was superior to most of the systems then on the market. The 500 was followed in 1953 by the 693cc Metor twin and a 750 version was later manufactured. Enfield twins were also sold in the US under the Indian name *(see pp.242–43)*. The bike on this page is a 1951 model.

Specifications

Model Royal Enfield 500
Capacity 495cc
Power Output 25 bhp @ 5500 rpm
Weight 390 lb (177 kg)
Top Speed 78 mph (125 km/h)
Country of Origin UK

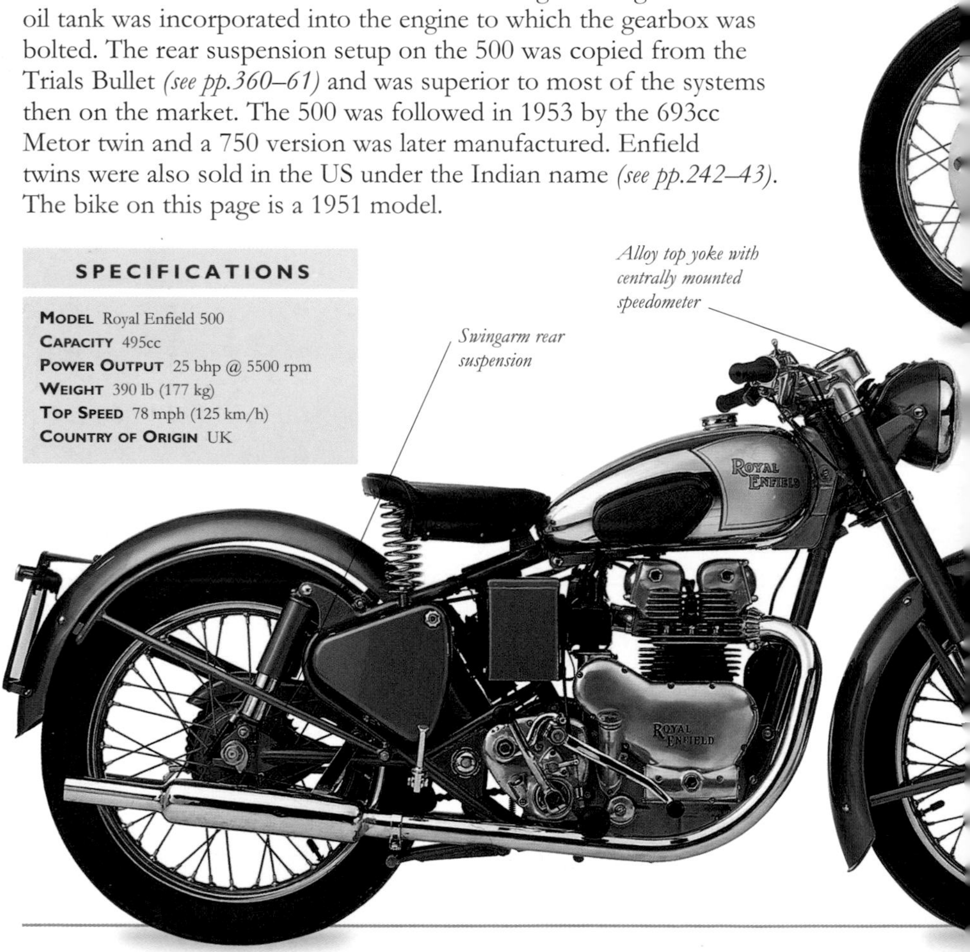
Alloy top yoke with centrally mounted speedometer
Swingarm rear suspension
ROYAL ENFIELD
ROYAL ENFIELD

Front drum brake with cooling discs

Large-diameter crankcase breather pipe

SPECIFICATIONS

MODEL Royal Enfield Continental GT
CAPACITY 248cc
POWER OUTPUT 26 bhp @ 7500 rpm
WEIGHT 300 lb (136 kg)
TOP SPEED 86 mph (138 km/h)
COUNTRY OF ORIGIN UK

Royal Enfield *Continental GT*

The Continental GT gained great popularity in the 1960s for its good looks and "café racer" image. Enfield began producing the practical, unit-construction Crusader 250 in 1956. This evolved into the Crusader Sports, Continental, and finally the Continental GT in 1964. The clip-on handlebars, racing fuel tank, and loud pipe were successful in attracting young buyers. However, the uncomfortable riding position and unreliability of its five-speed gearbox meant that its appeal was short-lived. Production stopped in 1967 when Enfield collapsed and Norton-Villiers took over the company.

RUDGE *TT Replica*

AFTER NEARLY 16 YEARS OF successful racing, the Rudge-Whitworth factory began production of the TT Replica in 1931. The bike had an open cradle frame with the engine as a stressed member and a four-speed gearbox with chain drive. The four-valve cylinder head had radial exhaust valves and parallel inlet ports. Alongside this 1933 machine, there were 348cc and, later, 248cc engine options, contributing to widespread use by British and Continental racers. A further advantage was the comparatively low cost of these machines for the private buyer. Under the guidance of George Hack, the racing team of Graham Walker, H.G. Tyrell Smith, and Ernie Nott won many Grand Prix, TT, and other races before the factory withdrew from official racing in the early 1930s.

SPECIFICATIONS

MODEL Rudge TT Replica
CAPACITY 499cc
POWER OUTPUT 32 bhp (est.)
WEIGHT 290 lb (131 kg)
TOP SPEED 100 mph (161 km/h)
COUNTRY OF ORIGIN UK

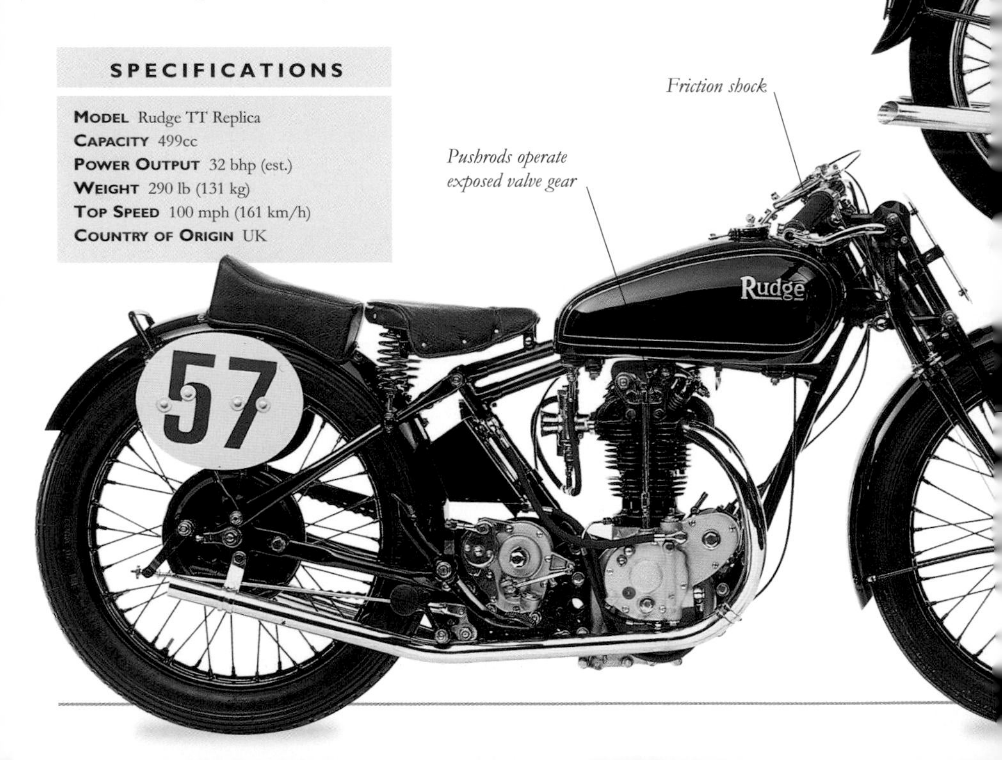

SPECIFICATIONS

MODEL Rudge Ulster
CAPACITY 499cc
POWER OUTPUT 35 bhp
WEIGHT 420 lb (191 kg)
TOP SPEED 95 mph (153 km/h)
COUNTRY OF ORIGIN UK

Design of the cylinder head is essentially the same as on the 1933 TT Replica

License plate

RUDGE *Ulster*

ADVERTISED IN 1937 AS "probably the fastest 500cc motorcycle in normal production," the Rudge Ulster offered a degree of comfort and refinement that was unusual for such a high-performance machine. Distinguished from two other 500cc models in the line by its aluminum-bronze-alloy cylinder head, other improvements to the famous Rudge four-valve engine included enclosed valvegear and pumped oil-feed to the top end. Coupled brakes and a hand-operated center stand added to the Ulster's luxury specification. The bike shown on this page is a 1937 model.

RUMI

THE RUMI 125 IS A NEAT DESIGN that features a twin-cylinder two-stroke engine in which the cylinders are horizontal. There is a four-speed gearbox and the simple frame features plunger rear suspension units. Moto Rumi of Bergamo first introduced the model in 1949, and it continued in production until Rumi quit the motorcycle business in 1962. The market for small capacity machines was very important in Italy after World War II. Large and small manufacturers produced many interesting designs and good quality machines. Rumi also used this engine in its famous "Formichino" ("Little Ant") scooter. Rumi are among the most desirable of small machines among collectors of classic motorcycles.

Four-speed gearbox driven by primary gear

SPECIFICATIONS

MODEL Rumi
CAPACITY 125cc
POWER OUTPUT 15 bhp
WEIGHT 187 lb (85 kg)
TOP SPEED 93 mph (150 km/h)
COUNTRY OF ORIGIN Italy

CONFIGURATION
The engine's twin cylinders are horizontal, and the clutch is mounted on the end of the crankshaft. The Rumi's engine unit is suspended from the machine's tubular-frame structure.

SCHÜTTOFF

SCHÜTTOFF ENTERED THE motorcycle market in the mid-1920s with a line of four-stroke singles of 246cc, 348cc, and 496cc with three-speed gearboxes. The model shown here is a 1926 o.h.v. machine with a four-valve cylinder head and a twin-port exhaust layout. It was a unit-construction engine in which the gearbox was housed in the same casings as the engine—an advanced feature for the period. Cycle parts were conventional and most machines appear to have been finished in the distinctive maroon paint job for the fuel and oil tanks. The frame is a tubular construction with the chain-stays and seat-post bolted to the back of the gearbox. Schüttoff later made two-stroke machines using engines supplied by DKW, which was also based in Saxony. The sporting 350 model was raced successfully in the late 1920s. DKW took complete control in the early 1930s.

SPECIFICATIONS

MODEL Schüttoff
CAPACITY 348cc
POWER OUTPUT 2.5 bhp
WEIGHT 232 lb (105 kg)
TOP SPEED Not known
COUNTRY OF ORIGIN Germany

Clutch can be operated by hand or foot controls

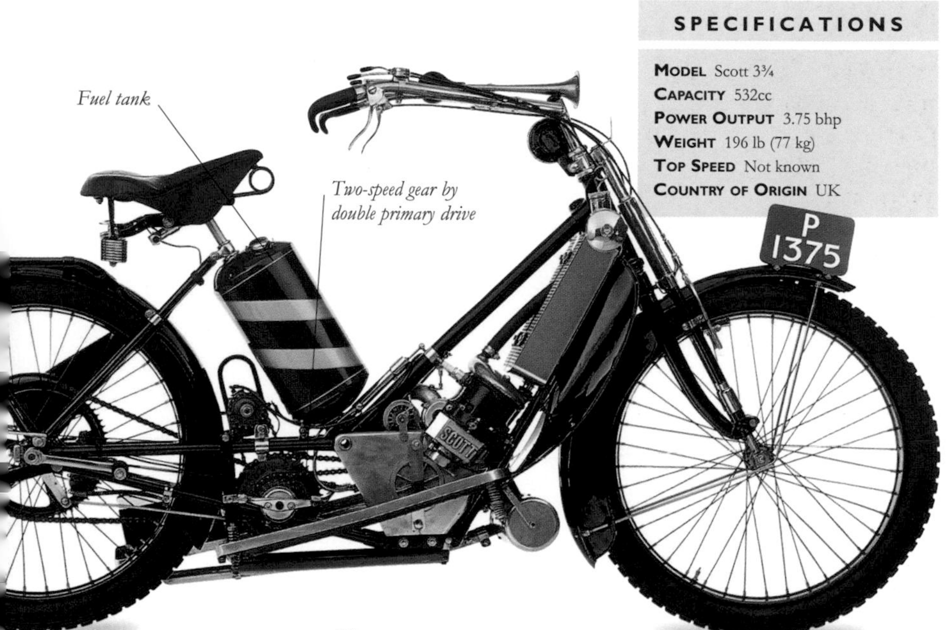

Fuel tank

Two-speed gear by double primary drive

SPECIFICATIONS
Model Scott 3¾
Capacity 532cc
Power Output 3.75 bhp
Weight 196 lb (77 kg)
Top Speed Not known
Country of Origin UK

Scott *3¾*

Alfred Angus Scott, a true innovator, was responsible for nearly 60 separate motorcycle patents. His first production Scott motorcycles appeared in 1908. The advanced design included a two-speed gear, all-chain drive, a triangulated frame, telescopic forks, and a kick starter. The unusual open-frame layout meant that the fuel tank was positioned on the seat-post. The whole machine had an integrated design that was far superior to most of the opposition, which was still little more than powered bicycles. The novel 333cc engine was a water-cooled, two-stroke twin; later models such as this had a 532cc capacity.

SCOTT *Super Squirrel*

ALTHOUGH A.A. SCOTT LEFT the motorcycle company he founded in 1915, subsequent models produced by the Shipley company followed the pattern that he had established. The first Squirrel models appeared in the early 1920s and were sporty 486cc versions of the existing 532cc machines. The Super Squirrel was introduced in 1924 and came equipped with a new 498cc or 596cc engine that incorporated water cooling of the cylinder head as well as the barrel. The Super Squirrel remained in production for seven years and was one of the company's most popular models. The bike on these pages dates from 1929.

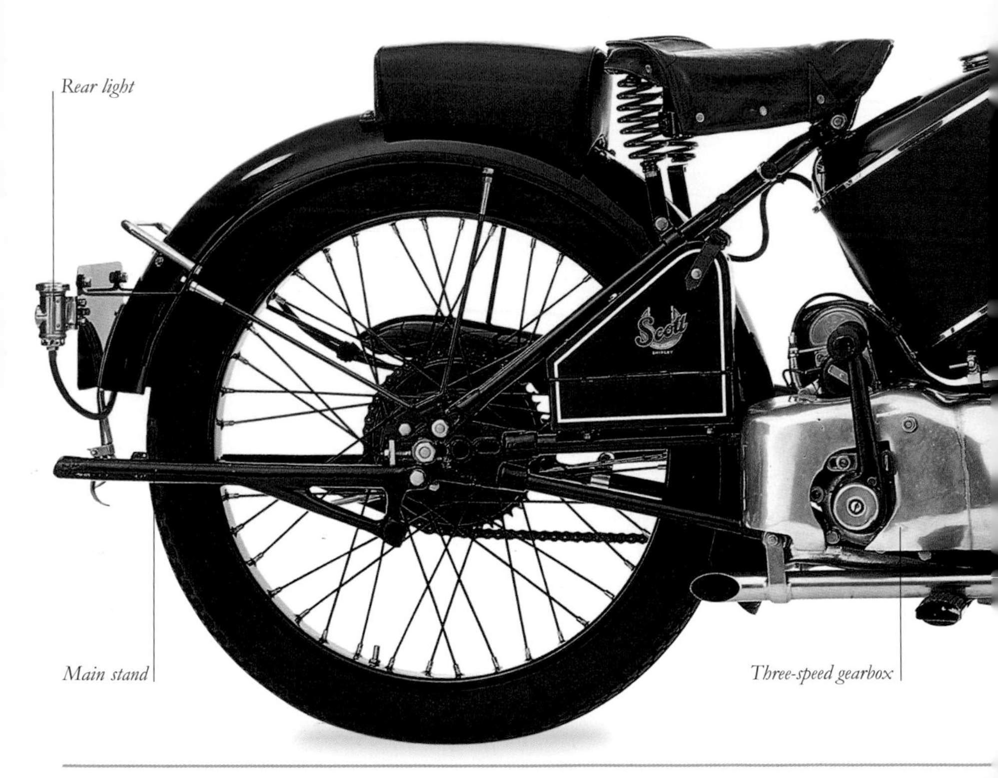

SPECIFICATIONS

Model Scott Super Squirrel
Capacity 498cc
Power Output Not known
Weight 235 lb (107 kg)
Top Speed 65 mph (105 km/h)
Country of Origin UK

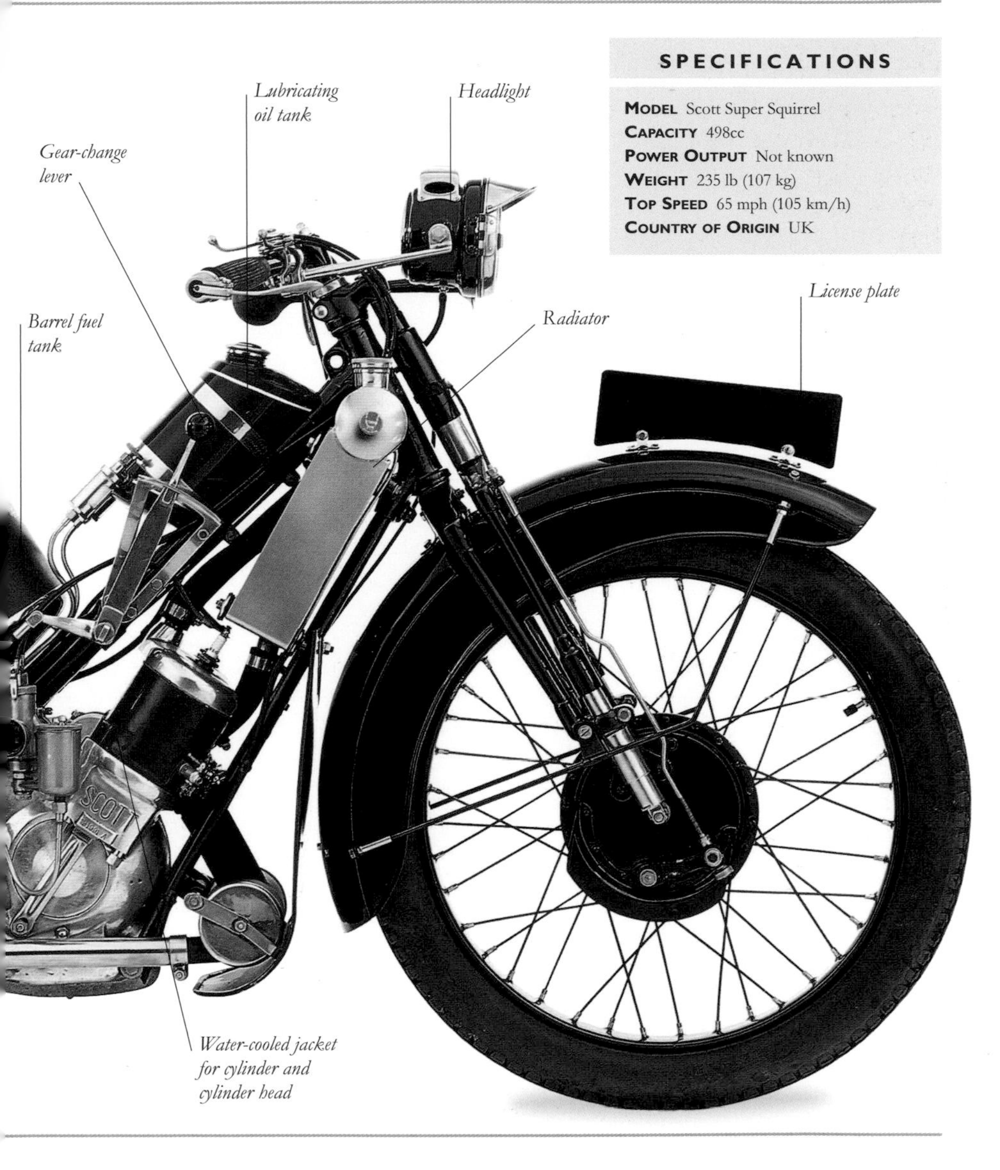

SCOTT *Flying Squirrel TT Replica*

IMPROVEMENTS IN FOUR-STROKE opposition meant that Scott never equaled its pre–World War I successes at the Isle of Man TT, although it did win the manufacturer's prize in 1922. A third place in the 1928 Senior TT was sufficient justification to produce a TT Replica model for 1929 based on the Flying Squirrel. It used the duplex frame that had first appeared in 1927 and incorporated the TT full-frame fuel tank. The company's telescopic forks had additional bracing for extra strength, but, sadly, this Scott innovation was replaced in 1931 by proprietary girders.

Triangulated frame was a Scott trademark

Lubrication system feeds oil directly to the crankshaft

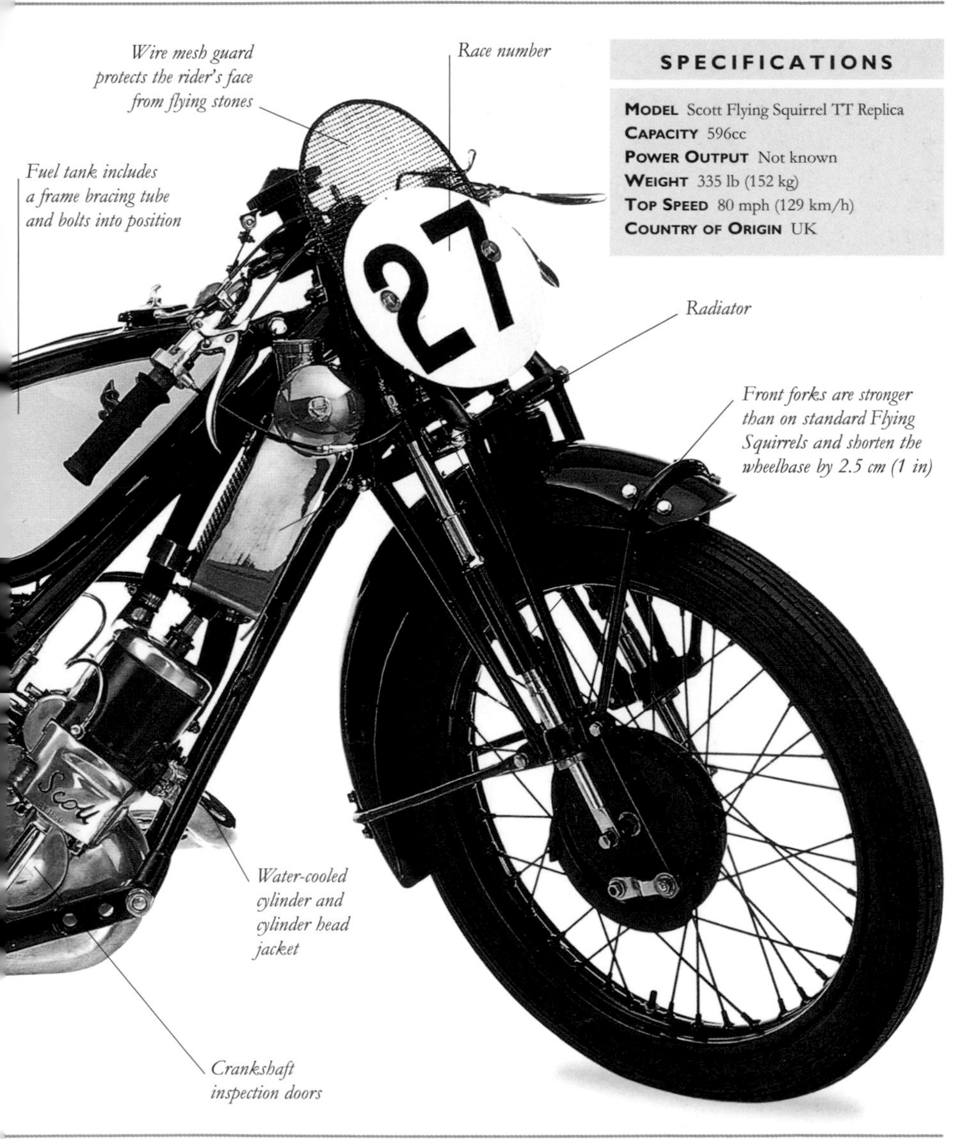

SPECIFICATIONS

MODEL Scott Flying Squirrel TT Replica
CAPACITY 596cc
POWER OUTPUT Not known
WEIGHT 335 lb (152 kg)
TOP SPEED 80 mph (129 km/h)
COUNTRY OF ORIGIN UK

SERTUM *250*

SERTUM BEGAN BUILDING motorcycles in Milan in 1932 and was among the first to resume production after World War II. The model shown here was built in 1947. The side-valve, single-cylinder machine had unit-construction of the engine and gearbox, with battery and coil ignition charged by dynamo. The steel frame uses the engine/gearbox unit as a structural member. Front suspension is by pressed-steel girder forks, and the rear has a swingarm with concealed springs. A 500cc model was also produced at this time. Despite the obvious quality of construction and advanced features, Sertum production ended in 1951.

SPECIFICATIONS

MODEL Sertum 250
CAPACITY 250cc
POWER OUTPUT 11 bhp (est.)
WEIGHT 280 lb (127 kg) (est.)
TOP SPEED 55 mph (89 km/h) (est.)
COUNTRY OF ORIGIN Italy

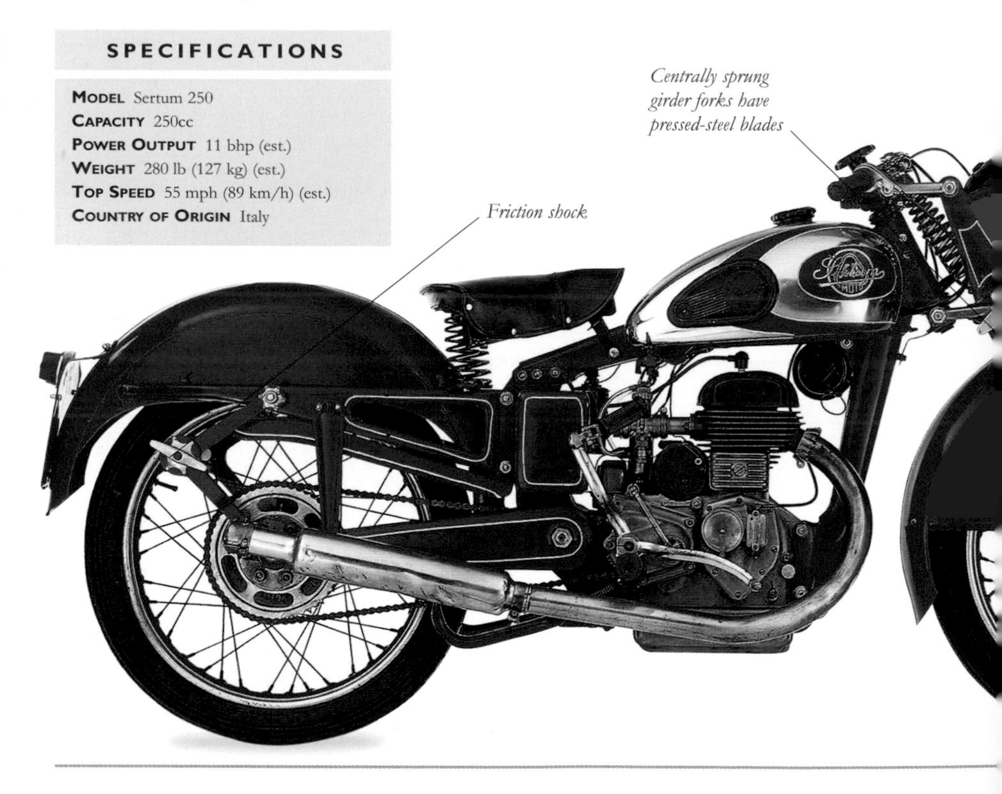

Centrally sprung girder forks have pressed-steel blades

Friction shock

SPECIFICATIONS

MODEL Spiegler
CAPACITY 348cc
POWER OUTPUT 12 bhp @ 3800 rpm
WEIGHT 264 lb (120 kg)
TOP SPEED 56 mph (90 km/h)
COUNTRY OF ORIGIN Germany

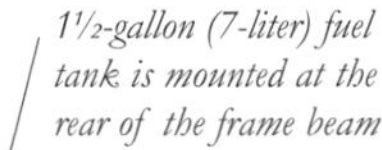

Dummy rim front brake

SPIEGLER

THE ADVANCED Spiegler frame design comprised a main beam running from the steering head to the rear axle. This was constructed of tubing with a steel skin. The engine and gearbox were supported by a twin-loop tubular cradle bolted to the main beam of the frame. The engine's overhead valves could be inspected through a lift-up panel. The lighting kit used on this machine was an accessory added after purchase. Fuel and oil tanks were within the frame structure. The machine pictured here was built in 1924 and uses Spiegler's own single-cylinder engine. Most machines were built using JAP or Motosacoche engines. The bike has conventional girder forks and dummy rim brakes.

Standard *Rex Sport*

Standard had used single-cylinder and V-twin engines supplied by JAP and MAG before the company began making its own engines around 1930. O.h.c. Rex models appeared later. These advanced machines were produced in 348cc and 493cc capacities. The engine had an inclined cylinder, and oil was contained in an integral reservoir. The barrel and head were cast-iron, and there were two exhaust ports. The four-speed Hurth gearbox was bolted to the back of the engine to create a compact power unit. The leading-link forks were made under license from Brough in England, based on a Harley-Davidson design. The overhead camshaft Standard Rex design was a very advanced machine for the 1930s and the machines were successful in competition. Motorcycle production at its factory in Stuttgart did not resume after World War II. The model shown here was built in 1935.

SPECIFICATIONS

Model Standard Rex Sport
Capacity 491cc
Power Output 22 bhp @ 4800 rpm
Weight Not known
Top Speed 81 mph (130 km/h)
Country of Origin Germany

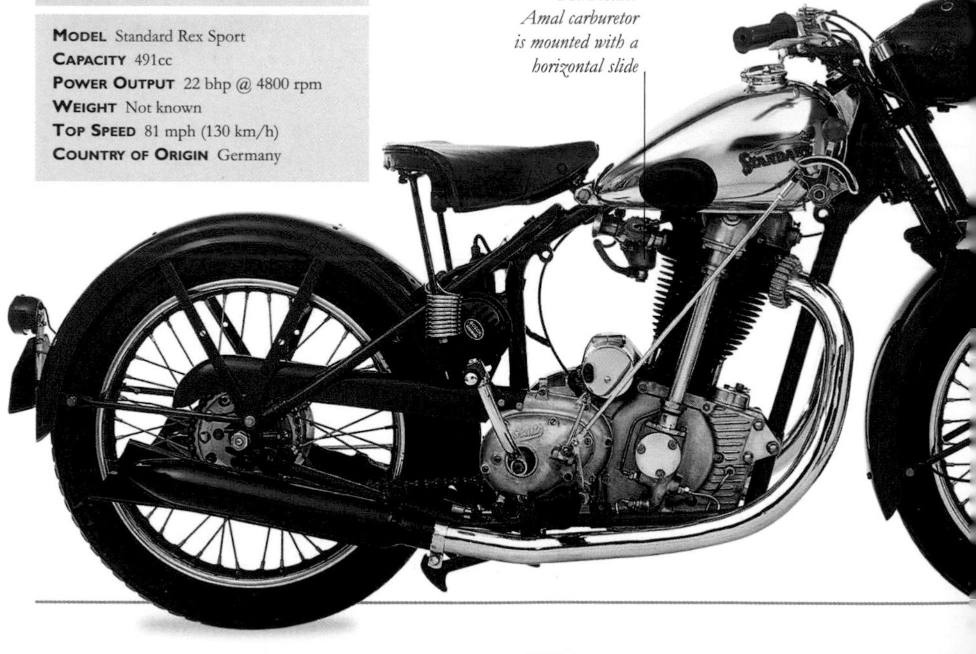

The Fischer Amal carburetor is mounted with a horizontal slide

SPECIFICATIONS

Model Sunbeam Sporting Model
Capacity 499cc
Power Output Not known
Weight Not known
Top Speed Not known
Country of Origin UK

SUNBEAM *Sporting Model*

IN ADDITION TO ROAD machines, Sunbeam also made competition machines based on the road bikes. Extra power was obtained through careful engine tuning and preparation, and weight was lost by removing auxiliary components. These Sunbeams achieved excellent results and enhanced the firm's reputation. George Dance was almost unbeatable in sprint and hill-climb competitions, while other riders won the 1920 and 1922 Senior TTs, the French Grand Prix, the Italian TT, and the Austrian championship. This machine dates from 1923.

SUNBEAM *Model 90*

SUNBEAM INTRODUCED O.H.V. singles during the 1923 racing season, and they appeared in its catalog alongside the very successful side-valve machines for 1924. The 347cc model 80 and the 493cc Model 90 were originally just listed as competition models and were not equipped with kick starter or other road features. By the time the 1928 model pictured here was built, they were available in road trim with kick-starter, optional lighting kit, and other extras to make life easier for the road rider. The twin-port cylinder head was an option on the Model 90 during 1926 and became a standard feature the following year. At this time, the twin-port layout was a popular feature on British single-cylinder machines. With two exhaust pipes splayed out into the passing air, it offered improved cooling as well as better gas flow.

The "little oilbath" final drive chaincase was not mounted on competition bikes

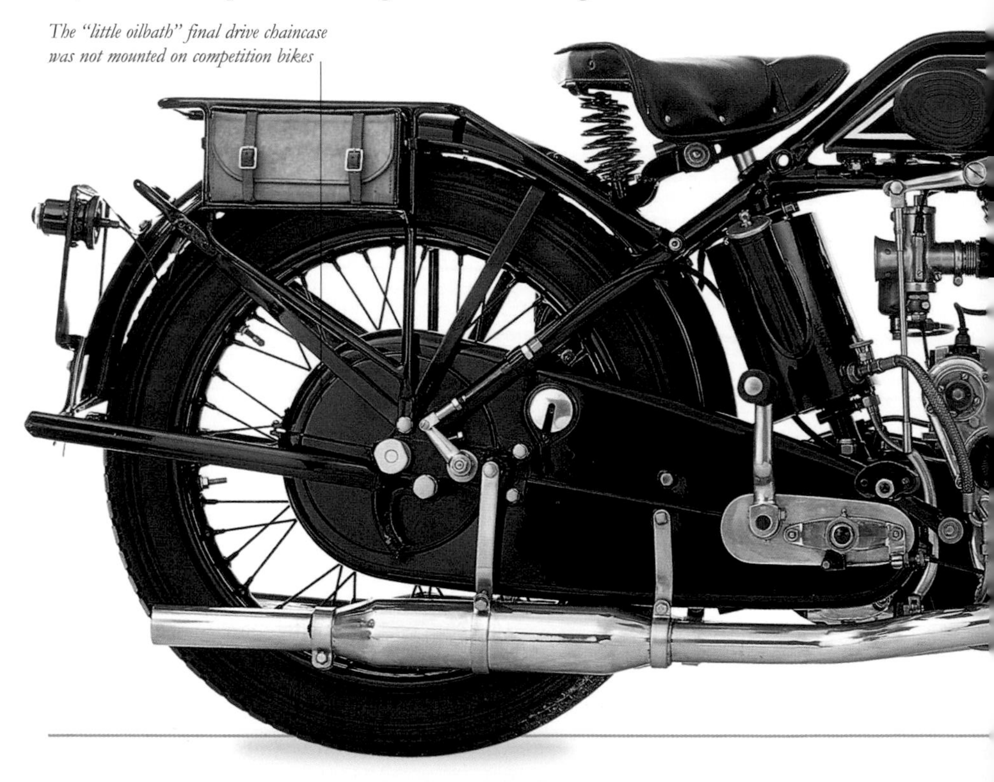

SPECIFICATIONS

MODEL Sunbeam Model 90
CAPACITY 493cc
POWER OUTPUT Not known
WEIGHT 300 lb (136 kg)
TOP SPEED Not known
COUNTRY OF ORIGIN UK

TANK CHANGE
For 1927–28 a "bullnose" flat tank was provided, differing from the earlier style with concave corners at the front. The flat tank was dropped in 1929 when more modern-looking saddle tanks were introduced.

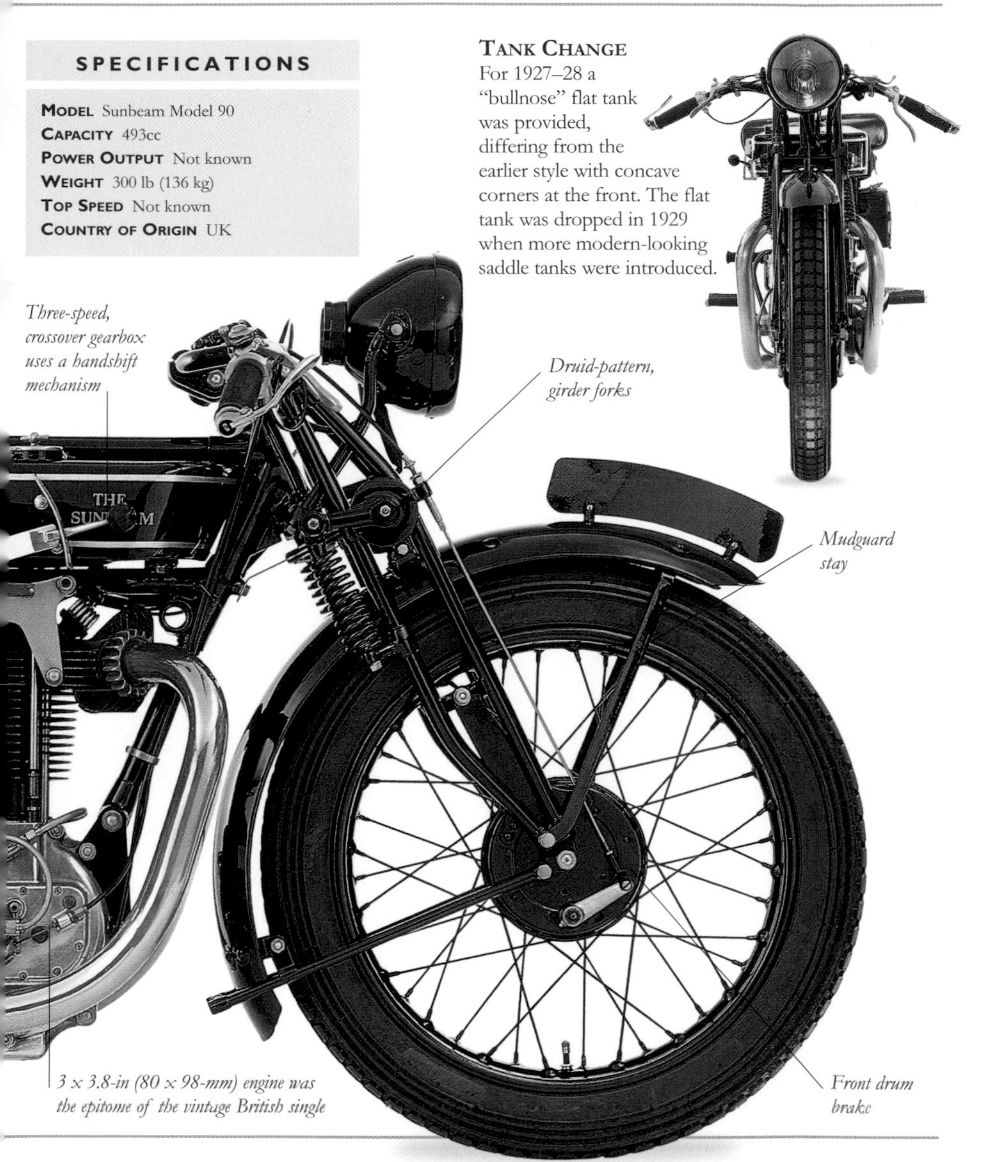

Three-speed, crossover gearbox uses a handshift mechanism

Druid-pattern, girder forks

Mudguard stay

3 x 3.8-in (80 x 98-mm) engine was the epitome of the vintage British single

Front drum brake

SUNBEAM *Model S7 De Luxe*

THE SUNBEAM S7 APPEARED IN 1946 and was a radical departure from conventional British design both in style and technology. The Sunbeam brand name had been bought by BSA during the war years to affix to a new luxury motorcycle. The Sunbeam uses an overhead camshaft, in-line, twin-cylinder engine with a four-speed gearbox and shaft final drive. This was to be the British equivalent of a BMW. Unfortunately, there were teething pains, and it also proved to be underpowered and overweight. The S7 De Luxe was a heavily revised version of the original design introduced in 1949 (this is a 1950 model). A lighter S8 version, with narrow wheels and a BSA front brake, was also produced.

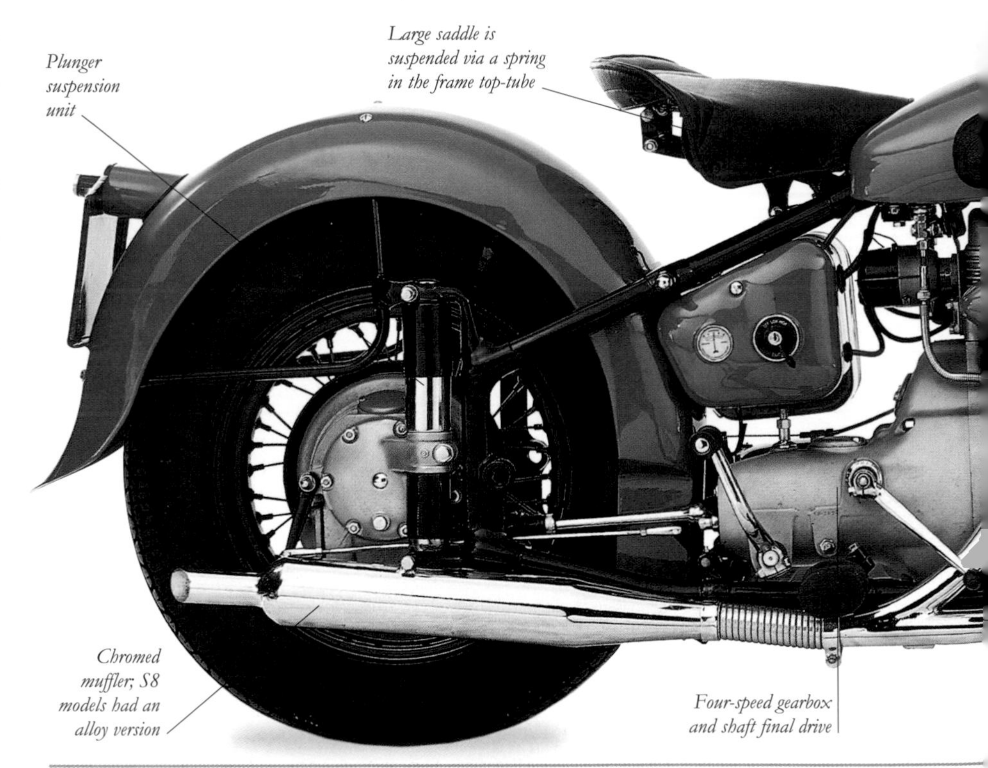

SPECIFICATIONS

MODEL Sunbeam Model S7 De Luxe
CAPACITY 487cc
POWER OUTPUT 25 bhp
WEIGHT 435 lb (197 kg)
TOP SPEED 72 mph (116 km/h)
COUNTRY OF ORIGIN UK

DE LUXE SPEC

The De Luxe had BSA A7 front forks and rear plunger units, although the duplex cradle frame was exclusive to the Sunbeam line, as were the balloon tires and drum brakes.

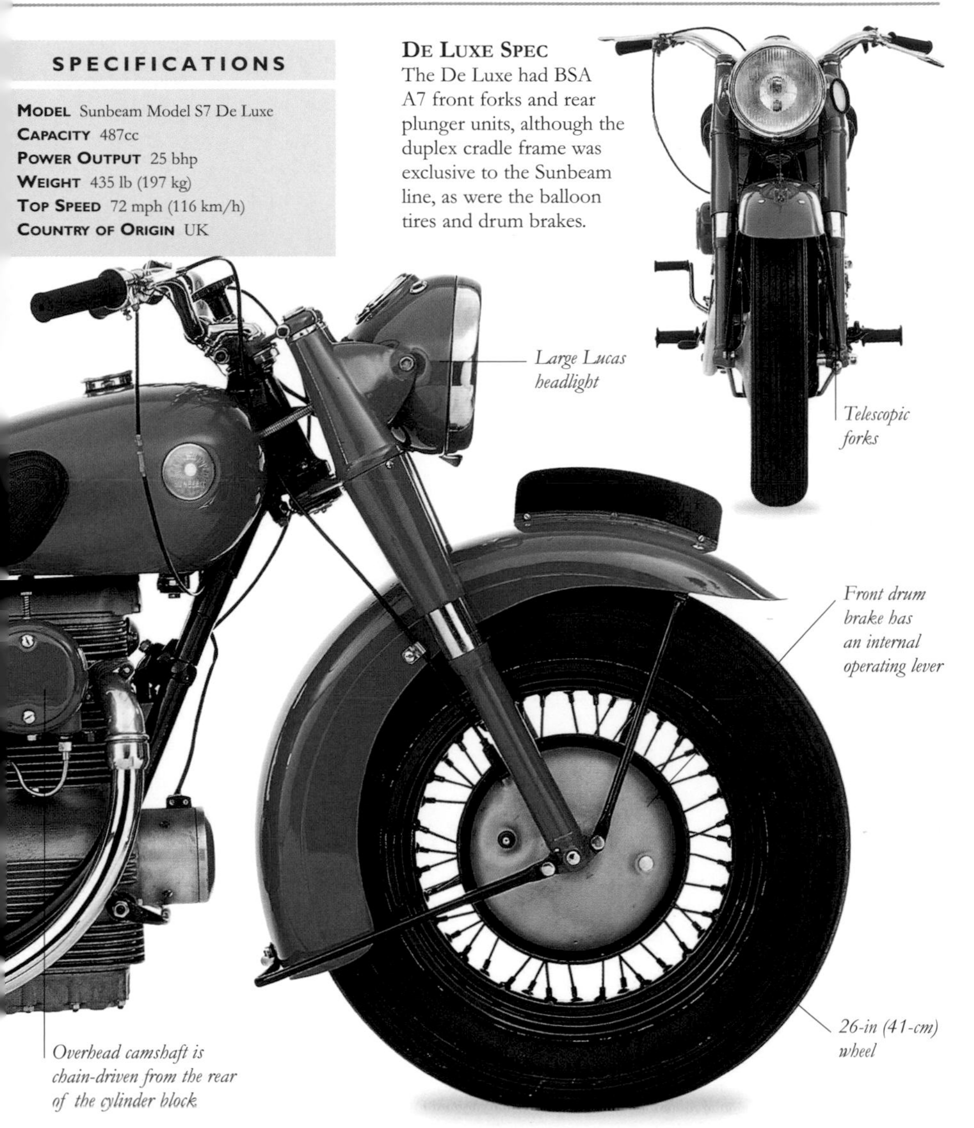

Large Lucas headlight

Telescopic forks

Front drum brake has an internal operating lever

26-in (41-cm) wheel

Overhead camshaft is chain-driven from the rear of the cylinder block

SUZUKI *RT63*

MZ RIDER ERNST DEGNER JOINED Suzuki at the end of 1961, bringing with him the secrets of advanced two-stroke design; even so, the 1962 125 Suzuki single was beaten by the four-stroke Hondas. For 1963, Suzuki built an all-new twin. The RT63 rotary-valve air-cooled twin was originally produced with both rear- and forward-facing exhausts, but was competitive for one season only. In 1964, the four-cylinder Hondas were dominant, and in the 125cc championship Suzuki rider Hugh Anderson could only finish third.

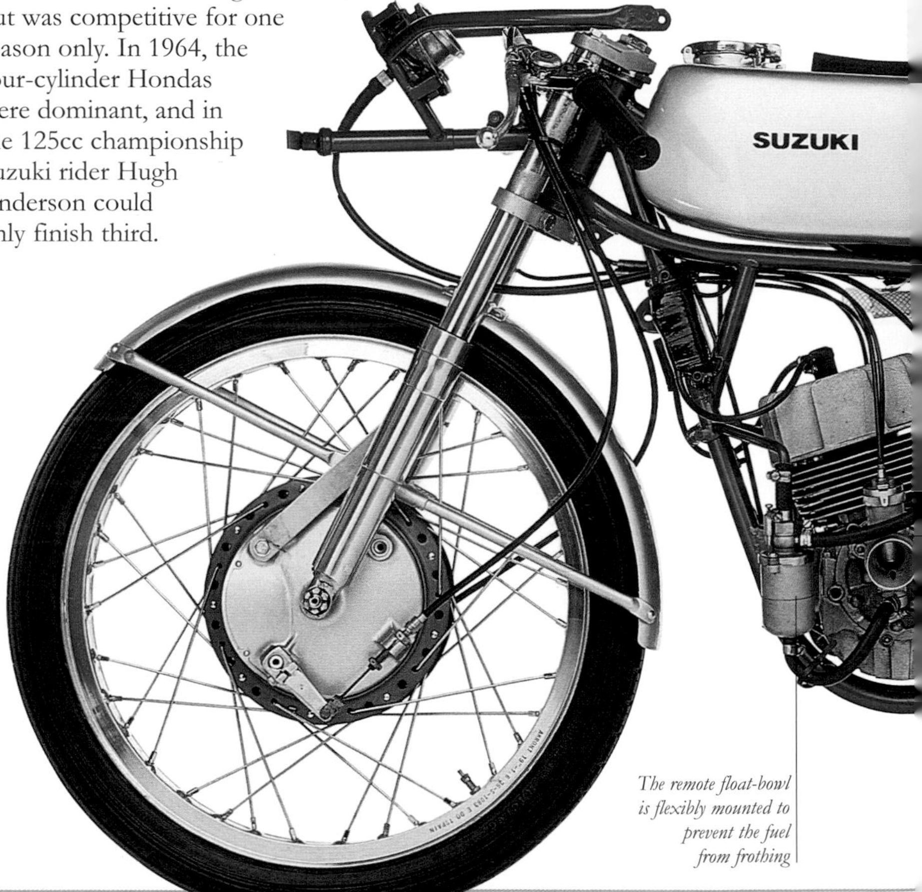

SUZUKI

The remote float-bowl is flexibly mounted to prevent the fuel from frothing

SPECIFICATIONS

MODEL Suzuki RT63
CAPACITY 124cc
POWER OUTPUT 25.5 bhp @ 12,000 rpm
WEIGHT 207 lb (94 kg)
TOP SPEED 114 mph (184 km/h)
COUNTRY OF ORIGIN Japan

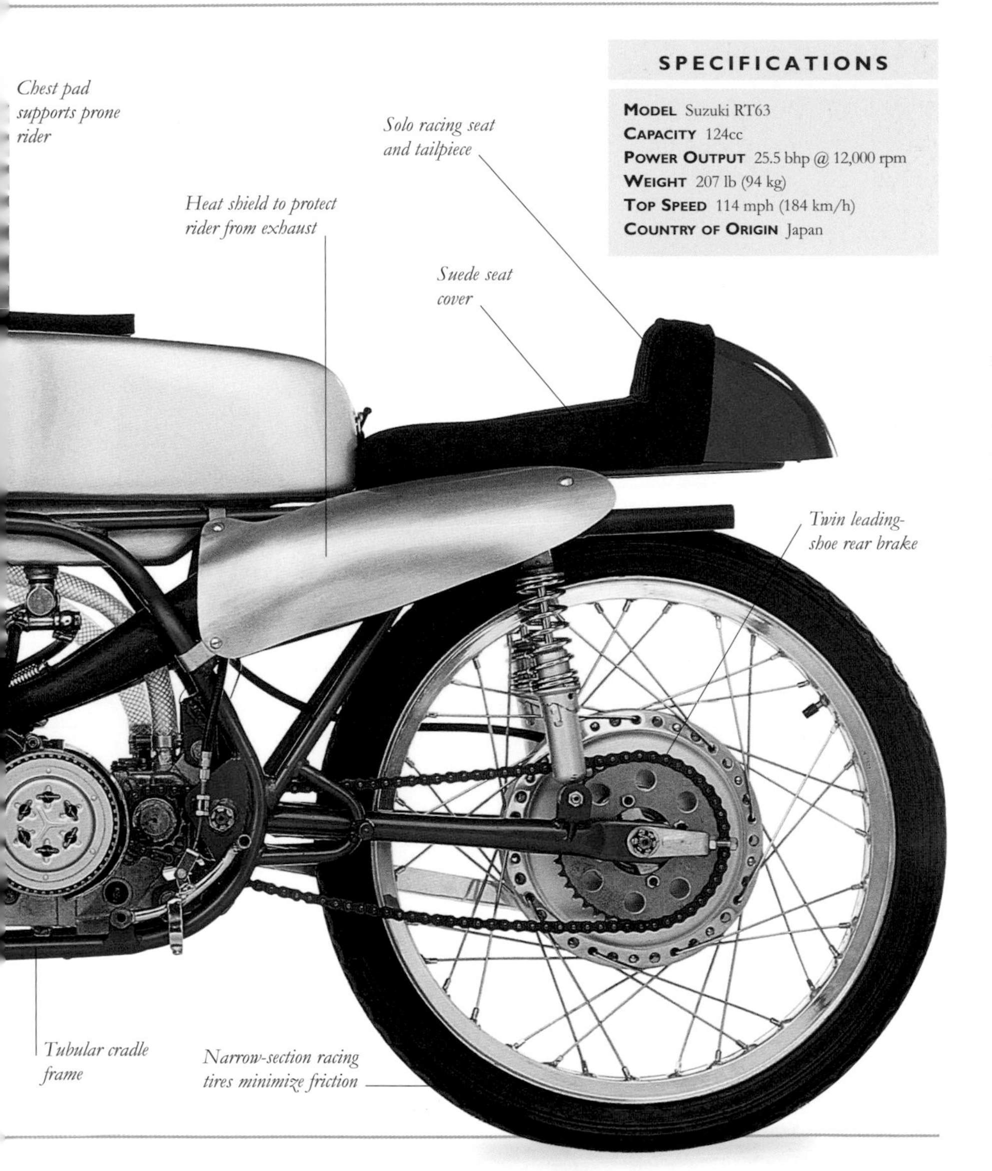

SUZUKI *X6 Hustler*

WHEN SUZUKI INTRODUCED the X6 Hustler in 1965 it caused an immediate stir, especially among lovers of high-performance lightweight machines. Despite its small, 247cc air-cooled, two-stroke engine, the X6, known as the T20 Super Six in Britain, was capable of 90 mph (145 km/h) and offered a sophistication rarely seen then on such small motorcycles. The X6 Hustler remained in production for three years, and at the height of its popularity, Suzuki produced 5,000 machines per month.

Rearview mirror

Externally sprung telescopic forks

Twin leading-shoe drum brake

Unusually for a Suzuki, the kick-starter is on the left

SPECIFICATIONS

MODEL Suzuki X6 Hustler
CAPACITY 247cc
POWER OUTPUT 29 bhp @ 7500 rpm
WEIGHT 316 lb (143 kg)
TOP SPEED 90 mph (145 km/h)
COUNTRY OF ORIGIN Japan

GEARBOX NAME

The Hustler had a six-speed gearbox (hence the "six"), an 8-in (20.3-cm) twin leading-shoe front brake, and a separate oil pump instead of the usual premix, gas/oil system.

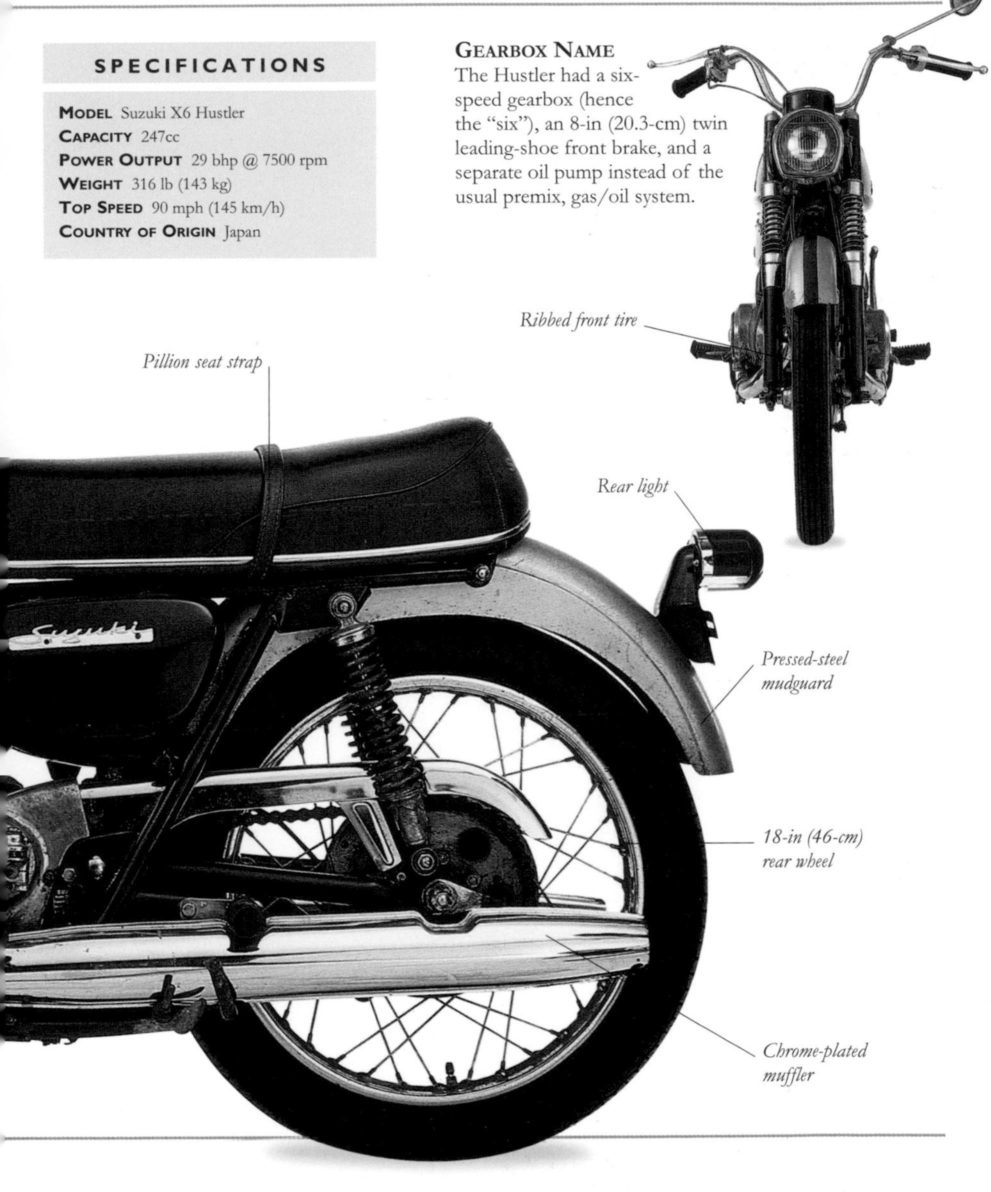

Suzuki *TR500*

Suzuki quit Grand Prix racing at the end of 1967, the same year it launched the T500 Cobra road bike—the biggest bike Suzuki had ever made. The TR500 racing version of the Cobra appeared in 1968 and went on to gain considerable racing success, especially in the US. The bike illustrated was ridden by Art Baumann to the company's first US National Championship win at Sears Point on September 7, 1969. In 1970, larger 1⅓-in (34-mm) carburetors were used and compression was raised to 7.34:1. It wasn't long before Suzuki was back competing in the Grand Prix.

Suspension front and rear is by Ceriani

Alloy wheel rim

Double-sided, Fontana, twin, leading-shoe front brake is vented to aid cooling

SPECIFICATIONS

MODEL Suzuki TR500
CAPACITY 500cc
POWER OUTPUT 64.5 bhp @ 8000 rpm
WEIGHT 298 lb (135 kg)
TOP SPEED 145 mph (233 km/h)
COUNTRY OF ORIGIN Japan

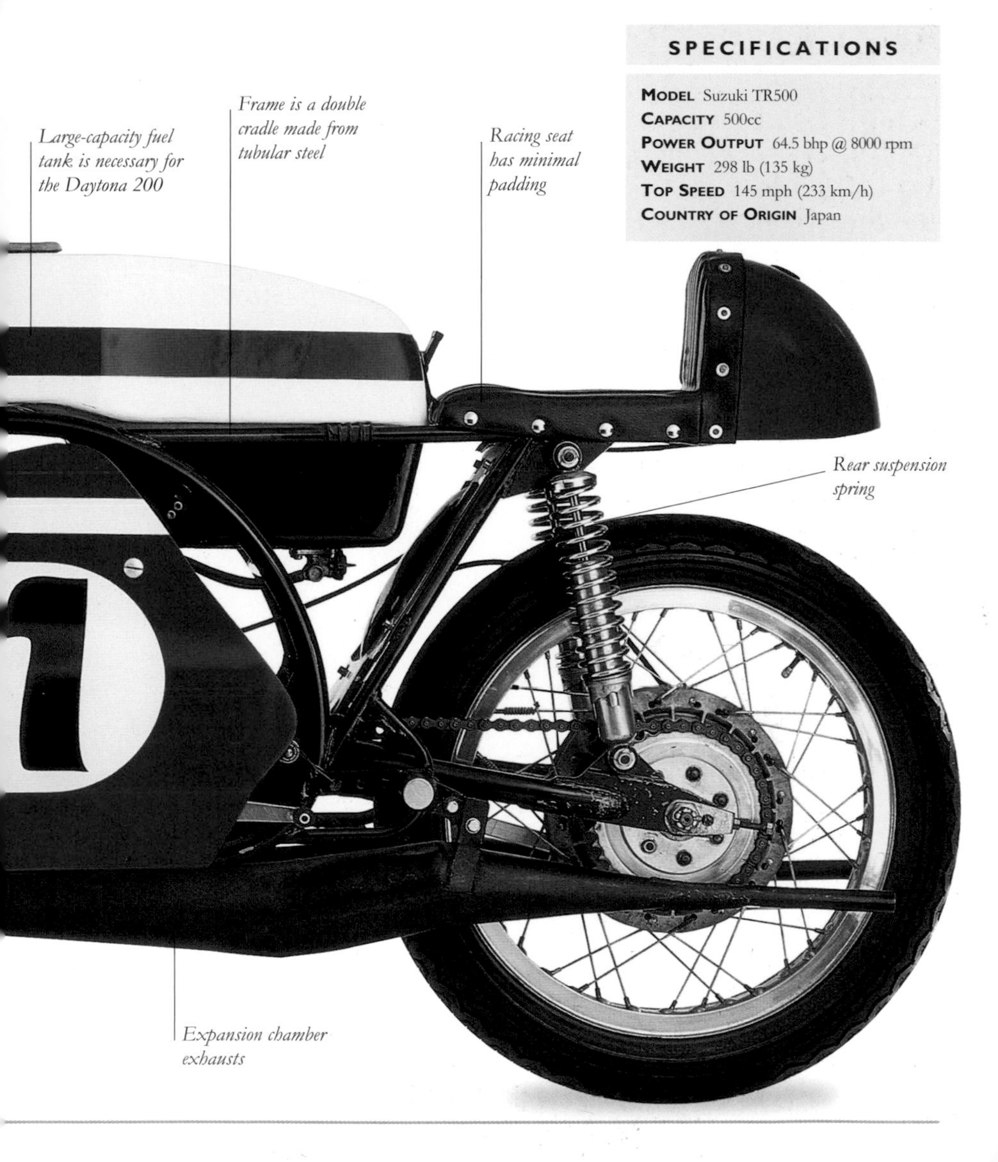

SUZUKI *GT750*

LAUNCHED IN 1971, Suzuki's GT750 made an immediate impact, thanks to its radical design and rapid acceleration. Building on its success in the 1960s with racing two-strokes, Suzuki decided to create a large-capacity, multicylinder sports bike for road riders. Suzuki's first venture into the superbike market was largely successful; the GT stayed in production for five years. The three cylinders were arranged in-line across the frame, which could cause the middle cylinder to overheat, being flanked by a cylinder on each side and shielded from the cooling breeze by the frame downtubes and front forks, hence the use of water cooling.

The four leading-shoe drum brakes were replaced in 1974 by twin discs

The water-cooling jacket means no need for cooling fins

ADVANCED COOLING

Relatively sophisticated for its time, the cooling system had a four-stage operation in which a thermostat blocked the flow of coolant when the engine was started so that the optimum operating temperature could be reached quickly.

SPECIFICATIONS

MODEL Suzuki GT750
CAPACITY 738cc
POWER OUTPUT 67 bhp @ 6500 rpm
WEIGHT 524 lb (238 kg)
TOP SPEED 108 mph (174 km/h)
COUNTRY OF ORIGIN Japan

Hard-ridden, the bike could only put out 25 mpg (8.86 km/l)

The bright, candy-pink paint was typical of the period, but later models were more subdued

Instruments include a centrally mounted temperature gauge

Passenger handrail

WATER COOLED 750

Three cylinders exit into four exhaust pipes

SUZUKI *Katana*

BY THE EARLY 1980s Suzuki had developed an extremely powerful and attractively compact four-cylinder engine, prefixed by the letters GSX. The company now needed a new design to match the modern performance of this d.o.h.c., 16-valve engine. It turned to former BMW car designer, Jan Fellstrom, and the Katana was the result. Characterized by its integral fuel tank and seat, half-fairing, and two-tone seat, the Katana was an immediate hit with riders wanting reliability and high performance. The 1982 model is shown here.

The 16-valve, four-cylinder engine was developed from an eight-valve

Anti-dive mechanism reduces fork movement when braking

TSCC logo advertises Suzuki's twin swirl combustion chamber

19-in (48-cm) front wheel

SPECIFICATIONS

Model Suzuki Katana
Capacity 997cc
Power Output 108 bhp @ 8500 rpm
Weight 554 lb (251 kg)
Top Speed 140 mph (225 km/h)
Country of Origin Japan

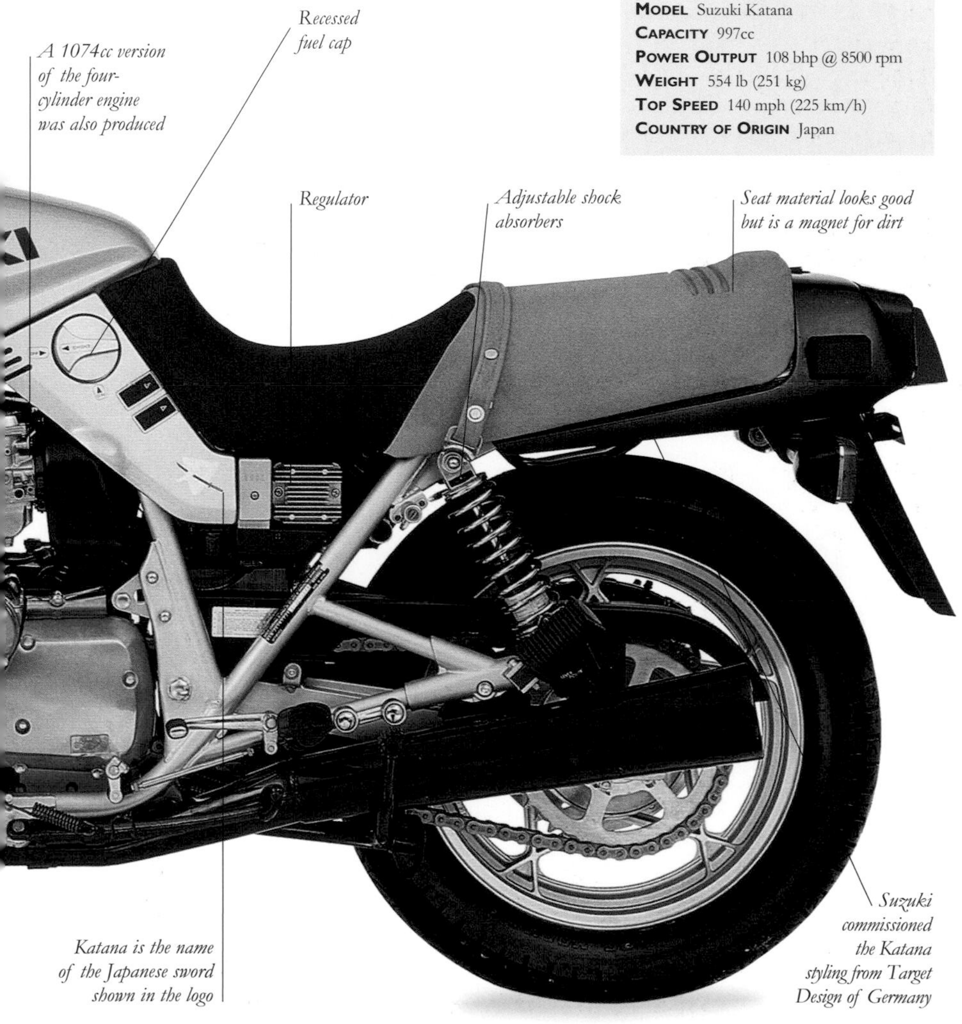

Recessed fuel cap

A 1074cc version of the four-cylinder engine was also produced

Regulator

Adjustable shock absorbers

Seat material looks good but is a magnet for dirt

Katana is the name of the Japanese sword shown in the logo

Suzuki commissioned the Katana styling from Target Design of Germany

Suzuki *RGV500*

In the competitive world of Grand Prix racing, manufacturers strive to give their products the edge over the opposition. With 170 bhp available on modern Grand Prix machines such as this Suzuki RGV500, lack of power is not a problem. Most modifications are intended to harness this level of energy and make the bikes easier to control. The RGV500 had its most successful season in 1993 when American rider Kevin Schwantz won the World Championship. Suzuki had to wait until 2000 for their next World Title, won by another American, Kenny Roberts Jr.

The swingarm is arched to provide clearance for exhausts

The mufflers are made from carbon fiber

Low Bike
The RGV's low frontal area keeps wind resistance down, aiding acceleration and maximizing the top speed. The slick tires help the bike to reach 185 mph (298 km/h).

SPECIFICATIONS

MODEL Suzuki RGV500
CAPACITY 498cc
POWER OUTPUT 170 bhp @ 13,000 rpm
WEIGHT 287 lb (130 kg)
TOP SPEED 185 mph (298 km/h)
COUNTRY OF ORIGIN Japan

Racing windshield

Fuel tank breather

The fairing is made from lightweight, Kevlar-reinforced carbon-fiber

Upside-down telescopic fork

Carbon-fiber brakes are most effective when they are hot; in very cold or wet conditions, covers are used

SUZUKI *GSX-R1100WR*

SUZUKI'S GSX-R1100, launched in 1986, heralded the start of the true racer-replica era. Its four-cylinder, four-stroke, d.o.h.c. engine, race-type features, and outstanding power made it an immediate classic. The GSX-R also became a favorite with racers, and there were many track successes. It is still in production and, though tall and heavy by modern standards and no longer the fastest in its field, it is revered by enthusiasts. This bike is a 1994 WR model.

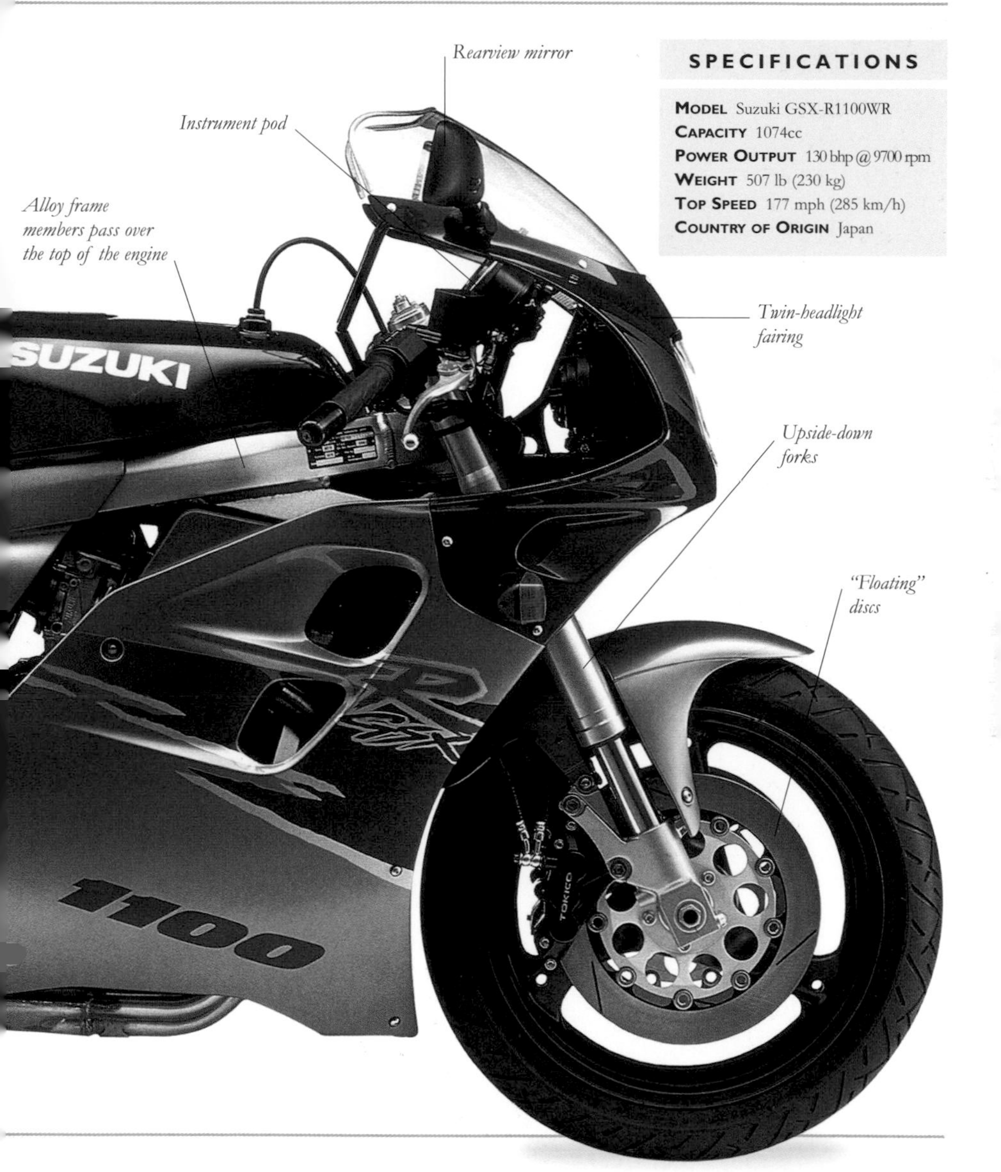

SPECIFICATIONS

MODEL Suzuki GSX-R1100WR
CAPACITY 1074cc
POWER OUTPUT 130 bhp @ 9700 rpm
WEIGHT 507 lb (230 kg)
TOP SPEED 177 mph (285 km/h)
COUNTRY OF ORIGIN Japan

SUZUKI *GSXR1000*

THE THEORY BEHIND THE ultimate sports bike is simple—it needs to have more power, less weight, and better handling and brakes than the opposition. The trouble is that opposition models are already at the optimum level in these areas, so progress tends to be by small degrees only. The new-for-2001 GSXR1000 was then just slightly more powerful, slightly lighter, and handled slightly better than the previous class leader, the Yamaha R1 *(see pp.462–63)*. Based on the previous year's 750 model, but with a bigger capacity engine and other detail changes, it moved to the top of the performance pecking order. The slick parts are the super efficient fuel injection and the balancer, which runs at twice engine speed.

SPECIFICATIONS

MODEL Suzuki GSXR1000
CAPACITY 998cc
POWER OUTPUT 139.9 bhp
WEIGHT 375 lb (170 kg)
TOP SPEED 176 mph (283 km/h)
COUNTRY OF ORIGIN Japan

FRONT VIEW
The front of the GSXR1000 looks no different than the smaller engined 600 and 750cc models. Air scoops below the headlight feed the fuel injection system via a large airbox located behind the steering head.

SUZUKI *GSX1300R Hayabusa*

WHEN THE SUZUKI HAYABUSA was introduced in 1999, it was the fastest production motorcycle in the world, with a top speed approaching 200 mph (322 km/h). Japanese manufacturers later agreed to a voluntary 186 mph (299 km/h) speed "restriction," and the 'Busa's performance was capped accordingly. The engineering behind the bike was conventional. It has an across-the-frame four-cylinder engine and a twin-spar alloy beam frame, but the strange looks are a unique consequence of the quest for improved aerodynamics.

SPECIFICATIONS

MODEL Suzuki GSX1300R Hayabusa
CAPACITY 1299cc
POWER OUTPUT 160 bhp
WEIGHT 474 lb (215 kg)
TOP SPEED 186 mph (299 km/h)
COUNTRY OF ORIGIN Japan

Rearview mirror

Ducts carry air away from the radiators

This is a 2001 model

Deeply faired front mudguard offers improved aerodynamics

AERODYNAMIC FRONT
Weird Hayabusa looks are all about achieving ultimate speed. The narrow headlight allows a narrower and more elongated front to the fairing. Indicators are faired in, and the mudguard is deep.

Hayabusa is the name of a superfast bird of prey

Forks are steeply angled for quicker steering

SUZUKI *GSX-R750*

THE INTRODUCTION OF THE ORIGINAL Suzuki GSX-R750 in 1985 helped to create and define the sports bike market. The GSX-R brand remains synonymous with focused, sporting riders' machines. Early bikes featured a distinctive frame that wrapped over and under an oil-cooled engine, but continual evolution meant that the engine was given water-cooling in 1992, a more conventional twin-spar frame in 1996, fuel injection in 1998, and a new shape for 2004 (seen here). But always the goal was to make a superior, performance-oriented motorcycle that would have better handling, and become lighter and more user-friendly with every reincarnation.

Pillion seat is hidden under cowl

Yoshimura exhaust system is a typical aftermarket modification

Substantial bracing reduces swingarm flex

Three-spoke alloy wheel is fitted with 17 inch, 180 x 55 tire

SPECIFICATIONS

Model Suzuki GSX-R750
Capacity 749cc
Power Output 146 bhp @ 12,750 rpm
Weight 359 lb (163 kg)
Top Speed 175 mph (282 km/h)
Country of Origin Japan

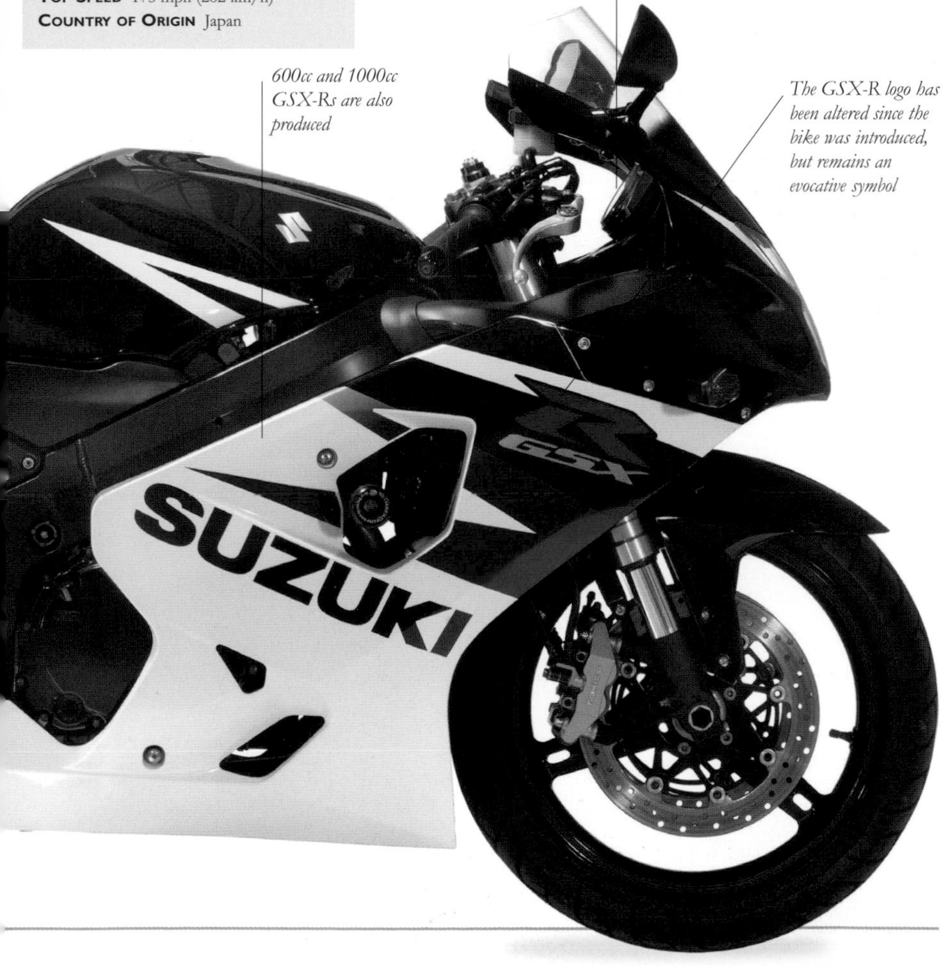
Dash features digital speedometer and analog rev-counter
600cc and 1000cc GSX-Rs are also produced
The GSX-R logo has been altered since the bike was introduced, but remains an evocative symbol
SUZUKI
GSX

TEMPLE-ANZANI

THIS AMAZING MACHINE WAS THE first motorcycle to cover more than 100 miles (161 kilometers) in one hour. The record was achieved at the banked Montlhery circuit near Paris in September 1925. The bike made its public debut at Brooklands two years earlier when rider/designer Claude Temple covered a flying mile at 112.53 mph (181.06 km/h). In October 1926, Temple raised the motorcycle speed record to 121.41 mph (195.34 km/h)—apparently the first time a motorcycle had traveled at over 120 mph (193 km/h). The fearsome machine was powered by an Anzani V-twin engine, which featured shaft-driven overhead camshafts and twin carburetors. Apparently the bike was constructed for Temple by OEC, which produced motorcycles in a variety of styles and sizes from 1901–54.

SPECIFICATIONS

MODEL Temple-Anzani
CAPACITY 116cc
POWER OUTPUT Not known
WEIGHT Not known
TOP SPEED 121 mph (195 km/h)
COUNTRY OF ORIGIN UK

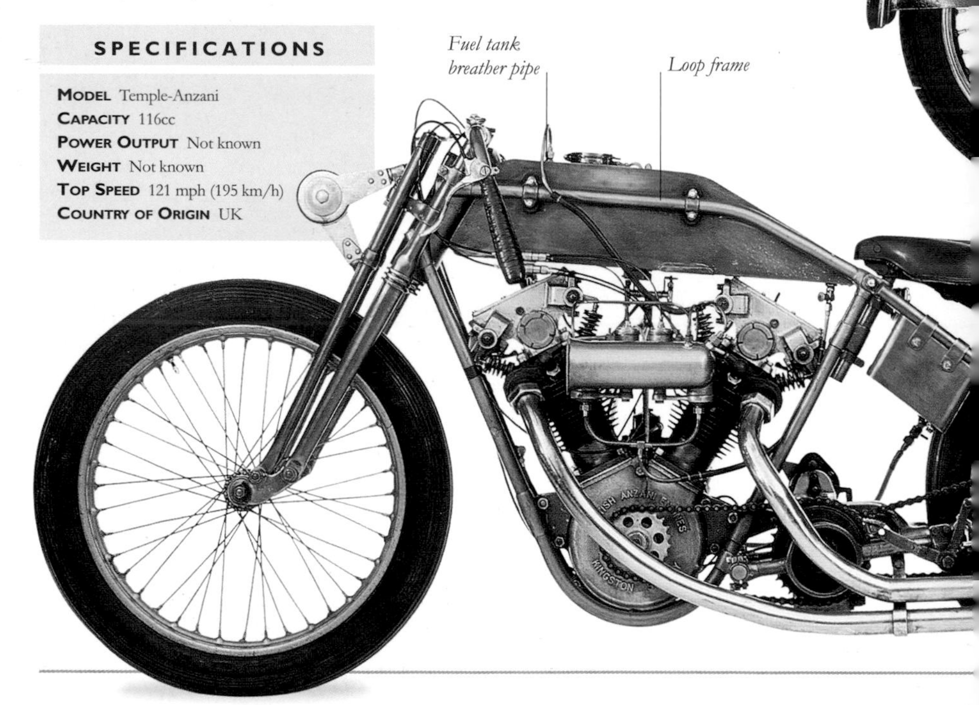

SPECIFICATIONS

MODEL Tornax Simplex Sport
CAPACITY 592cc
POWER OUTPUT 22 bhp
WEIGHT 392 lb (178 kg)
TOP SPEED 75 mph (120 km/h)
COUNTRY OF ORIGIN Germany

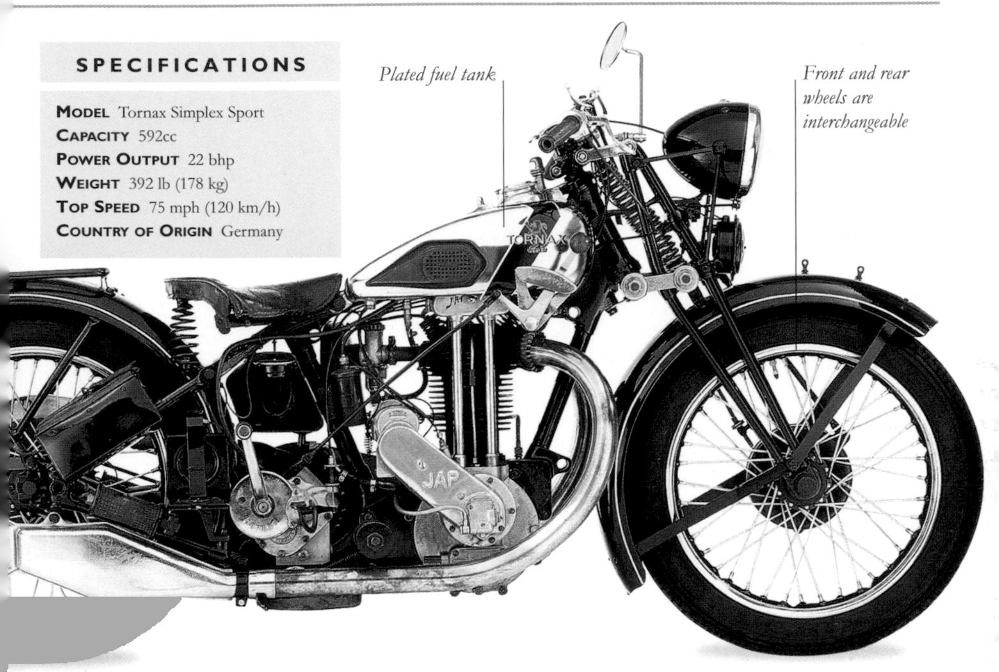

TORNAX *Simplex Sport*

ESTABLISHED IN 1925, Tornax soon established a reputation for high-quality sports machines. It used proprietary engines during its time as a manufacturer. This 1934 Simplex Sport model was typical, using a 592cc twin-port o.h.v. engine supplied by the British firm JAP. The gearbox was a three-speed Hermes unit. The construction is conventional, with a diamond-pattern frame and girder forks. Instruments were recessed into the plated fuel tank. This model was the last of the JAP-engined Tornax machines after government restrictions prohibited the import of foreign components.

TRIUMPH *Model R Fast Roadster*

THE MODEL R WAS BASED ON Triumph's popular Model H. Many of the cycle parts were the same, and the only major changes were the new cylinder and head designs by engineer Harry Ricardo. He changed the bike from a 550cc side valve to a 499cc with advanced four-valve cylinder head and centrally positioned spark plug. The result was that performance increased greatly. The Model R, which was nicknamed "Riccy," made its debut in the 1921 Senior TT and remained in its catalog until 1927. Triumph concentrated on a two-valve model from 1924; 35 years later Honda realized the potential of four valves. This 1923 bike predates the use of drum brakes by one year.

SPECIFICATIONS

MODEL Triumph Model R Fast Roadster
CAPACITY 499cc
POWER OUTPUT 20 bhp
WEIGHT 240 lb (109 kg)
TOP SPEED 70–75 mph (113–121 km/h)
COUNTRY OF ORIGIN UK

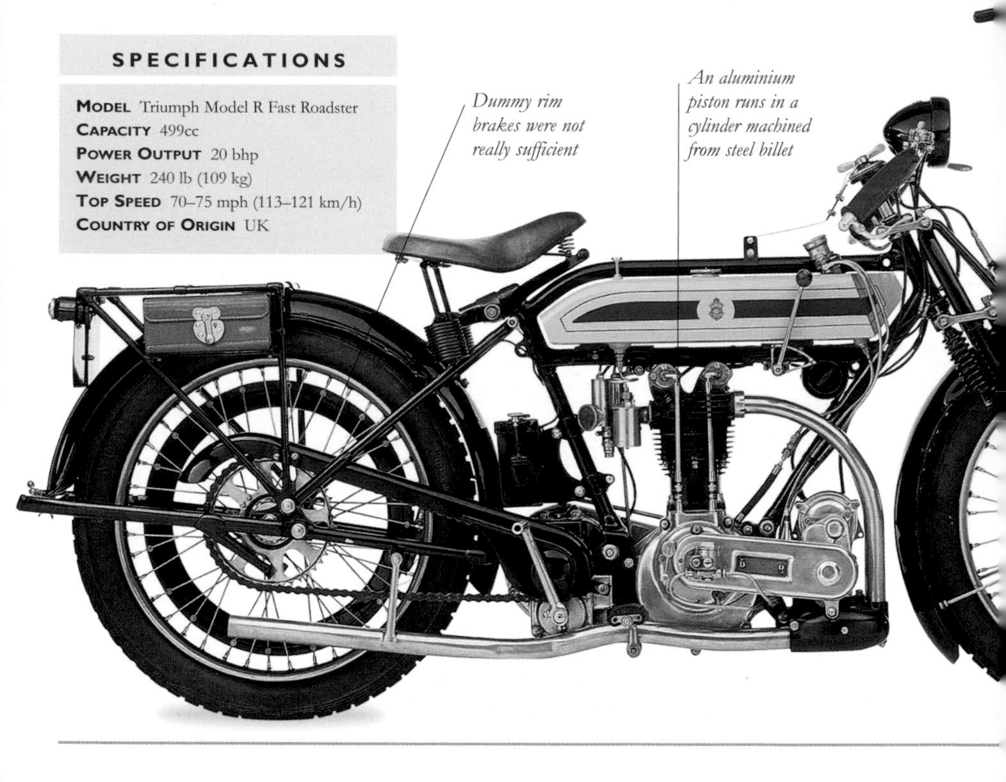

Dummy rim brakes were not really sufficient

An aluminium piston runs in a cylinder machined from steel billet

Luggage rack

Coupled brake linkage, with locking facility for sidecar parking

SPECIFICATIONS

MODEL Triumph 6/1
CAPACITY 649cc
POWER OUTPUT 25 bhp @ 4500 rpm
WEIGHT 435 lb (197 kg)
TOP SPEED 85 mph (137 km/h) (est.)
COUNTRY OF ORIGIN UK

TRIUMPH *6/1*

WHILE MANY MANUFACTURERS began building V-twins, Triumph stuck to its single-cylinder engines until 1933, when it unveiled a novel parallel twin-cylinder machine. The 649cc 6/1 was developed by Triumph's design chief, Valentine Page, and was intended mainly for sidecar use. The innovative design had helical gear drive to the gearbox, which was bolted to the back of the engine. Although the 6/1 won prestigious awards in factory tests, and the parallel twin configuration was to become a firm favorite with British manufacturers, the 6/1 was discontinued in 1936.

TRIUMPH *Tiger 80*

TRIUMPH WAS TAKEN OVER IN 1936, and one of the first things the new owners did was install Edward Turner as design chief. His main priority was to revamp the capable, but staid, 250cc, 350cc, and 500cc overhead-valve singles. Chrome, polished alloy, and new paintwork improved the looks, and revised engines improved the performance. The bikes were relaunched as the Tiger 70, Tiger 80, and Tiger 90—the numbers after the name were an optimistic indication of the speed of each model in miles per hour. The new models were better looking, faster, and cheaper. The success of these new bikes helped ensure the survival of the Triumph brand.

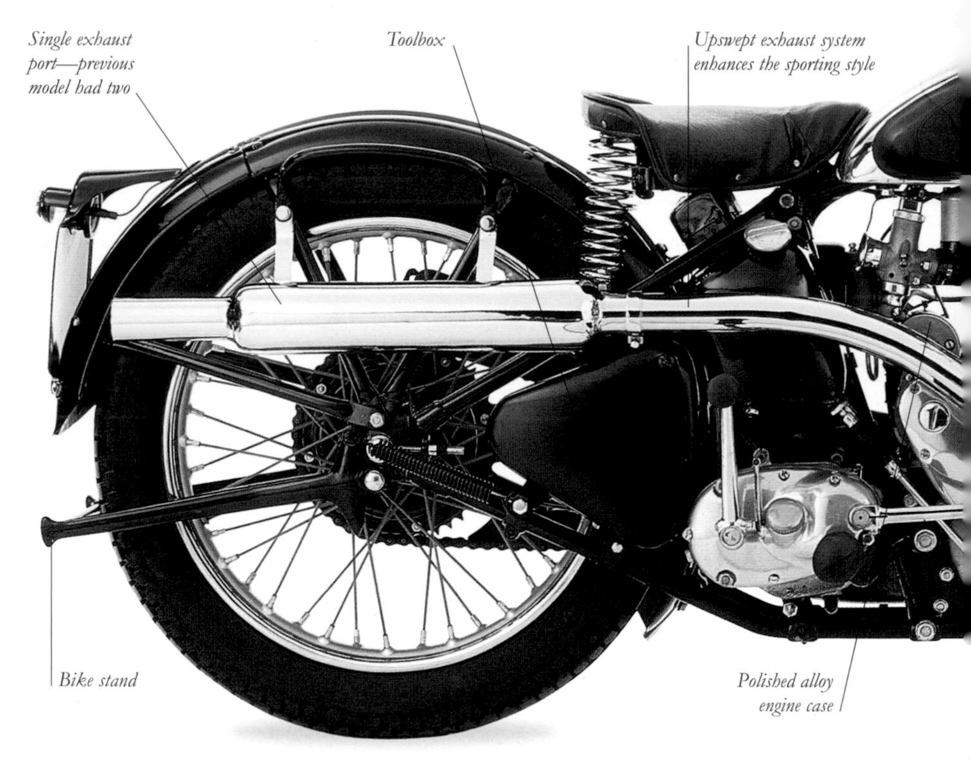

Specifications

Model Triumph Tiger 80
Capacity 343cc
Power Output 20 bhp @ 5700 rpm
Weight 320 lb (145 kg)
Top Speed 75 mph (121 km/h)
Country of Origin UK

Tiger Specs
The o.h.v. single-cylinder engine employed 70 x 89-mm bore and stroke dimensions with a 7.5:1 compression ratio. As with other Tigers, each engine was tuned and tested prior to despatch.

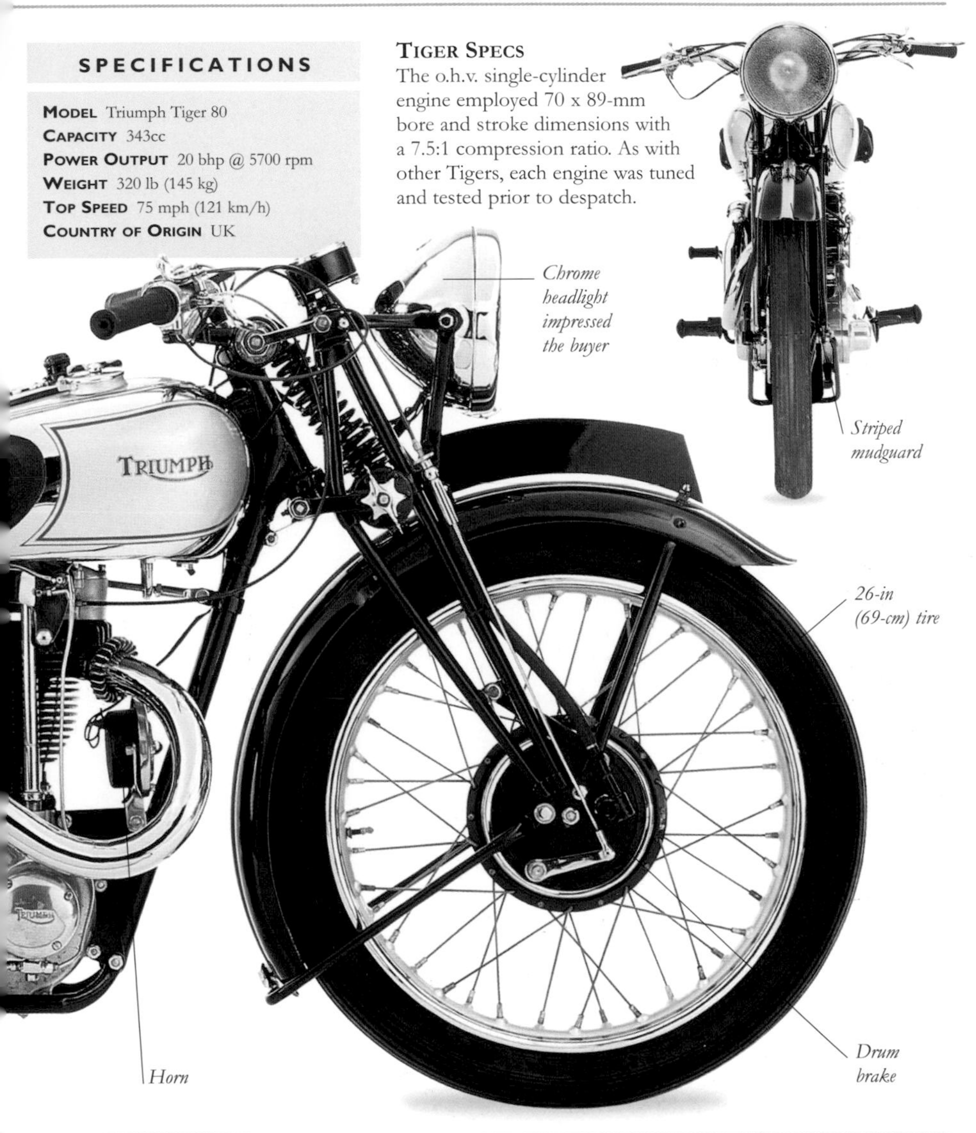

TRIUMPH *Speed Twin*

INTRODUCED IN 1937, THIS WAS the most influential British bike ever made. Edward Turner's talent for marketing style within a tight budget was epitomized by this parallel twin: it was 5 lb (2.2 kg) lighter than the Tiger 90 sports single but cost only fractionally more. The parallel twin-cylinder engine had a two-bearing crankshaft with a central flywheel and also had two camshafts (behind and in front of the cylinders) and enclosed valvegear. The new engine was mounted in the chassis of the Tiger 90. It offered a new level of power, performance, and sophistication to motorcycle buyers and was a template that other British manufacturers soon attempted to copy. The outbreak of war in 1939 delayed the arrival of a swarm of copycat parallel twins from other British factories. When they finally arrived, the Triumph outlived them all. A sports version appeared for 1939 in the form of the Tiger 100.

Amaranth red paintwork became a Speed Twin hallmark

Separate gearbox

SPECIFICATIONS

MODEL Triumph Speed Twin
CAPACITY 498cc
POWER OUTPUT 27 bhp @ 6300 rpm
WEIGHT 378 lb (171 kg)
TOP SPEED 93 mph (150 km/h)
COUNTRY OF ORIGIN UK

CLASSIC LINES
Viewed from any angle, Edward Turner's designs are always characterized by graceful lines. Even the 8-in (20-cm) Lucas headlight does not detract from the Speed Twin's overall slimness and elegance.

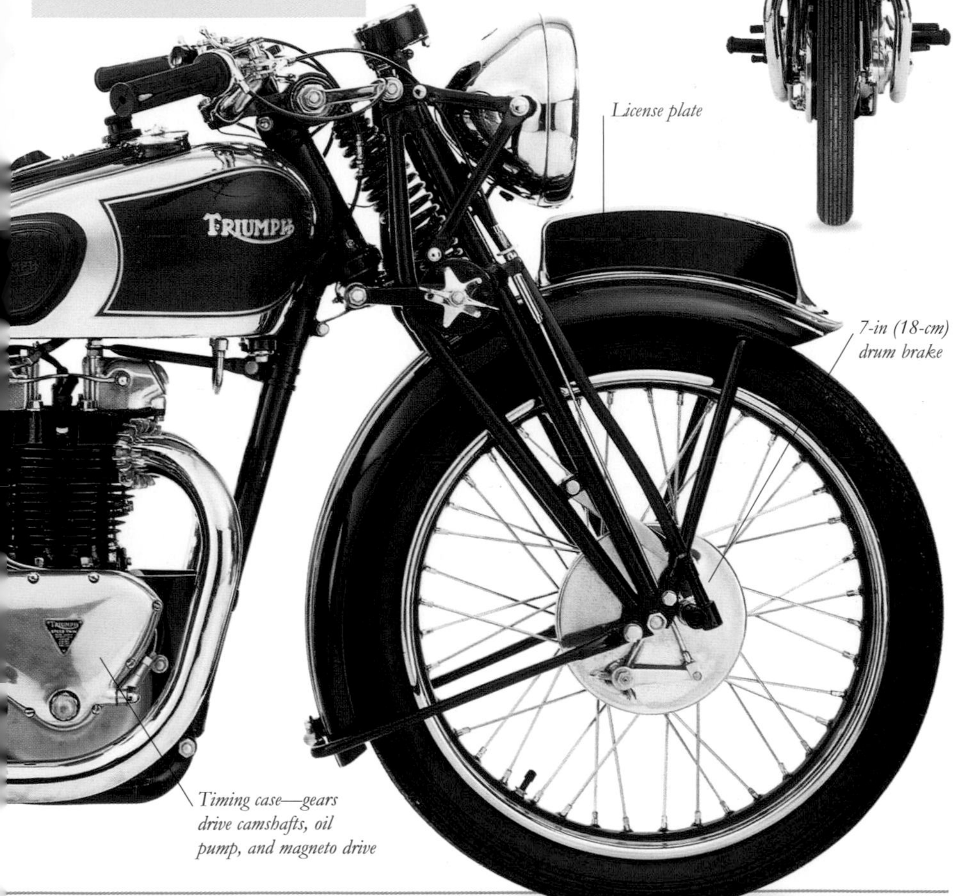

License plate

7-in (18-cm) drum brake

Timing case—gears drive camshafts, oil pump, and magneto drive

TRIUMPH *Record Breaker*

TRIUMPHS SOLD IN large numbers in postwar America and soon became popular with racers and mechanics. This cigar-shaped projectile played an essential part in establishing the Triumph legend. It was built by a team from Texas to beat the world land-speed record for motorcycles. The frame is a multitubular structure and the souped-up 649cc Thunderbird engine runs on alcohol. At the Bonneville Salt Flats in 1956, Johnny Allen sped to a new world-record speed of 214 mph (345 km/h). However, despite the fact that the timekeeper and his equipment had been approved before the record attempt, the FIM (Federation of International Motorcyclists) refused to recognize the new record. Triumph had the last laugh when the company named its most famous model *(see pp.411–13)* after the salt flats.

SPECIFICATIONS

MODEL Triumph Record Breaker
CAPACITY 649cc
POWER OUTPUT 65 bhp (est.)
WEIGHT Not disclosed
TOP SPEED 214 mph (345 km/h)
COUNTRY OF ORIGIN UK

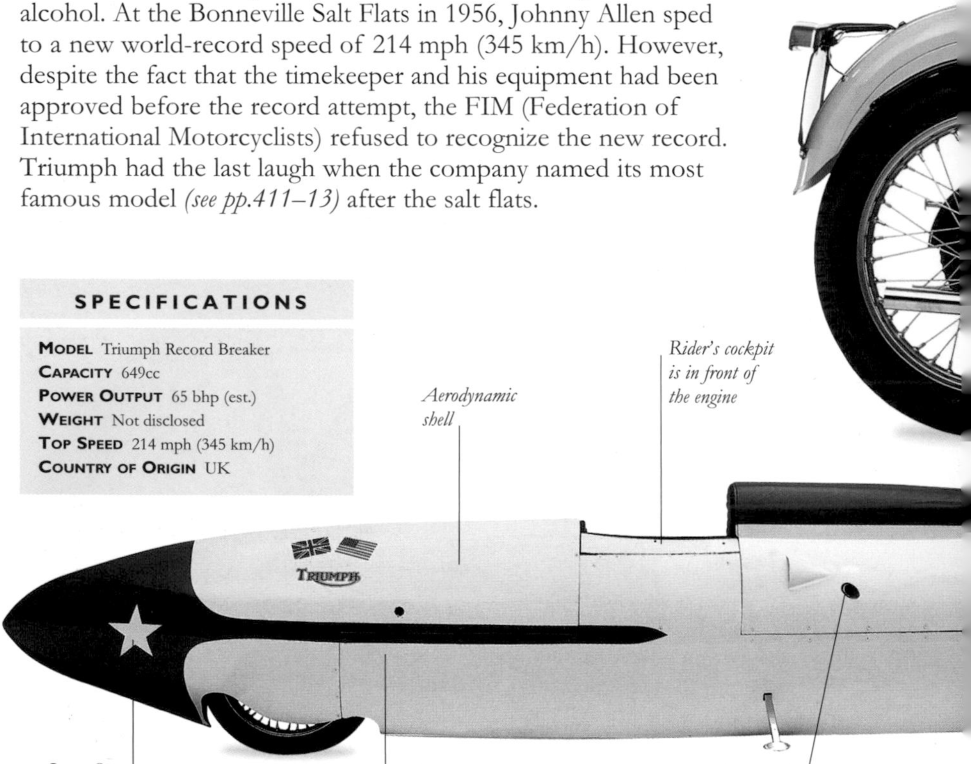

Aerodynamic shell

Rider's cockpit is in front of the engine

Lone Star logo honors team's home state

Alloy bodywork mounted on a tubular frame

Exhaust outlet

SPECIFICATIONS

MODEL Triumph T120 Bonneville
CAPACITY 649cc
POWER OUTPUT 46 bhp @ 6500 rpm
WEIGHT 404 lb (183 kg)
TOP SPEED 110 mph (177 km/h)
COUNTRY OF ORIGIN UK

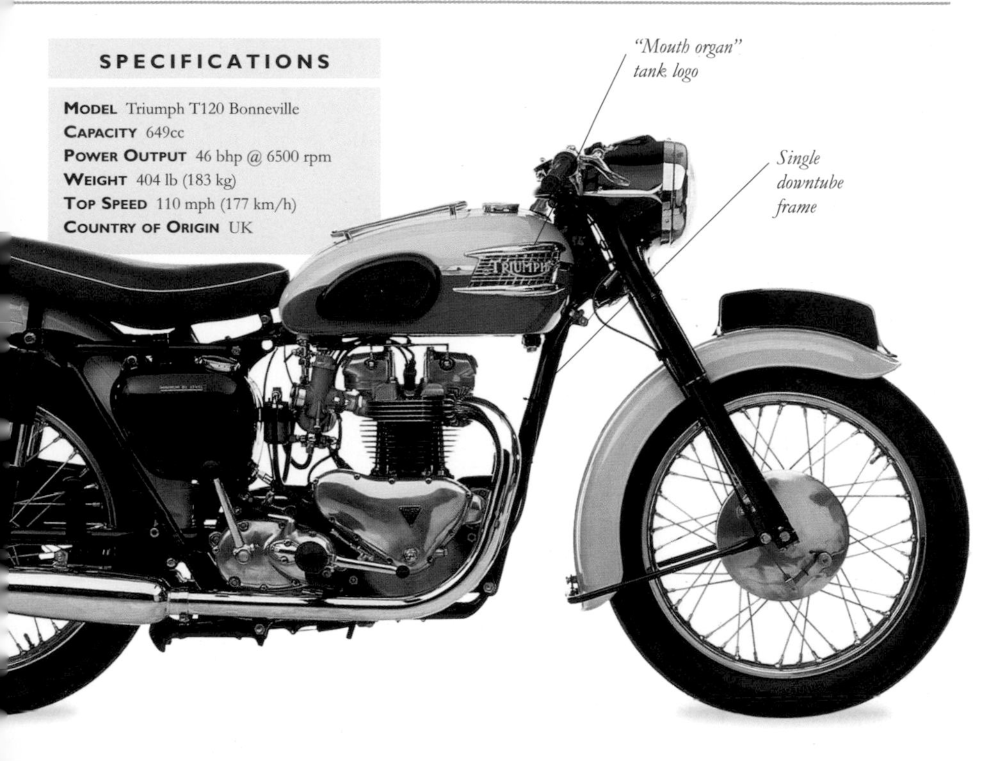

TRIUMPH *T120 Bonneville*

NAMED AFTER THE Utah salt flats on which a Triumph twin achieved a world speed-record *(see left)*, the T120 Bonneville was a high-performance version of Triumph's 650 twin introduced at the London Earl's Court Show in 1958. Power was increased by attaching twin carburetors to an improved cylinder head (thus increasing compression ratio) and using performance camshafts. The single downtube frame on this rare 1959 model was replaced by a duplex version in 1960.

TRIUMPH *T120TT Bonneville Special*

TRIUMPH AMERICA DEVELOPED THE TT Special for the West Coast market, where desert and beach racing were very popular. To save the weight of a battery, the problematic Energy Transfer ignition was used, while the duplex frame gave way to a single downtube running into a duplex cradle. The rear end was stiffened by plates supporting the ends of the swinging fork pivot. The T120TT was made for only two seasons, and few survived the rigors of sand racing and the stresses imposed on the bearings. The advent of genuine scramble machines made the TT Special obsolete.

Rear mudguard

MINIMUM OIL LEVEL

SPECIFICATIONS

Model Triumph T120TT Bonneville Special
Capacity 649cc
Power Output 54 bhp @ 6500 rpm
Weight 350 lb (159 kg)
Top Speed 130 mph (209 km/h)
Country of Origin UK

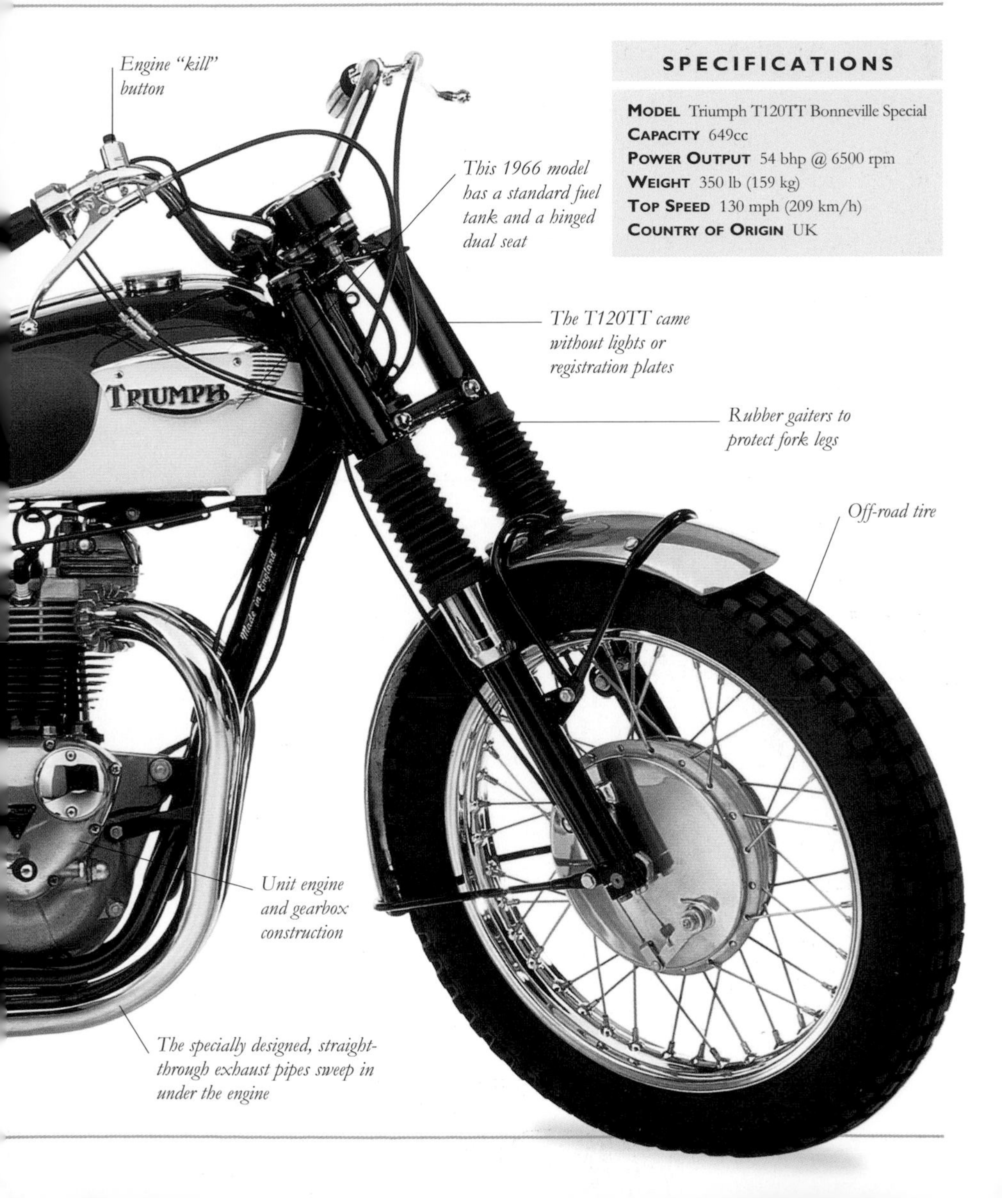

TRIUMPH *T150 Trident*

DEMAND FOR MORE POWER led motorcycle designers to build multi-cylinder machines. To avoid the costs of developing an all-new machine, Triumph effectively added another cylinder to its 500cc twin to create a 740cc triple. A 120° crankshaft made it smoother and better-sounding than the twin. The Trident was fast and handled well, but it couldn't compete with the glitzy and reliable four-cylinder Honda that appeared at the same time. It wasn't introduced into the UK until 1969, a year after its launch in export markets. Later versions of the model came with electric starters, disc brakes, and five-speed gearboxes, but the strength of the Japanese competition meant production ceased in 1975.

Passenger handrail

"Ray gun" muffler

Three Amal Concentric carburetors

SPECIFICATIONS

MODEL Triumph T150 Trident
CAPACITY 740cc
POWER OUTPUT 58 bhp @ 7250 rpm
WEIGHT 482 lb (219 kg)
TOP SPEED 122 mph (196 km/h)
COUNTRY OF ORIGIN UK

RACE PEDIGREE

Racing versions of the Trident dominated the 750 class. In 1972, Ray Pickrell won the Formula 750 TT, with an average speed over the five laps of 104 mph (168 km/h).

Gearchange is on the right

Zener diode charging regulator

Twice-into-two exhaust system

Twin leading-shoe ventilated front drum brake

Points cover

TRIUMPH *Trident Racer*

TRIUMPH SHOWED LITTLE INTEREST in motorcycle racing until a production-based class became popular in Europe and the US. In 1970 and '71, the factory race shop assembled special race bikes using souped-up 750cc Trident engines attached to lightweight racing frames. These fantastic machines outperformed and outhandled all opposition for two seasons, with big wins recorded on both sides of the Atlantic. In 1971, the Triumph on these pages—ridden by Ray Pickrell and Percy Tait—won the grueling Bol d'Or at Le Mans for the second year running.

Ergonomic tank- and seat-design are essential as riders spend long periods on the track

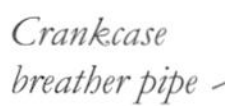

Crankcase breather pipe

In the early 1970s, endurance racers used road tires

The rear wheel must be easily removable for rapid pit stop tire changes

SPECIFICATIONS

MODEL Triumph Trident Racer
CAPACITY 749cc
POWER OUTPUT Over 70 bhp
WEIGHT 395 lb (179 kg)
TOP SPEED 164 mph (264 km/h)
COUNTRY OF ORIGIN UK

Tachometer

The headlight is angled to give maximum visibility when cornering or braking

Twin cast-iron drum brakes

A light above the racing number assists the pit crew and the scorers recording lap times

TRIUMPH *X75 Hurricane*

UNVEILED IN NOVEMBER 1972, the X75 Hurricane was based on the discontinued BSA Rocket 3 *(see p.77)*, inheriting its duplex frame and inclined cylinders. BSA-Triumph had commissioned US designer Craig Vetter to put together a limited edition of chopper-style triples to cash in on the chopper craze caused by the movies *Wild Angels* and *Easy Rider*. By 1972 (the date of the bike shown on these pages), the BSA name had been dropped, so a Triumph logo was used. The Hurricane was more brash than efficient: the 2-gallon (9-liter) fuel tank restricted distance, the extended forks impaired handling, and the wide handlebars made high speeds difficult. Japanese manufacturers took up the idea of producing "factory custom" machines a few years later.

SPECIFICATIONS

MODEL Triumph X75 Hurricane
CAPACITY 740cc
POWER OUTPUT 58 bhp @ 7250 rpm
WEIGHT 444 lb (201 kg)
TOP SPEED 105 mph (169 km/h)
COUNTRY OF ORIGIN UK

For style rather than handling, the forks were extended by 1 in (2.5 cm)

A small steel fuel tank is hidden inside the fiberglass seat and dummy tank unit

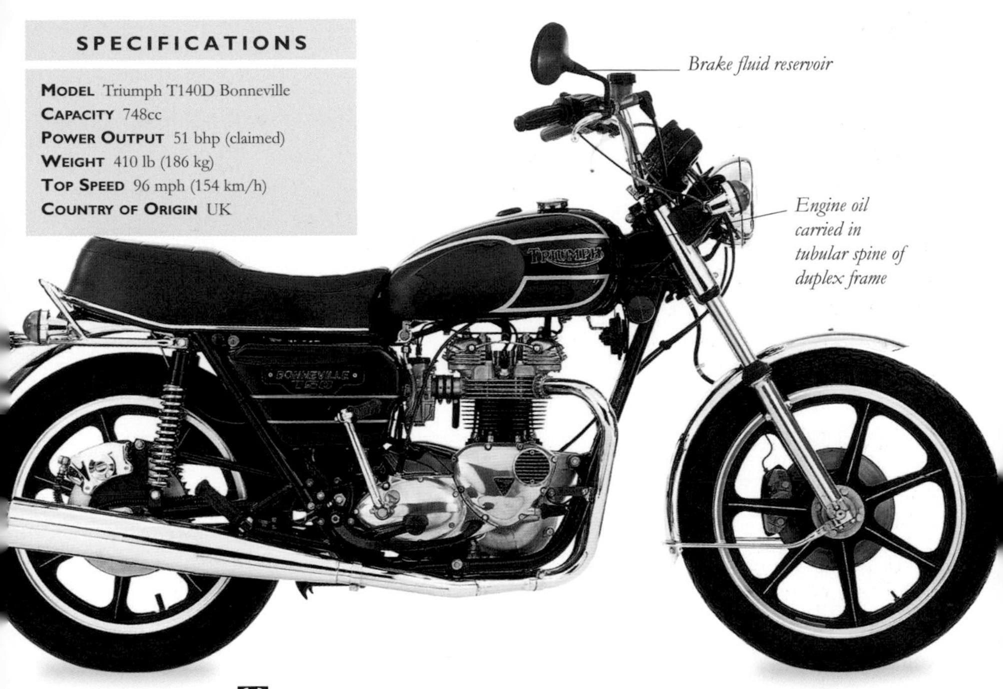

SPECIFICATIONS

MODEL Triumph T140D Bonneville
CAPACITY 748cc
POWER OUTPUT 51 bhp (claimed)
WEIGHT 410 lb (186 kg)
TOP SPEED 96 mph (154 km/h)
COUNTRY OF ORIGIN UK

TRIUMPH *T140D Bonneville*

BASED ON A DESIGN dating from 1937, the Triumph Bonneville was, by the 1970s, well past its expiration date. However, its charisma and handling kept it popular. By the end of the decade, capacity had grown to 750cc, and there were five speeds in the gearbox. The T140D model shown was built in 1979 and featured detail changes, including cast wheels and the black-and-gold paint job. It may have looked good, but performance was poor—the engine was strangled by noise restrictions. This was the last Triumph twin to be exported to the US in significant numbers. Bonneville production continued at Triumph's Meriden factory until 1983.

TRIUMPH *T595 Daytona*

SIX YEARS AFTER TRIUMPH'S 1991 relaunch, the company entered the lucrative supersports market with the T595, built to compete with the market-leading Honda Fireblade *(see pp.212–13)* and charismatic Ducati 916 *(see pp.110–11)*. The bike used Triumph's trademark three-cylinder setup and the Daytona name from earlier models, but almost everything else was new. Though lighter, more powerful, and better-looking than its predecessor, the T595's reputation suffered when some early bikes were recalled because of a frame problem.

Removable cowl conceals passenger seat

Passenger footrest

Engine unit is a structural part of the frame

SPECIFICATIONS

MODEL Triumph T595 Daytona
CAPACITY 955cc
POWER OUTPUT 114 bhp @ 9500 rpm
WEIGHT 437 lb (198 kg)
TOP SPEED 160 mph (257 km/h)
COUNTRY OF ORIGIN UK

DAYTONA POWER
The Daytona had a six-speed gearbox, was able to push out 114 bhp, and could hit 160 mph (257 km/h). Stopping power was provided by 13½-in (320-mm) disc brakes with Nissin calipers.

TRIUMPH *Street Triple R*

IN 1994, INSPIRED BY THE STREETFIGHTER custom trend, Triumph created a large-capacity naked bike by stripping the fairing off their three-cylinder 900cc Daytona sports bike to make the Speed Triple model. The Street Triple, a smaller capacity naked bike, was made in 2007 by stripping the fairing from the 675cc three-cylinder sports bike. Competitively priced, the new bike was light, agile, and great fun, and became a global sales success. Styling details followed the bigger bike, with underseat exhaust and a cowl for the radiator. A higher specification model, the Street Triple R, with improved brakes and suspension, appeared in 2008. Both bikes were substantially changed for 2013.

Dual seat offers minimalist passenger accommodation

Underseat exhausts were fitted, but dropped for 2013

Cast-alloy swingarm

SPECIFICATIONS

MODEL Triumph Street Triple R
CAPACITY 675cc
POWER OUTPUT 107 bhp @ 11,700 rpm
WEIGHT 368 lb (167 kg)
TOP SPEED 153 mph (246 km/h)
COUNTRY OF ORIGIN UK

Radiator is hidden behind plastic shroud

Bug-eyed twin headlights are fitted

Suspension features adjustable damping and pre-load settings

VELOCETTE *Model K*

THE 1925 MODEL K WAS THE first Velocette to feature the Percy Goodman-designed overhead camshaft engine that became synonymous with the Birmingham, England, factory. The new engine was the first four-stroke that Velocette had produced since converting to war production during the 1914–18 conflict. It was the basis for successful race bikes that were produced up to 1953, and for sports road bikes that continued until 1947. It also inspired the overhead camshaft Norton CS1 *(see pp.318–19)*, produced in another Birmingham factory in 1927. The first racing KTT models appeared in 1928, but the K remained essentially unchanged until the early 1930s. The machine shown on these pages is a 1925 Model K.

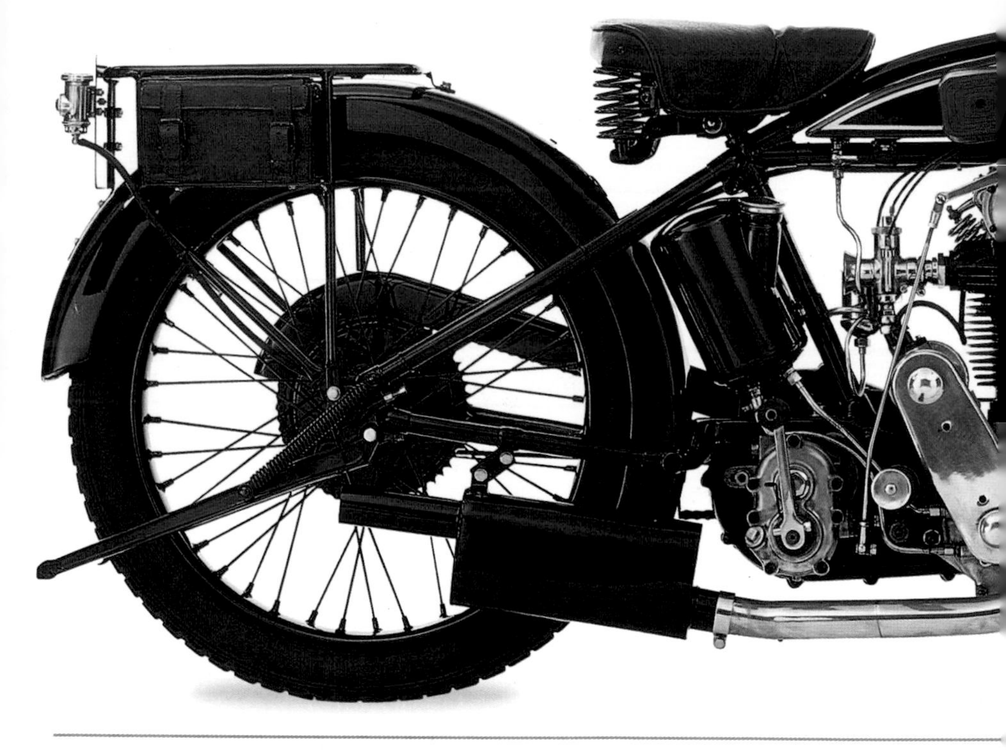

SPECIFICATIONS

MODEL Velocette Model K
CAPACITY 348cc
POWER OUTPUT Not known
WEIGHT 260 lb (118 kg)
TOP SPEED 65 mph (105 km/h)
COUNTRY OF ORIGIN UK

ROLLER LINK
Much of Velocette's wartime work involved making parts for Rolls-Royce armored cars; the association proved to be a positive influence on Velocette's engineering quality.

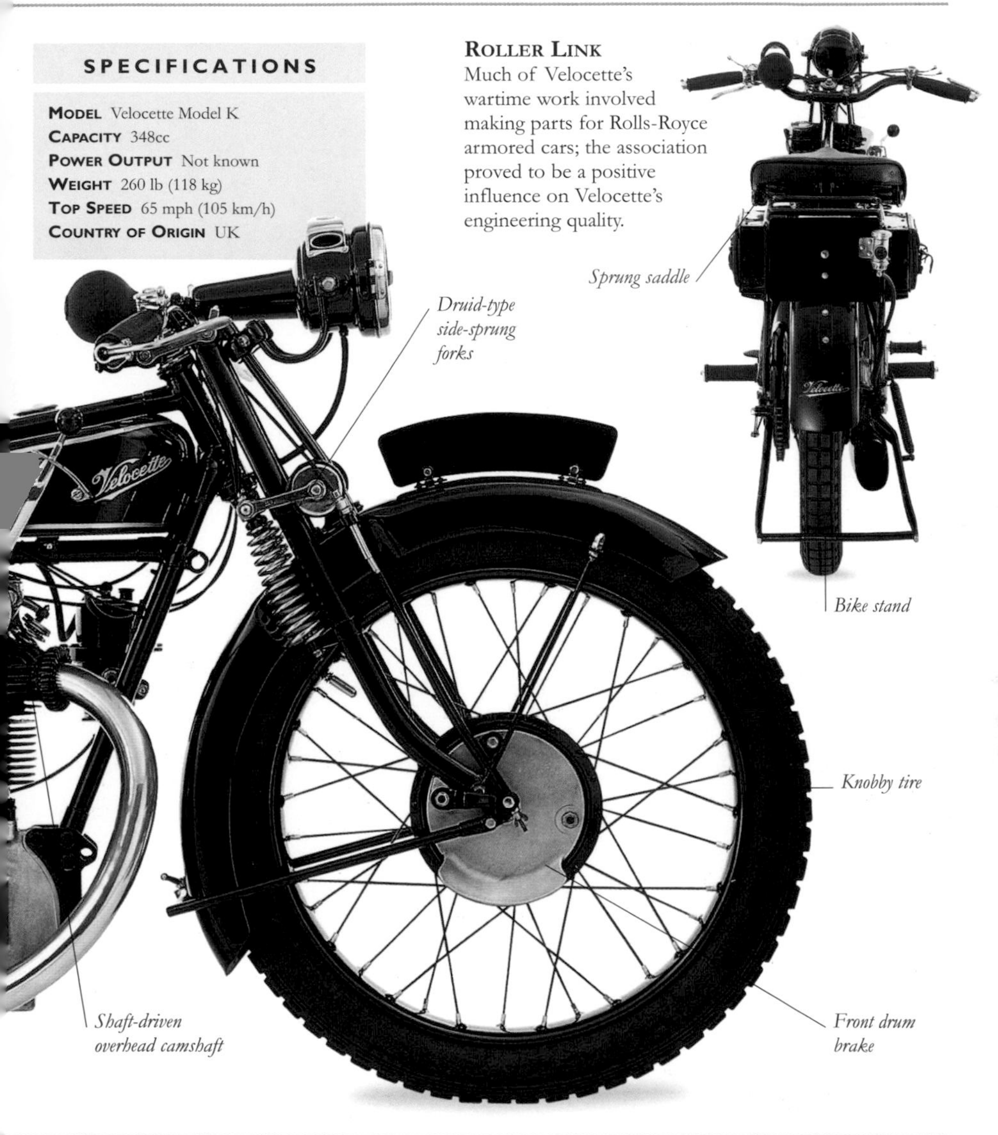

VELOCETTE *KTT MkVIII*

DESPITE A PREWAR HERITAGE, variants of the o.h.c. Velocette engine designed by Percy Goodmann were built up until the 1950s, thus signaling the strength of the initial design. The last model to bear the famous moniker, the KTT MkVIII (1947 model pictured), was introduced a year after the Velocette factory team had cleaned up at the 1938 TT, and it soon established a name for reliability that distinguished it among highly temperamental production racers of the era. Modifications after the war were limited, but the winning World Championship machines ridden by Freddie Frith in 1949 and 1950 used double overhead camshafts.

SPECIFICATIONS

MODEL Velocette KTT MkVIII
CAPACITY 348cc
POWER OUTPUT 34 bhp
WEIGHT 320 lb (145 kg)
TOP SPEED 115 mph (185 km/h)
COUNTRY OF ORIGIN UK

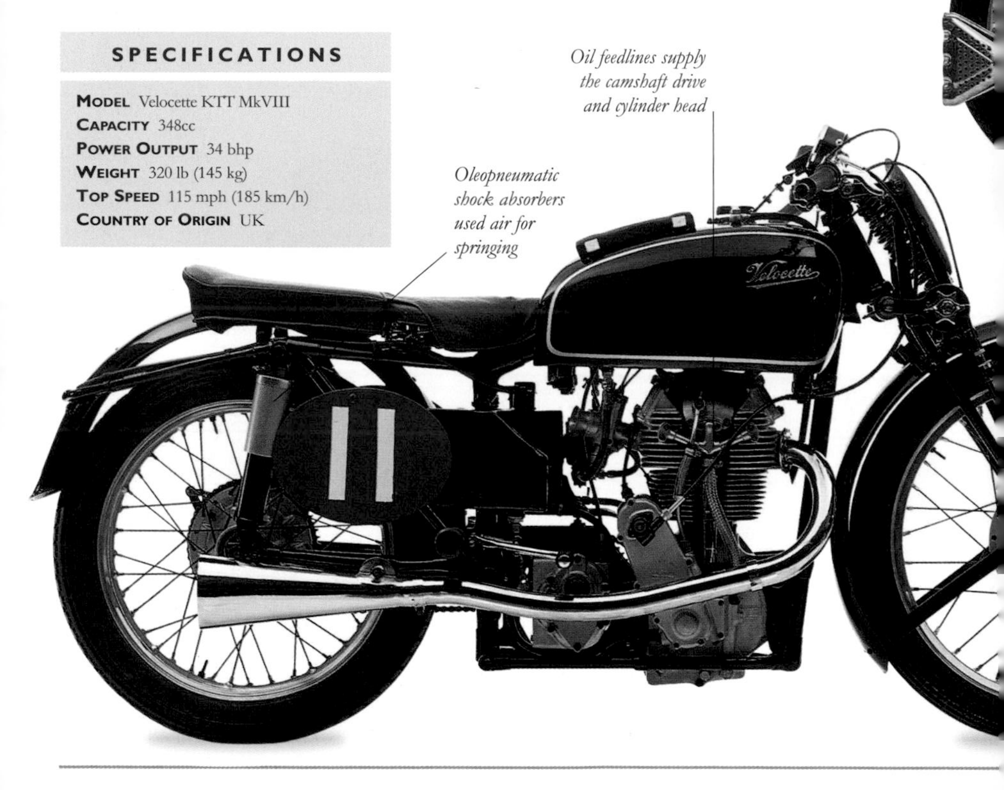

Oil feedlines supply the camshaft drive and cylinder head

Oleopneumatic shock absorbers used air for springing

Rear springing is adjusted by moving the upper mounting bolts of the rear shock unit through an arc

License plate

The fuel tank is cut away to accommodate the downdraft carburetor

SPECIFICATIONS

MODEL Velocette Thruxton Venom
CAPACITY 499cc
POWER OUTPUT 40 bhp @ 6200 rpm
WEIGHT 390 lb (177 kg)
TOP SPEED 105 mph (169 km/h)
COUNTRY OF ORIGIN UK

VELOCETTE *Thruxton Venom*

IN ESSENCE A SOUPED-UP version of Velocette's Venom model, the big single engine proved ideally suited to the Thruxton Nine Hour production race, after which later developments of the sports Venom were named. Prior to the introduction of the Thruxton in 1965, the Velocette factory sold a high-performance kit to prospective racers. Ironically, the 1965 800-km (500-mile) race moved to Castle Combe in Wiltshire, England, but Velocettes still dominated the 500cc class. The Thruxton Venom illustrated here is a 1967 model.

VICTORIA *KR50S*

VICTORIA INTRODUCED A LINE of single-cylinder machines in 1928 to supplement the flat-twin models already produced by the German company. The new machines were of conventional construction with tubular frames, girder forks, and British-built Sturmey-Archer engines that ranged from 198cc to 495cc. The KR50S, a top-of-the-line sports model with three-speed gearbox, was introduced in 1931. Victoria's Sturmey-Archer-engined machines were phased out in the early 1930s: the KR50S was dropped in 1933, while the 198cc side-valve version lasted until 1934. The bike on this page is a 1931 model.

SPECIFICATIONS

MODEL Victoria KR50S
CAPACITY 495cc
POWER OUTPUT 18 bhp @ 5000 rpm
WEIGHT 330 lb (150 kg)
TOP SPEED 70 mph (113 km/h)
COUNTRY OF ORIGIN Germany

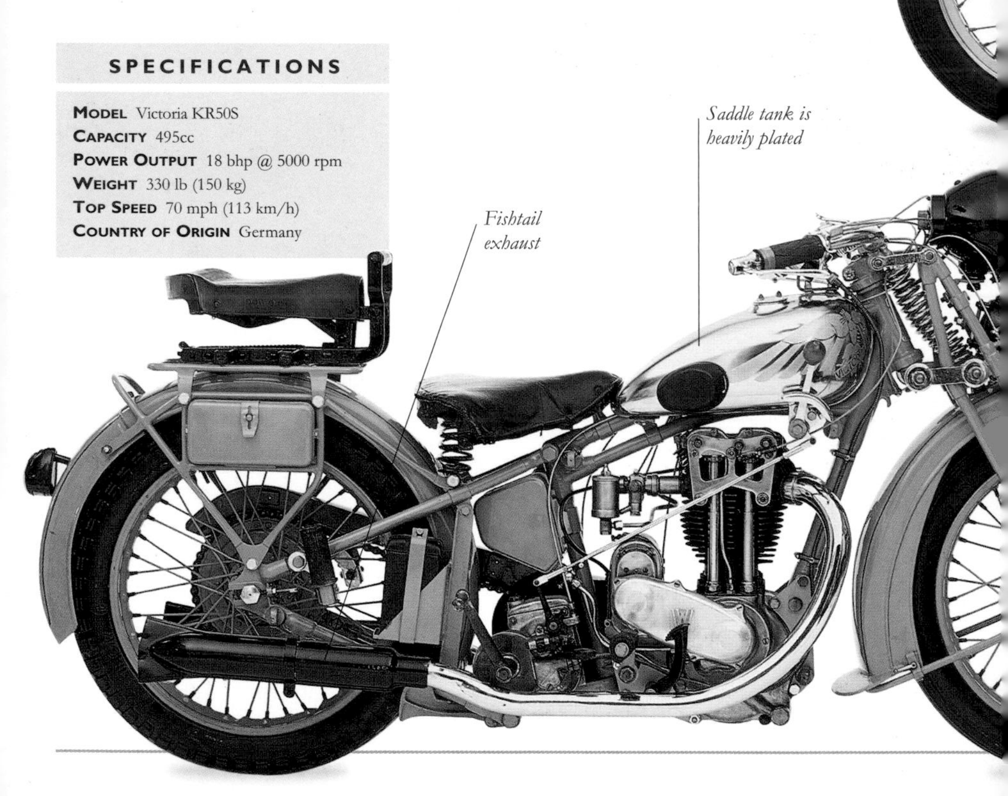

Saddle tank is heavily plated

Fishtail exhaust

Plunger suspension at the rear

SPECIFICATIONS

MODEL Victoria Bergmeister
CAPACITY 347cc
POWER OUTPUT 21 bhp @ 6300 rpm
WEIGHT 389 lb (177 kg)
TOP SPEED 81 mph (130 km/h)
COUNTRY OF ORIGIN Germany

Carburetor cowl

Gray was a traditional color on Victoria machines

VICTORIA *Bergmeister*

THE BERGMEISTER WAS VICTORIA'S first postwar four-stroke. It was introduced in 1951, but production versions were not available until 1953. The bike was designed by Richard Küchen, and the four-speed gearbox used chains rather than pinions. Final drive was by shaft and the bike foreshadowed the Moto Guzzi layout, which followed 15 years later. In 1956, production ended and Victoria returned to the lightweight two-stroke. The Bergmeister's heavy development costs damaged the firm's finances, and it merged with the Zweirad Union in 1958. The Bergmeister shown here dates from 1954.

VINCENT-HRD *Comet*

PHILIP VINCENT HAD DEVELOPED his ideas for producing a spring-frame motorcycle while studying at university. His father bought the defunct HRD marque and a few spares to start him in business in 1928. After a disastrous showing at the 1934 TT, Philip Vincent dispensed with engines bought from JAP. With less than four months to go to the 1934 Motorcycle Show, he commissioned Phil Irving to develop a 500cc single-cylinder engine. Irving succeeded, and the 1935 model was exhibited with the new engine, even though it had never run. The new design, the basis of subsequent Vincent machines, featured a high camshaft and widely splayed pushrods. The valves ran in double guides. This model dates from 1938. Production of the Vincent at the Stevenage factory never reached large numbers and the bikes were always expensive machines. The HRD part of the name was dropped after World War II.

SPECIFICATIONS

MODEL Vincent-HRD Comet
CAPACITY 498cc
POWER OUTPUT 25 bhp @ 5300 rpm
WEIGHT 385 lb (175 kg)
TOP SPEED 92 mph (148 km/h)
COUNTRY OF ORIGIN UK

SUSPENSION
The Comet was fully sprung with Brampton girder forks and included a cantilever rear suspension system, patented by Vincent in 1927 after buying the defunct HRD company.

Gear pedal

License plate

Grooved tire

The front wheel has two brake drums for effective high-speed braking

Mudguard stay doubles as front wheel stand

3.3 x 3.5-in (84 x 90-mm) o.h.v. engine drives a four-speed Burman gearbox

VINCENT-HRD *Series C Black Shadow*

THE BEST-KNOWN OF ALL Vincent-HRD models, the Black Shadow began as a souped-up version of the Series B Rapide, which superseded the prewar Series A Rapide in 1946. The Series C version appeared in 1949 and was in production until 1954. The 84 x 90-mm (3.3 x 3.5-in), V-twin, unit-construction engine formed an integral part of the rolling chassis, suspended from a spine frame that doubled as an oil tank. The rear subframe pivoted from the rear of the gearbox with twin spring boxes and, from 1949, a shock absorber was mounted between its apex and the rear of the oil tank. A large 150 mph (241 km/h) speedometer emphasized that, at the time, the Vincent was the world's fastest standard motorcycle.

Friction shocks for the rear suspension are located at the junction of the seat stays and the swingarm

Stainless steel mudguard

A six-volt 40-watt Miller dynamo is driven from the clutch

SPECIFICATIONS

MODEL Vincent-HRD Series C Black Shadow
CAPACITY 998cc
POWER OUTPUT 55 bhp @ 5700 rpm
WEIGHT 458 lb (208 kg)
TOP SPEED 122 mph (196 km/h)
COUNTRY OF ORIGIN UK

VERSION DIFFERENCES

The postwar Black Shadows retained many of the best features of the Series A line: cantilever rear suspension, detachable twin-braked wheels, and, on Series Bs, Brampton girder forks. Series C models had the firm's own Girdraulic forks.

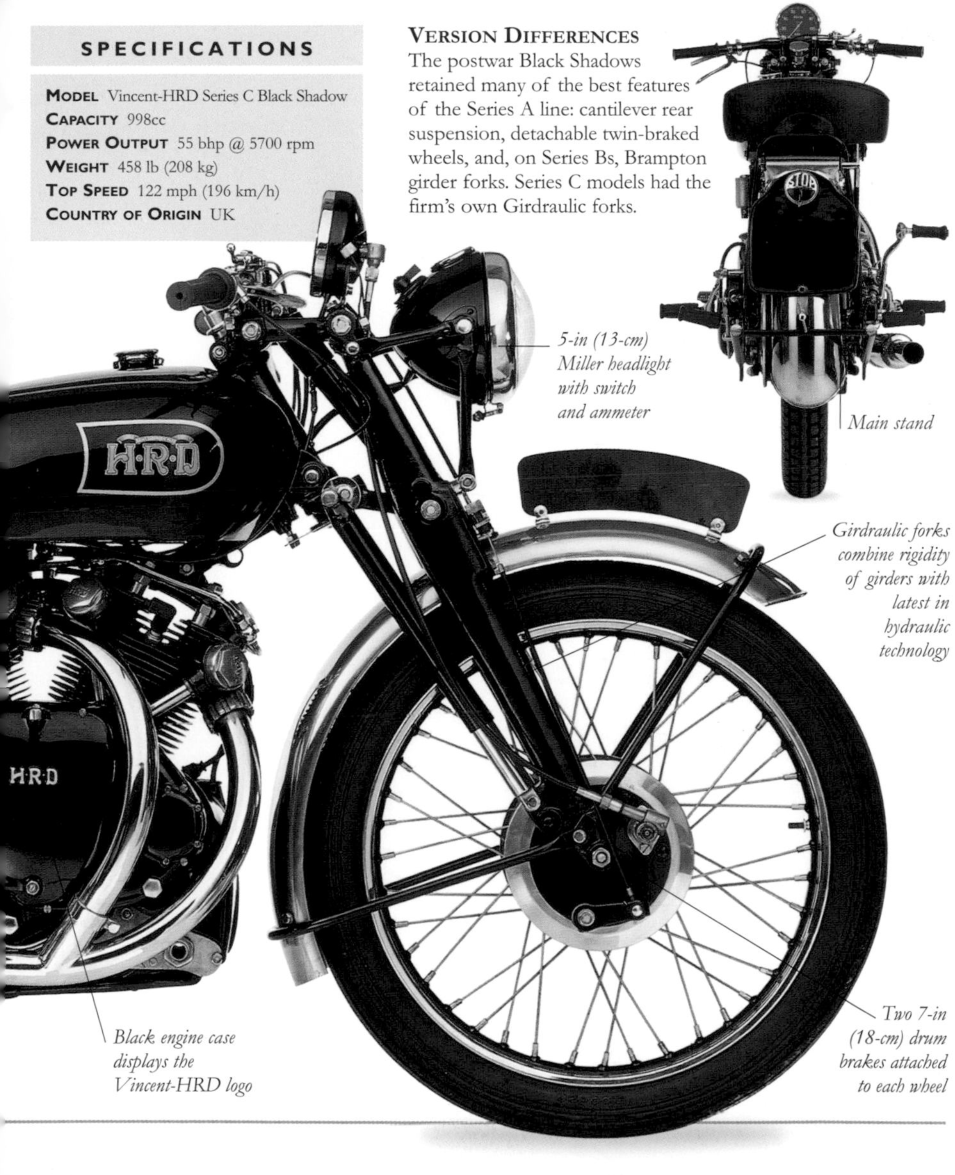

5-in (13-cm) Miller headlight with switch and ammeter

Main stand

Girdraulic forks combine rigidity of girders with latest in hydraulic technology

Two 7-in (18-cm) drum brakes attached to each wheel

Black engine case displays the Vincent-HRD logo

VINCENT *Series D Rapide*

VINCENT INTENDED THAT ALL the Series D models should have enclosed bodywork. These streamlined machines were announced in late 1954. A revised frame layout meant that the oil tank moved to a position under the seat. Coil ignition was used and suspension was controlled by Armstrong shock absorbers. But the bodywork, which was made by an outside supplier, could not be delivered in time and so the first 1955 models, like the Rapide on this page, got the improvements without the bodywork. The Series D was the last range produced by Vincent before production ended in 1955. Although the bikes were expensive, they cost so much to make that the company's profits were negligible.

SPECIFICATIONS

MODEL Vincent Series D Rapide
CAPACITY 998cc
POWER OUTPUT Not known
WEIGHT 447 lb (202 kg)
TOP SPEED 110 mph (177 km/h)
COUNTRY OF ORIGIN UK

The dual seat and oil tank are mounted on a tubular subframe

Girdraulic forks have alloy legs

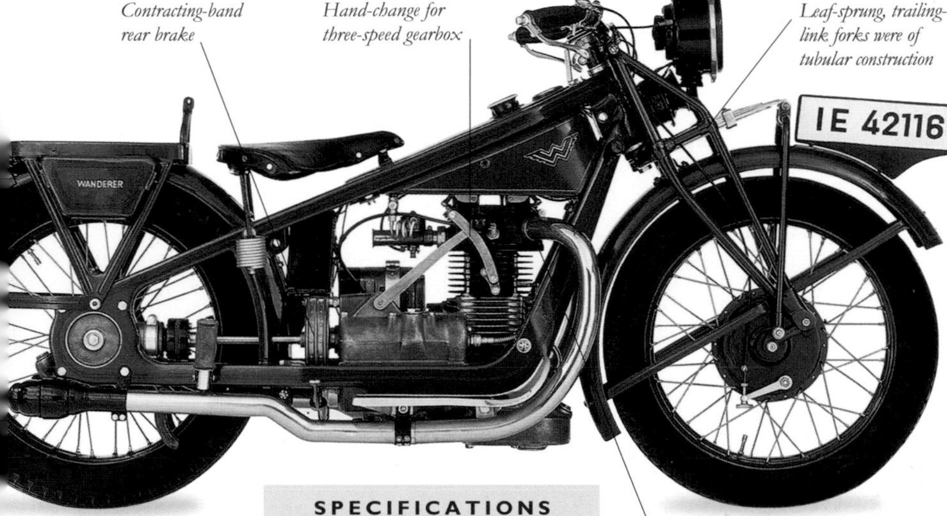

Double-sided front drum brake

SPECIFICATIONS

Model Wanderer K500
Capacity 498cc
Power Output 18 bhp
Weight 364 lb (165 kg)
Top Speed 75 mph (120 km/h)
Country of Origin Germany

WANDERER *K500*

DESIGNED BY ALEXANDER NOVIKOFF, this radical machine was introduced in 1928. Intended as a high-quality rival to BMW's shaft-drive twins, the K500's single-cylinder engine was mounted with its crank running longitudinally in the pressed-steel frame. There was a three-speed gearbox and shaft final drive. However, Wanderer staked too much on the new bike and the firm was taken over by NSU in 1930. The takeover ended the bike's production, but its design proved more durable.

WESLAKE

THE SPEEDWAY BIKE IS AMONG the most specialized of motorcycles, and its design has altered little in 60 years. The skeletal lightweight frames and vertically mounted, single-cylinder, four-stroke engines of modern machines differ little from their forebears. Until the four-valve Weslake engine appeared in 1974, two-valve engines, made first by JAP and then by Jawa, had dominated the sport since the 1930s. The new Weslake was more powerful than the two-valve engines and was to become the dominant engine in speedway racing during the 1970s and '80s. The bike on these pages was used by 1981 World Champion Bruce Penhall.

SPECIFICATIONS

MODEL Weslake
CAPACITY 500cc
POWER OUTPUT 48 bhp
WEIGHT 183 lb (83 kg)
TOP SPEED 60–80 mph (97–128 km/h)
COUNTRY OF ORIGIN US

Minimal handlebar controls operate clutch and throttle

Fuel tank holds enough methanol to last four laps

Forks are steeply angled, giving more responsive steering

Lightweight telescopic forks allow little movement

Front wheel is much larger than the rear and carries a narrower tire

Oil pump

WIMMER *GG35*

THE BAVARIAN FIRM OF B. Wimmer und Sohn began building motorcycles and clip-on bicycle engines in the early 1920s. These included small-capacity o.h.v. engines with unit construction and water cooling. The GG35 model shown here dates from 1932 and is typical of the company's neat design and high-quality construction. The three-speed Hurth gearbox was bolted to the back of the inclined-cylinder 344cc o.h.v. engine to make a rigid semi-unit-construction power unit. The compact frame was a twin-loop cradle structure equipped with conventional girder forks. The engine had a cast-iron cylinder and head with twin exhaust ports. These exit into high-level pipes which were both stylish and practical given Wimmer's trials heritage. Wimmer production ended in 1939.

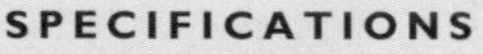

SPECIFICATIONS

MODEL Wimmer GG35
CAPACITY 344cc
POWER OUTPUT 16 bhp
WEIGHT Not known
TOP SPEED 70 mph (113 km/h) (est.)
COUNTRY OF ORIGIN Germany

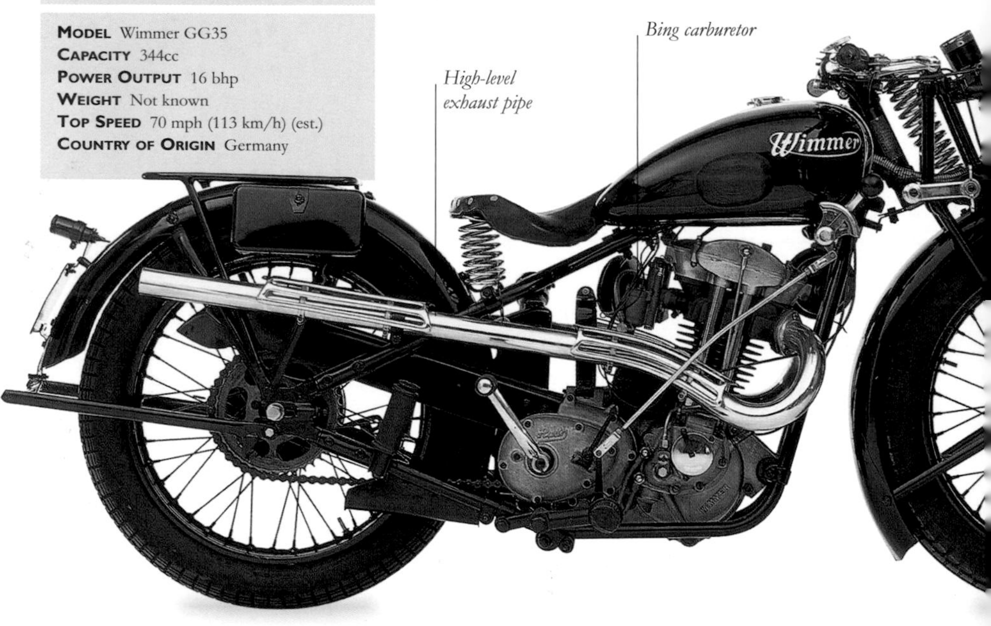

Transmission is by three-speed gearbox and shaft final drive

Trailing-link forks have two leaf springs

Bi.1932

SPECIFICATIONS

MODEL Windhoff
CAPACITY 748cc
POWER OUTPUT 22 bhp @ 4000 rpm
WEIGHT 441 lb (200 kg)
TOP SPEED 80 mph (129 km/h)
COUNTRY OF ORIGIN Germany

WINDHOFF

THE EXTRAORDINARY four-cylinder Windhoff was shown for the first time at the 1927 Berlin Show. The futuristic design was based around a massive oil-cooled in-line engine unit. There was no frame—the gearbox and rear subframe were bolted to the back of the engine, the steering head assembly bolted to the front. The one-piece engine and cylinder block were made from cast alloy, and the steel cylinder liners were cooled by oil pumped from the massive sump. Oil capacity was 1.3 gallons (6 liters). The o.h.c. was gear driven. The model seen here is from 1928.

YALE *1910*

THE CONSOLIDATED MANUFACTURING COMPANY of Toledo, Ohio, began making motorcycles in 1902 after buying the rights to the California motorcycle. The early machines were called Yale-Californias, but the suffix had been dropped by the time this single-cylinder model was built in 1910. The Yale was typical of the rugged machines produced in the US at this time. Throttle and ignition were controlled by twist grips that operated via complicated linkage arrangements. The engine retained the atmospheric inlet valve, and battery and coil ignition. The other model built by Yale was a 61cu. in. (1000cc) V-twin with a two-speed gearbox, chain drive, and a power output of 6½ horsepower.

SPECIFICATIONS

MODEL Yale 1910
CAPACITY 39.5cu. in. (649cc)
POWER OUTPUT 3.5 bhp
WEIGHT 160 lb (72.6 kg)
TOP SPEED 45 mph (72 km/h)
COUNTRY OF ORIGIN US

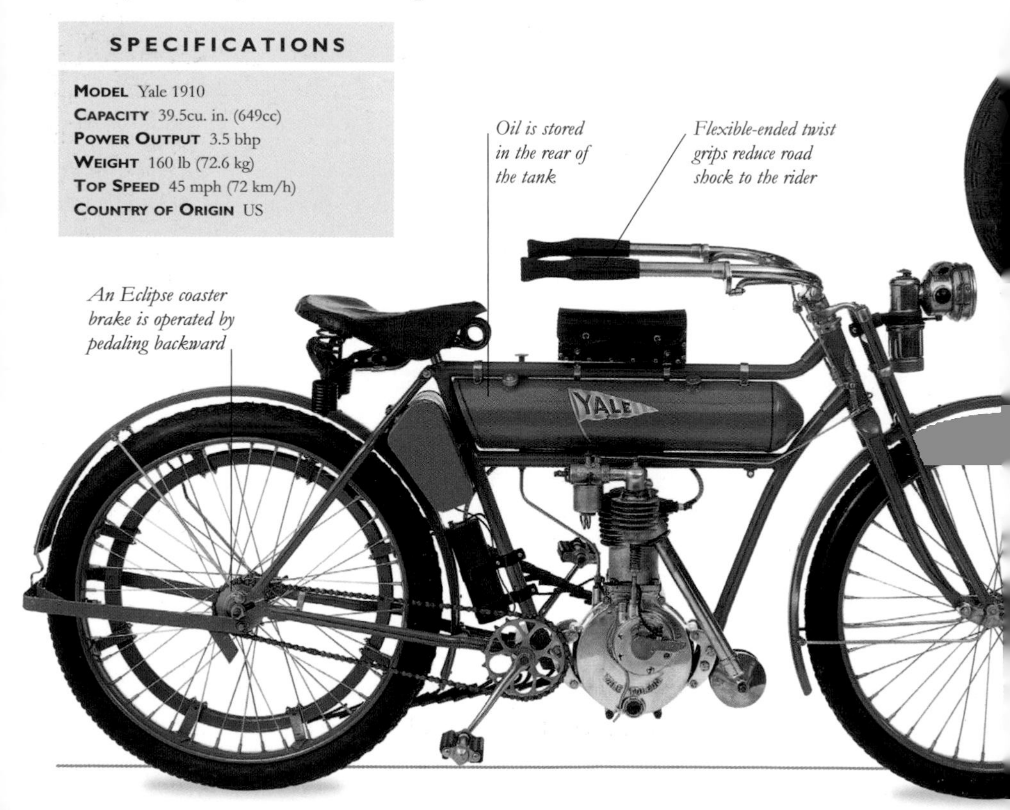

Oil is stored in the rear of the tank

Flexible-ended twist grips reduce road shock to the rider

An Eclipse coaster brake is operated by pedaling backward

SPECIFICATIONS

MODEL Yamaha YD2
CAPACITY 247cc
POWER OUTPUT 14.5 bhp @ 6000 rpm
WEIGHT 309 lb (140 kg)
TOP SPEED 70 mph (113 km/h) (est.)
COUNTRY OF ORIGIN Japan

Clutch mounted on the left end of the crankshaft

Single carburetor is concealed behind an alloy cowl

YAMAHA *YD2*

YAMAHA'S FIRST TWO-STROKE TWIN, the YD1, was introduced to the Japanese market in 1957. It was heavily inspired by the German Adler MB200 *(see pp.16–17)*. In 1959, the bike was replaced by the YD2 (the letter D was used on the model name to indicate a 250). The crankcases and cylinders were changed on the new model, and it was notable for Yamaha as the first of its bikes to be exported to the West. The YD2 was a practical motorcycle that featured an enclosed chain, a combined dynamo/electric starter unit, and deeply valanced mudguards. Styling was typical of Japanese motorcycles of the period, but the appearance of a more sporty stablemate meant that few YD2s were sold. This is a 1959 model.

YAMAHA *YDS3C Big Bear*

YAMAHA WAS QUICK TO DROP THE heavy, staid look that was typical of its, and other Japanese manufacturers', first export models in the late Fifties and early Sixties. Tubular frames, chrome mudguards, and a lighter styling touch helped to sell the bikes in Western markets. Yamaha produced its first trail bikes by adding high-level pipes to road models. The Big Bear, the first Yamaha "street scrambler," was introduced for 1965. Based on the YDS3, it was the first of Yamaha's two-strokes to be equipped with an automatic oiling system. Yamaha won its first road racing World Championship in 1964, and the YDS3's sales success was attributable partly to Yamaha's increasing reputation.

SPECIFICATIONS

MODEL Yamaha YDS3C Big Bear
CAPACITY 246cc
POWER OUTPUT 21 bhp @ 7500 rpm
WEIGHT 350 lb (159 kg)
TOP SPEED 88 mph (142 km/h)
COUNTRY OF ORIGIN Japan

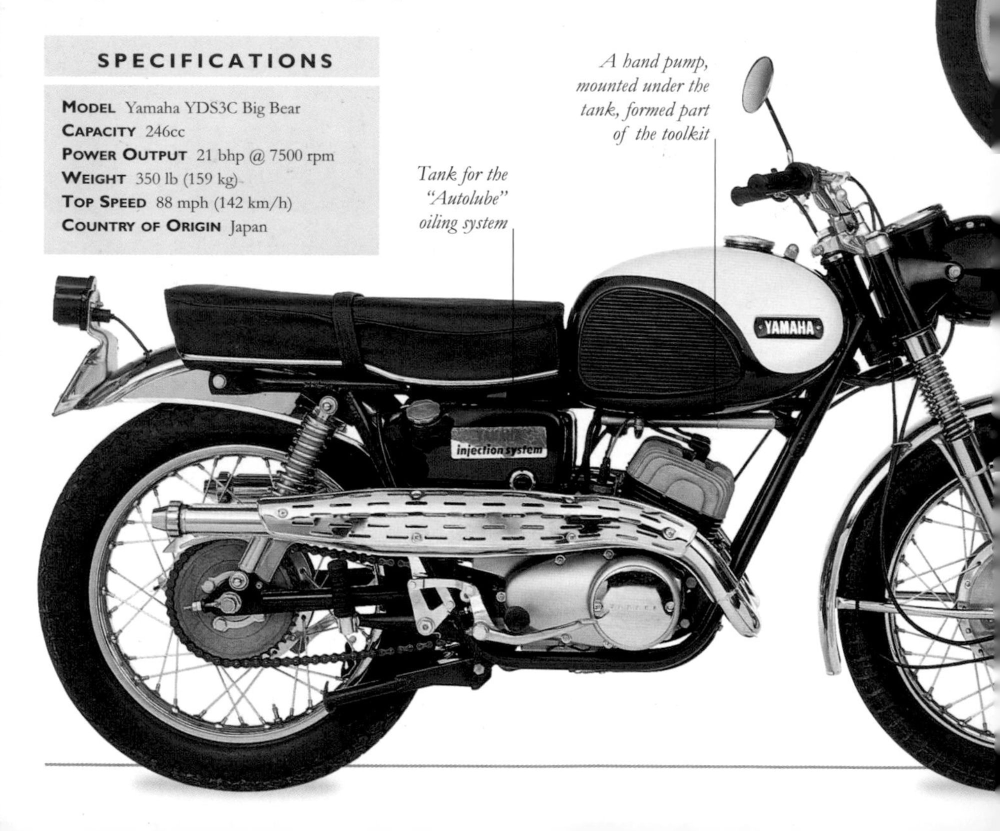

A hand pump, mounted under the tank, formed part of the toolkit

Tank for the "Autolube" oiling system

Electronic ignition control box mounted under fairing stay

Box-section swingarm

Four leading-shoe drum brakes

SPECIFICATIONS

MODEL Yamaha TR3
CAPACITY 347cc
POWER OUTPUT 58 bhp @ 9500 rpm
WEIGHT 235 lb (107 kg)
TOP SPEED Not known
COUNTRY OF ORIGIN Japan

YAMAHA *TR3*

YAMAHA HAD OFFERED racing versions of its two-stroke twin road bikes since the TD1 of 1961; these had competed successfully at club and national level with riders of varying abilities. The withdrawal, in the late 1960s, of many Japanese factories' teams from international competition meant Yamaha's two-stroke twins came to dominate that field too. The 1972 TR3, shown above, was ridden to victory by Don Emde in the Daytona 200 against full-blown 750s, becoming the smallest capacity machine ever to win the event.

YAMAHA *YZ250*

YAMAHA WON THE 250CC Motocross World Championship in 1973. The following year the company introduced the monoshock suspension system that has since become commonplace on both road and competition machinery and placed it on the YZ250. The system consisted of a single long-travel rear suspension unit mounted horizontally underneath the seat and fuel tank. As a result, the YZ250 was probably the best production motocross bike around at the time and became enormously popular with amateur riders.

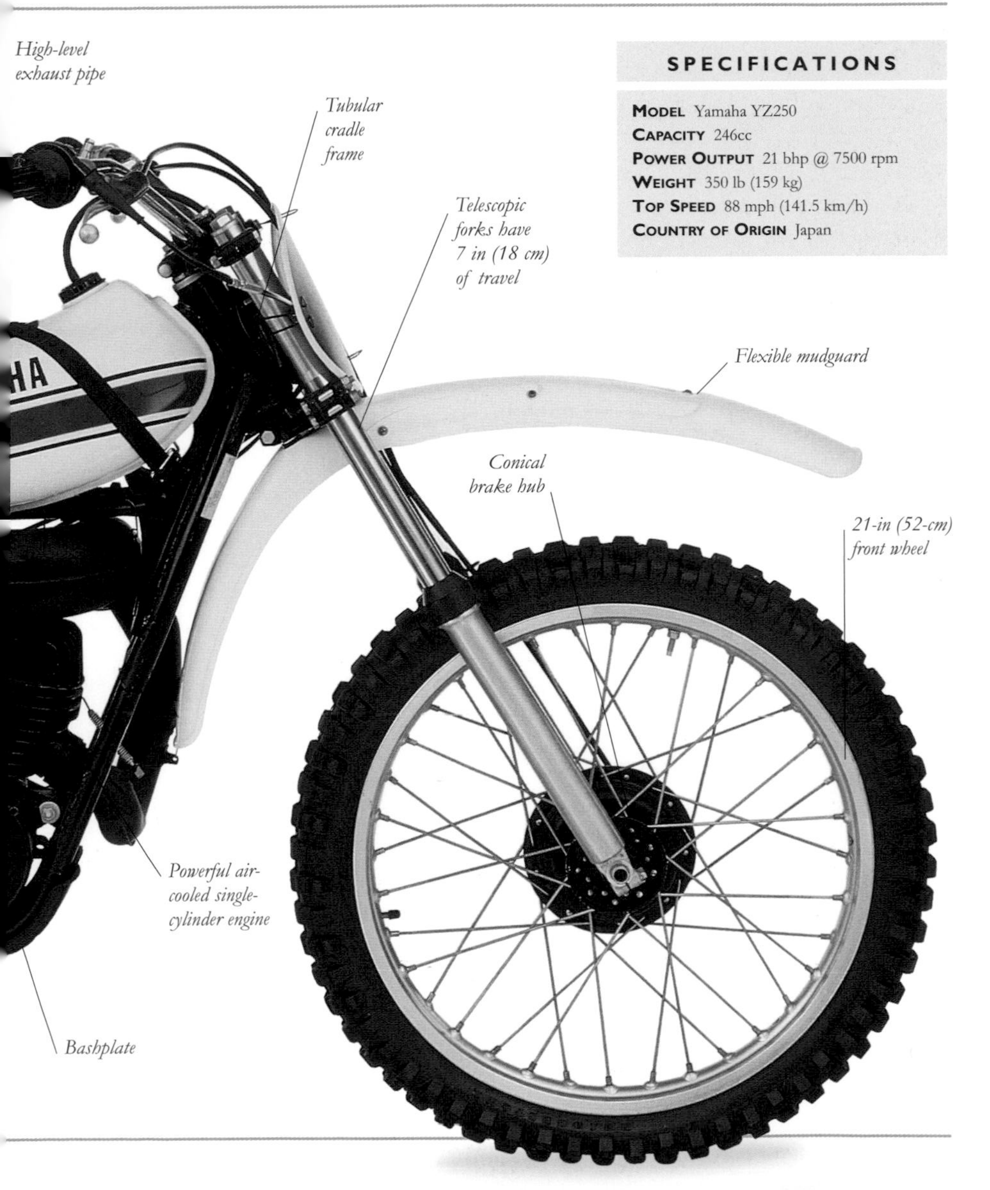

SPECIFICATIONS

Model Yamaha YZ250
Capacity 246cc
Power Output 21 bhp @ 7500 rpm
Weight 350 lb (159 kg)
Top Speed 88 mph (141.5 km/h)
Country of Origin Japan

YAMAHA *RD350B*

YAMAHA UPGRADED ITS RANGE of two-stroke twins for 1973, giving the revised line of 124cc to 347cc models the "RD" prefix. The most notable change was the adoption of reed valves throughout the line. Situated between the carburetor and the crankcase, the reed valve operated as a one-way valve. It allowed the motor to run a higher crankcase pressure without risking "blowback" through the carburetor. This increased performance, especially at low and medium revs. The 350 was replaced by the 398cc RD 400 in 1976 but reverted to the 350 engine size for the water-cooled RD350LC of 1980. The Yamaha two-stroke twin achieved something of a cult following among two-stroke enthusiasts.

One of two exhausts

Easily tunable twin-cylinder engine made this bike popular with racers

SPECIFICATIONS

MODEL Yamaha RD350B
CAPACITY 347cc
POWER OUTPUT 39 bhp @ 7500 rpm
WEIGHT 340 lb (154 kg)
TOP SPEED 105 mph (169 km/h)
COUNTRY OF ORIGIN Japan

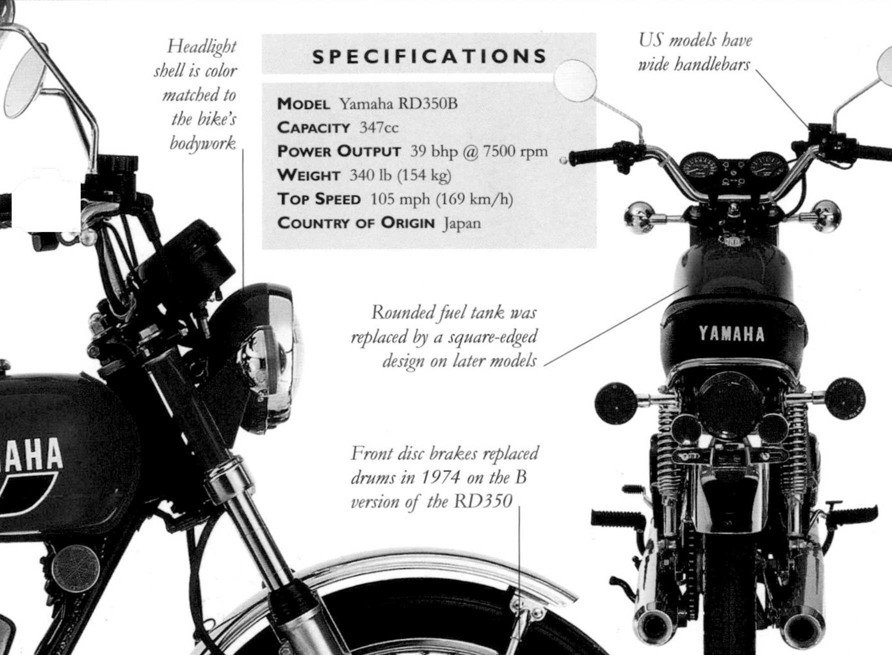

Headlight shell is color matched to the bike's bodywork

US models have wide handlebars

Rounded fuel tank was replaced by a square-edged design on later models

Front disc brakes replaced drums in 1974 on the B version of the RD350

Duplex cradle frame

Mudguard stay

Brake caliper

DEFINITIVE BIKE
Throughout the 1970s, Yamaha twins—such as the 1975 model shown here—were the definitive, sporty two-strokes, with performance capable of outstripping much larger machines.

YAMAHA *TZ250*

THE WATER-COOLED TZ RACERS replaced Yamaha's air-cooled TD models when they were introduced in 1973. In the following 15 years a TZ was essential for private racers looking for racing success. The engine layout followed the pattern of Yamaha's two-stroke production twins, with parallel cylinders and a 180° crankshaft. To maintain its competitive edge, Yamaha continually improved the TZ. Mechanics also worked hard to make them faster, and many machines, such as this 1979 model, were heavily modified. This bike was used by German rider Dieter Braun, who won the 250 World Championship in 1973 for Yamaha.

Monoshock frames were used on TZs from 1976

Telescopic front forks

Drilled front discs

SPECIFICATIONS

MODEL Yamaha TZ250
CAPACITY 247cc
POWER OUTPUT 53 bhp @ 10,500 rpm
WEIGHT 239 lb (108 kg)
TOP SPEED 140 mph (225 km/h)
COUNTRY OF ORIGIN Japan

LOW POSITION
A race bike like the TZ Yamaha is designed to allow the rider to tuck in so as to reduce wind resistance. On the straights the rider will crouch down to achieve maximum speed.

YAMAHA *OW48*

YAMAHA'S PISTON-PORTED TWO-STROKES were the most widely used and successful racing machines of the 1970s, in every class from 125 to 750cc. Giacomo Agostini won Yamaha's first 500cc world title in 1975, then Barry Sheene won in 1976 and 1977 on the superior disc-valve Suzuki. The riding skills of Kenny Roberts negated the Suzuki advantage, and he won the world title for Yamaha in 1978, 1979 (on the bike seen here), and 1980. The OW48 was an in-line, two-stroke, four-cylinder machine with twin Mikuni carburetors and a six-speed box. Its 130 horsepower capacity meant it was able to hit speeds of up to 180 mph (290 km/h).

Clip-on bars

REAR VIEW
The four expansion-chamber exhausts are given adequate ground clearance by contorting the exhaust pipes. The pipe exiting by the seat comes from the cylinder on the far left.

Morris seven-spoke alloy wheels

Shock absorber adjustment screws

SPECIFICATIONS

MODEL Yamaha OW48
CAPACITY 498cc
POWER OUTPUT 130 bhp
WEIGHT 320 lb (145 kg)
TOP SPEED 180 mph (290 km/h)
COUNTRY OF ORIGIN Japan

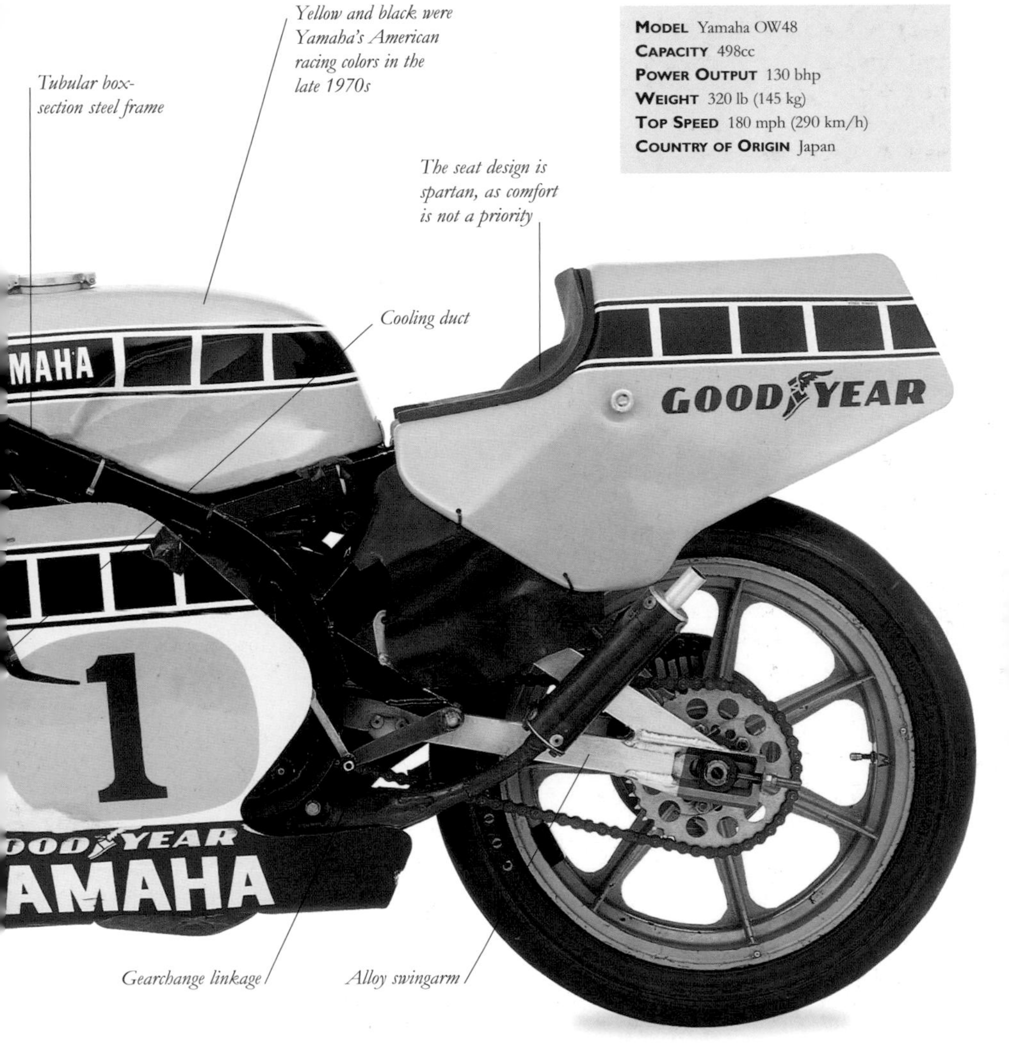

Tubular box-section steel frame

Yellow and black were Yamaha's American racing colors in the late 1970s

The seat design is spartan, as comfort is not a priority

Cooling duct

Gearchange linkage

Alloy swingarm

YAMAHA *RD250LC*

INTRODUCED IN 1980, the LC achieved cult status in the UK and elsewhere during a comparatively short model run. Technically the LC was a logical progression from earlier air-cooled road bikes, which were themselves developed from racers—RD allegedly stood for Race Developed. The LC (Liquid Cooled) took the monoshock suspension and liquid-cooled cylinders from the TZ series race bikes and put them on the road. An LC, in 250 or 350 capacity, was every Eighties' teenage hood's dream bike. Production racing around the world was dominated by LCs in the 1980s, and many future GP stars started their careers on LCs.

Kick starter—no electric starter was used

Expansion chamber-style exhausts

Carburetors fed the engine via reed valves

SPECIFICATIONS

MODEL Yamaha RD250LC
CAPACITY 247cc
POWER OUTPUT 35 bhp
WEIGHT 306 lb (139 kg)
TOP SPEED 106 mph (171 km/h)
COUNTRY OF ORIGIN Japan

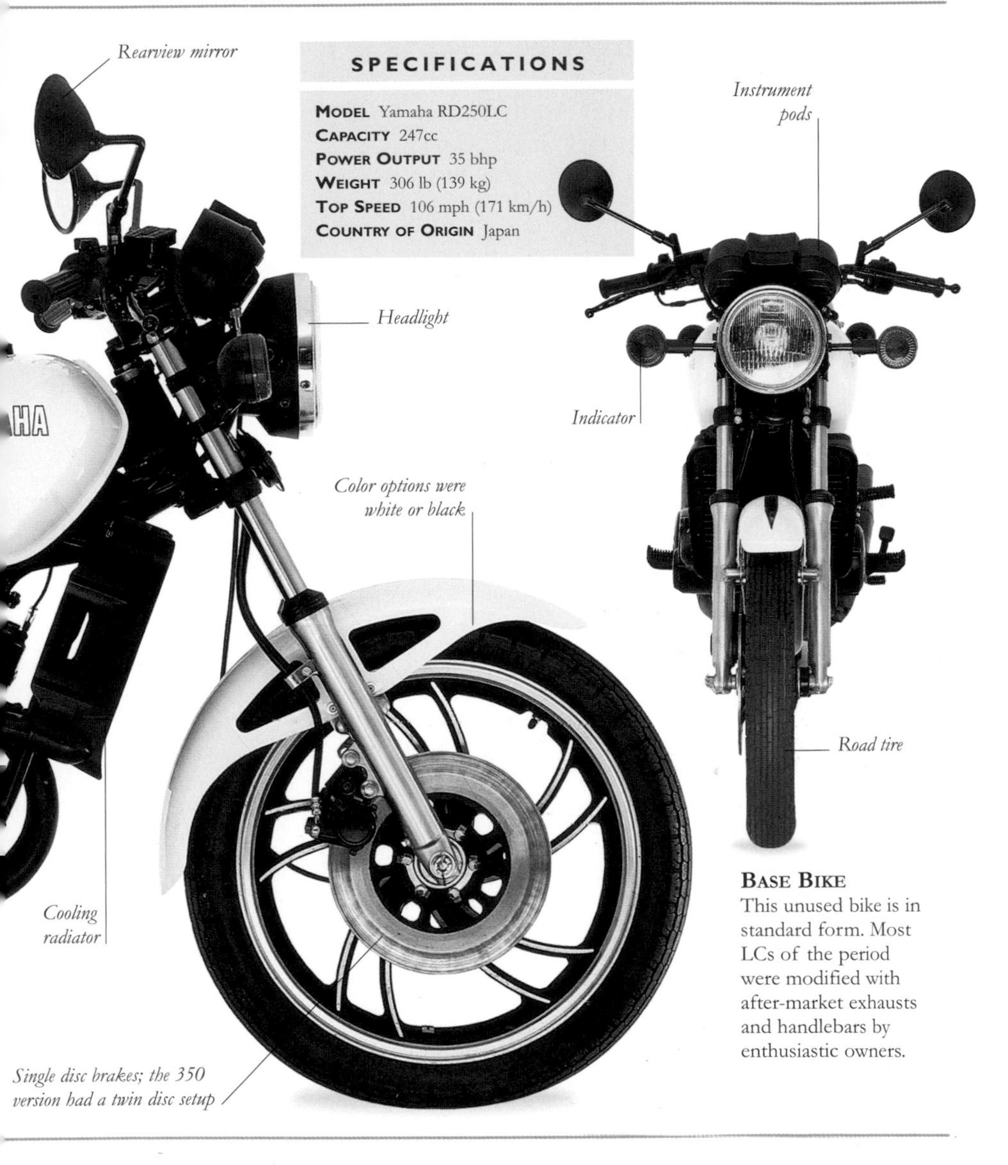

BASE BIKE
This unused bike is in standard form. Most LCs of the period were modified with after-market exhausts and handlebars by enthusiastic owners.

YAMAHA *RZ500*

AS EMISSION CONTROLS AND restrictive legislation closed in on large-capacity two-strokes, Yamaha produced the ultimate version of its race-developed road bikes. Produced from 1984 to 1987, the RZ was a water-cooled twin-crankshaft V4 machine whose engine layout derived from Yamaha's early 1980s' Grand Prix machines. Like many other machines that were manufactured in Japan, the V4 Yamaha was sold in various forms in different world markets. In North America it was known as the RZ500; in Japan it was the RZV500 with an alloy frame; and in Britain the steel-framed bike was sold as the RD500. The bike shown on these pages is a 1984 model.

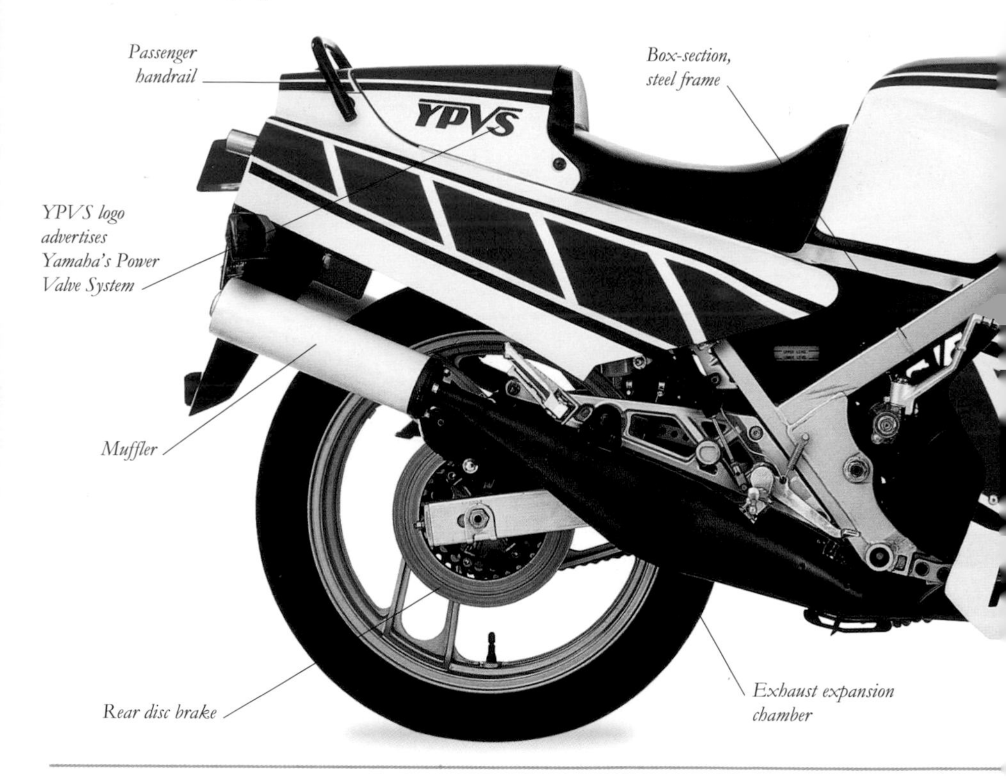

SPECIFICATIONS

MODEL Yamaha RZ500
CAPACITY 499cc
POWER OUTPUT 87 bhp @ 9500 rpm
WEIGHT 396 lb (180 kg)
TOP SPEED 135 mph (216 km/h)
COUNTRY OF ORIGIN Japan

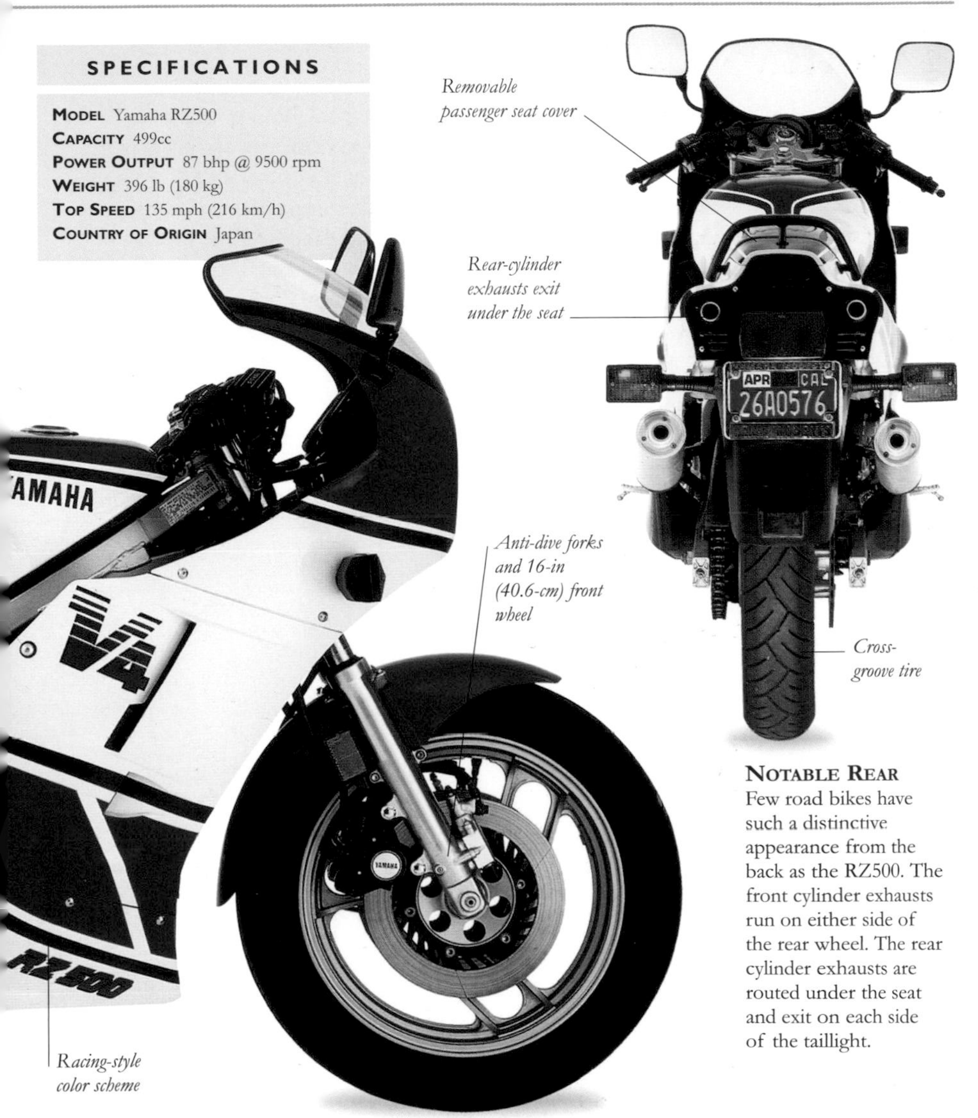

NOTABLE REAR
Few road bikes have such a distinctive appearance from the back as the RZ500. The front cylinder exhausts run on either side of the rear wheel. The rear cylinder exhausts are routed under the seat and exit on each side of the taillight.

YAMAHA *V-Max*

ENGLISHMAN JOHN REED WAS ASKED to design the V-Max when Yamaha wanted to build the ultimate custom bike. Since Reed lived in California, the outrageous styling unsurprisingly showed strong American influences. Instead of the usual long, low styling of custom bikes, the V-Max had a much denser mechanical look. Its massive water-cooled V4 motor was based on that of the Venture touring bike. Some markets, including the UK, were supplied with a restricted version of the V-Max, with power output cut to 95 bhp. In unrestricted form the V-Max is capable of stunning straight-line performance; the handling is less inspiring. This V-Max dates from 1985.

Clutch cover

15-in (38-cm) rear wheel

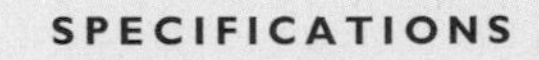

SPECIFICATIONS

MODEL Yamaha V-Max
CAPACITY 1198cc
POWER OUTPUT 145 bhp
WEIGHT 596 lb (270 kg)
TOP SPEED 144 mph (230 km/h)
COUNTRY OF ORIGIN Japan

Rearview mirror

Single speedometer instrument

Huge air intakes on each side of the tank dominate the styling

Small mudguard

Twin hydraulic disc front brake

Cooling radiator

Massive V-four engine

YAMAHA *FZR1000*

THE FZR1000 WAS YAMAHA'S largest capacity supersports bike when launched. Its history can be traced back to the 1985 FZ750—the first mass-produced five-valve-per-cylinder bike. The original d.o.h.c. FZR1000 became the FZR1000 Exup in 1989. The Exup's variable exhaust valve offered a wider spread of power than would otherwise be available, while the large frame offered a riding position that suited all. Transmission was through a six-speed gearbox, and suspension consisted of upside-down telescopic forks and a rear swingarm with rising rate single shock. The power output was phenomenal, particularly when de-restricted to 147 bhp; this was the standard in some markets. The FZR1000 illustrated dates from 1992.

SPECIFICATIONS

MODEL Yamaha FZR1000
CAPACITY 1002cc
POWER OUTPUT 125 bhp @ 10,000 rpm
WEIGHT 529 lb (240 kg)
TOP SPEED 167 mph (269 km/h)
COUNTRY OF ORIGIN Japan

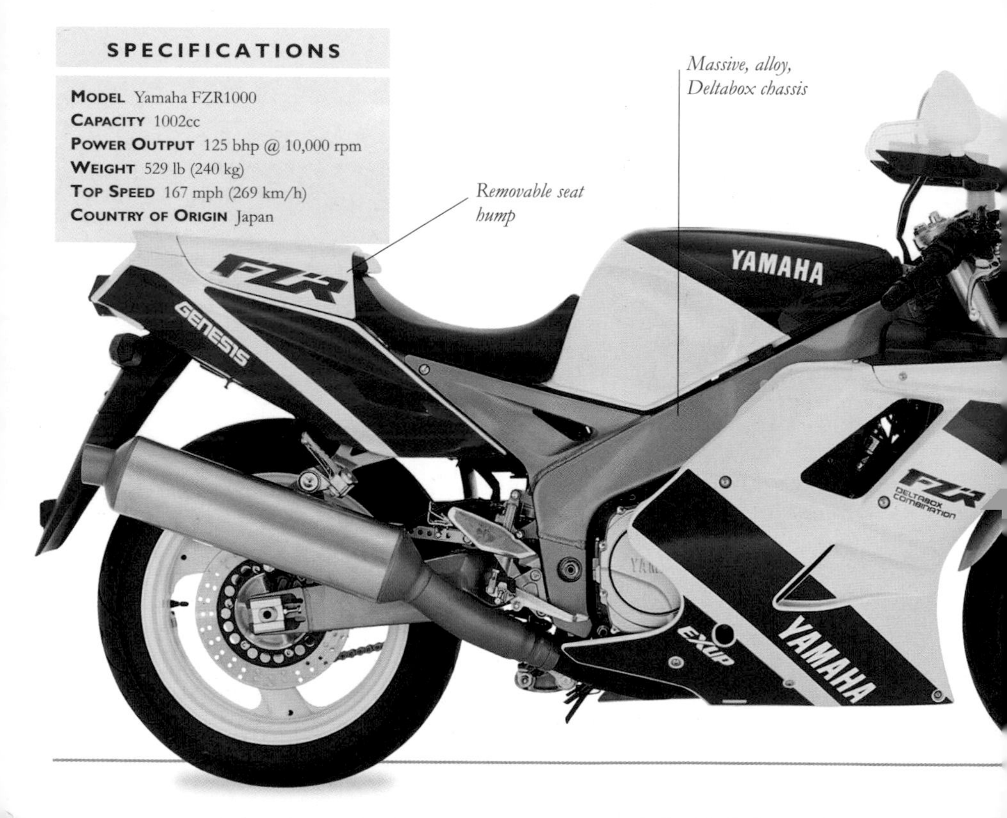

Removable seat hump

Massive, alloy, Deltabox chassis

Exhaust system features a catalytic converter

Alloy frame

SPECIFICATIONS

MODEL Yamaha GTS1000
CAPACITY 1002cc
POWER OUTPUT 100 bhp @ 9000 rpm
WEIGHT 553 lb (251 kg)
TOP SPEED 132 mph (213 km/h)
COUNTRY OF ORIGIN Japan

Massive, six-piston, front brake caliper

YAMAHA *GTS1000*

THE GTS1000 WAS THE FIRST Japanese mass-produced bike to have hub-center steering in place of the more traditional front fork. Yamaha's "Omega chassis," with single-sided front swingarm and a monoshock rear, gave it a rock-steady ride and a low center of gravity. The downside was heavy steering and a poor steering lock. Powered by Yamaha's FZR1000 Exup motor with added fuel injection and catalytic converter, the GTS1000 was aimed at the more experienced rider.

Yamaha *YZF600 R6*

With 150 mph (241 km/h) performance, more nimble handling, and more realistic running costs than 1000cc sports machines, the 600cc sports bike class had attained massive importance in the 1990s. But until the arrival of the R6 for 1999, Yamaha had been unable to make a serious impact on the category. The R6 was created with the same "no compromise" philosophy as the larger R1 model *(see pp.462–63)*. The result was a performance-focused bike that was lighter and more powerful than its rivals while having its own unique appearance. It was an almost perfectly balanced sporting machine with excellent handling and exceptional performance from its 100 bhp four-cylinder engine.

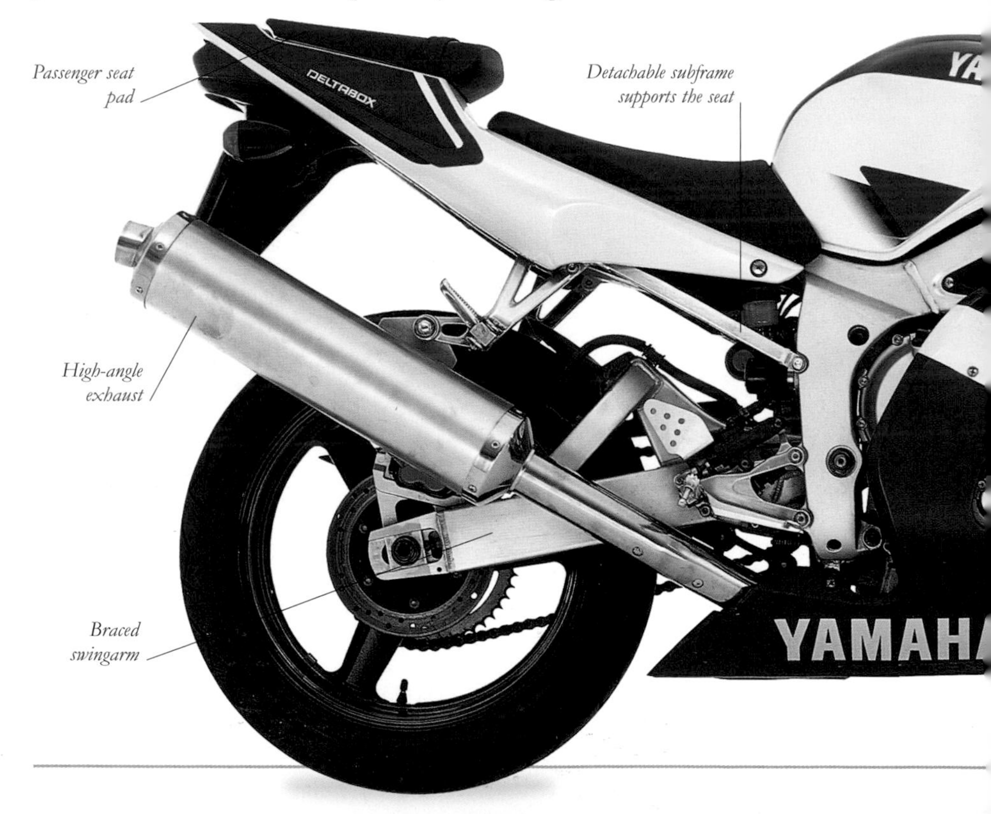

SPECIFICATIONS

MODEL Yamaha YZF600 R6
CAPACITY 599cc
POWER OUTPUT 98.7 bhp
WEIGHT 373 lb (169 kg)
TOP SPEED 155 mph (250 km/h)
COUNTRY OF ORIGIN Japan

FAMILY RESEMBLANCE
The R6 has a similar look to its bigger brother, the 1000cc R1, but is actually very different. The central air duct between the headlights directs air into the air box, which feeds the four carburetors.

YAMAHA *YZF1000 R1*

IT WAS ALWAYS GOING TO TAKE an extraordinary bike to topple the Honda Fireblade *(see pp.212–13)* from the top of the performance tree. And it was. The Yamaha YZF1000 R1 moved the performance perimeter on by several steps even if there was nothing radically new in the package. Yamaha went to great lengths to make everything lighter, stronger, and more efficient. The result was a bike that was faster, lighter, handled better, and was more radical looking than the Fireblade. Unfortunately, but almost inevitably, the R1's time at the top of the tree was also short-lived, with the arrival of the Suzuki GSXR1000 *(see pp.396–97)* in 2001 pushing the Yamaha off its perch.

Straight-edged rear "hugger" mudguard

Water-cooled, 16-valve four-cylinder engine

High-performance exhaust

Brake caliper

SPECIFICATIONS

MODEL Yamaha YZF1000 R1
CAPACITY 998cc
POWER OUTPUT 132.1 bhp
WEIGHT 375 lb (170 kg)
TOP SPEED 165 mph (266 km/h)
COUNTRY OF ORIGIN Japan

SHARP FRONT
The distinctive pointed snout of the R1 briefly set the standards for 21st century sports bikes. This is a 2001 model that sports a taller than standard windshield.

YAMAHA *MT-01*

ORIGINALLY SHOWN AS A PROTOTYPE to gauge public interest, the MT-01 became a production model for 2005. The concept combined a large, but comparatively low power, V-twin engine from one of Yamaha's cruisers, with a sports bike's unfaired, sit-up stance, and high-quality suspension and brakes. The easy delivery of the engine and the style of the bike were great fun at lower speeds, although at higher speeds its weight and lack of power were drawbacks. This brave attempt to do something different was largely misunderstood by the market; the bike never sold in big numbers, and was eventually discontinued in many markets.

SPECIFICATIONS

MODEL Yamaha MT-01
CAPACITY 1670cc
POWER OUTPUT 90 bhp @ 4750 rpm
WEIGHT 529 lb (240 kg)
TOP SPEED 131 mph (211 km/h)
COUNTRY OF ORIGIN Japan

4-gallon (15-liter) fuel tank gives a range of just 115 miles (185 km)

Large analog rev-counter also contains a digital speedometer and other instrument functions

Huge 1670cc engine delivers power t low revs with a 5500 rpm redline

ZÜNDAPP *KS601*

BETTER KNOWN AS THE "Green Elephant," the KS601 was Germany's fastest road machine when it was introduced in 1950. The horizontally opposed, twin-cylinder, o.h.v. engine and four-speed chain and sprocket "gearbox" were essentially those of prewar days, but they were now mounted in a tubular frame equipped with telescopic forks, plunger rear suspension, and interchangeable wheels. In typical Zündapp fashion, the engine was smoothly styled with its ancillaries enclosed as much as possible. It soon became popular with sporting sidecar drivers. Comparisons with BMW's flat-twin designs were inevitable and there were many who thought that the Zündapp was a superior machine. They had a committed following—the famous Elephant rally in Germany is named after them. Unfortunately, production of the flat-twins ended in 1957.

SPECIFICATIONS

MODEL Zündapp KS601
CAPACITY 597cc
POWER OUTPUT 28 bhp @ 4700 rpm
WEIGHT 472 lb (214 kg)
TOP SPEED 87 mph (140 km/h)
COUNTRY OF ORIGIN Germany

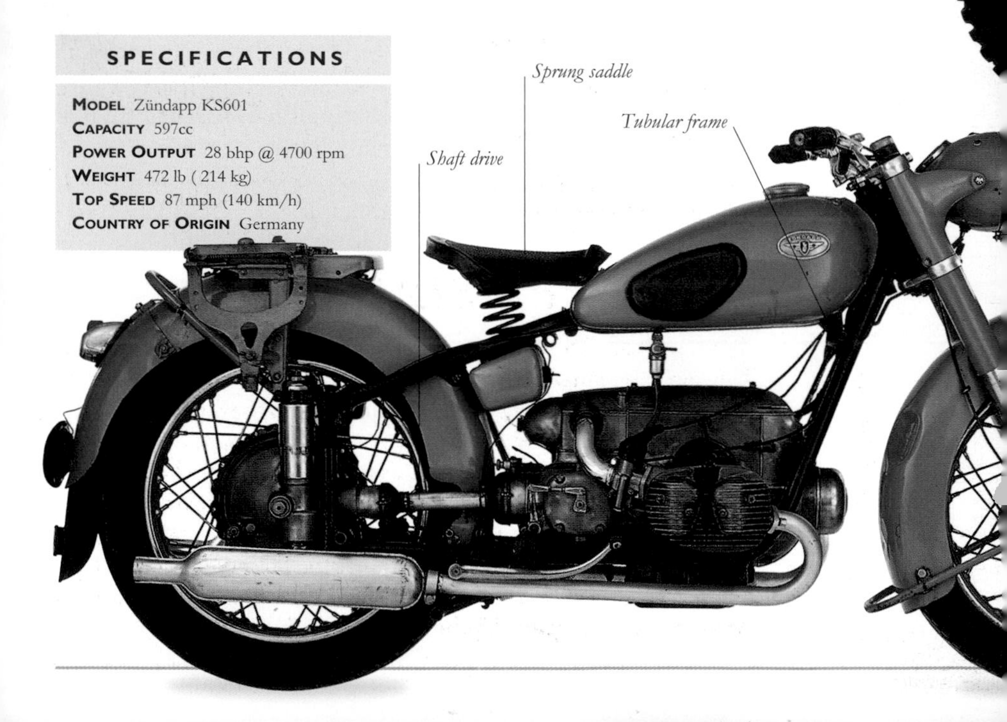

High-level exhaust system

Lights attached to fulfill legal requirements for road use

SPECIFICATIONS

MODEL Zündapp GS125
CAPACITY 124cc
POWER OUTPUT 18 bhp @ 7900 rpm
WEIGHT 220 lb (100 kg) (est.)
TOP SPEED 65 mph (105 km/h) (est.)
COUNTRY OF ORIGIN Germany

This Zündapp engine was also supplied to other firms as a proprietary unit

ZÜNDAPP *GS125*

THE LETTERS GS STAND FOR Gelände Sport—Gelände meaning "terrain." Enduro-type events were very popular in the early 1970s, especially in the US where cheap Japanese trail bikes had given many riders a taste for something more serious. The requirement was for a machine combining good off-road capabilities with high speed when necessary, in a form legal on the road. The GS125 fulfilled this, being built on the lines of a motocross machine with lights. The machine shown is a 1972 bike; developed versions of it won the world 125cc Motocross Championship in 1973 and 1974.

INDEX

A

B

C

D

E

F

G

H

N

O

P

R

S

ACKNOWLEDGMENTS

DORLING KINDERSLEY WOULD LIKE TO THANK THE FOLLOWING FOR THEIR ASSISTANCE:

Neil Lockley, Kathryn Hennessy, and Suparna Sengupta for additional editorial assistance; Shanker Prasad for additional DTP design assistance; Franziska Marking for picture research; Myriam Megharbi, David Saldanha, Melanie Simmonds, and Hayley Smith in the picture library; Dave King and Jonathan Buckley for photography; Tom Bedford for locating and delivering bikes; and Margaret McCormack for compiling the index.

DORLING KINDERSLEY WOULD ALSO LIKE TO THANK ALL THOSE WHO ALLOWED THEIR BIKES TO BE PHOTOGRAPHED, INCLUDING:

The Barber Vintage Motorsports Museum 18, 68–69, 79, 109, 136–37, 138–39, 145, 150–51, 152–53, 161, 165, 166–67, 170–71, 174–75, 176–77, 244–45, 350–51
F. W. Warr and Sons 80–81, 168
Steve Slocombe 144
Dave Griffiths 162–63
Harley-Davidson UK 180–81

PICTURE CREDITS

THE PUBLISHER WOULD LIKE TO THANK THE FOLLOWING FOR THEIR KIND PERMISSION TO REPRODUCE THEIR PHOTOGRAPHS:

(Key: a=above; b=below; c=center; l=left; r=right; t=top)

28-29 Dorling Kindersley: Palmers Motor Company. **62-63 Dorling Kindersley:** Phil Davies. **86–87:** Bike/Emap Automotive. **112-113 Dorling Kindersley:** Neil Mort, Mott Motorcycles. **182-183 Dorling Kindersley:** Alan Purvis. **120b:** The World's Motorcycles News Agency. **226-227 Honda (UK).** **270-271 Dorling Kindersley:** Pegasus Motorcycles. **273t:** The World's Motorcycles News Agency. **274-275 Dorling Kindersley:** Trevor Pope Motorcycles. **310-311 Dorling Kindersley:** Ken Small. **332-333 Dorling Kindersley:** John Pitts. **400-401 Dorling Kindersley:** Rob Johnson. **422-423 Dorling Kindersley:** David Riman. **464-465 Courtesy of Yamaha Motor**.

NOTE:
Every effort has been made to trace the copyright holders. Dorling Kindersley apologizes for any unintentional omissions and would be pleased, in such cases, to add an acknowledgment in future editions.